DEVELOPMENTS IN PLASTICS TECHNOLOGY—3

CONTENTS OF VOLUMES 1 AND 2

Volume 1

Volume 2

DEVELOPMENTS IN PLASTICS TECHNOLOGY—3

Edited by

A. WHELAN

London School of Polymer Technology,
The Polytechnic of North London, Holloway, London, UK

and

J. L. CRAFT

Technical Consultant, Plastics and Rubbers,
Enfield, UK

ELSEVIER APPLIED SCIENCE PUBLISHERS
LONDON and NEW YORK

ELSEVIER APPLIED SCIENCE PUBLISHERS LTD
Crown House, Linton Road, Barking, Essex IG11 8JU, England

Sole Distributor in the USA and Canada
ELSEVIER SCIENCE PUBLISHING CO., INC.
52 Vanderbilt Avenue, New York, NY 10017, USA

WITH 43 TABLES AND 158 ILLUSTRATIONS

Softcover reprint of the hardcover 1st edition 1986

British Library Cataloguing in Publication Data

Developments in plastics technology.
3
1. Plastics
I. Whelan, A. II. Craft, J.L. (Joan Letitia)
668.4 TP1120

The Library of Congress has cataloged this serial publication as follows:
Developments in plastics technology.—1-—London; New York : Applied Science Publishers, c1982–
v.:ill.; 23 cm.—(Developments series)

1. Plastics—Collected works. I. Series.
TP1101.D48 668.4′05—dc19 84-644560

ISBN-13: 978-94-010-8358-4 e-ISBN-13: 978-94-009-4183-0
DOI: 10.1007/978-94-009-4183-0

PREFACE

Because of the sheer size of the plastics industry, the title *Developments in Plastics Technology* now covers an incredibly wide range of subjects or topics. No single volume can survey the whole field in any depth and so what follows is therefore a series of chapters on selected topics. The topics were selected by us, the editors, because of their immediate relevance to the plastics industry.

When one considers the materials produced and used by the modern plastics industry, there is a tendency to think of the commodity thermoplastics (such as poly(vinyl chloride) or polyethylene); the thermosetting materials are largely ignored. Because of this attitude we are very pleased to include in this volume a chapter which deals with the processing of a thermosetting material, i.e. the pultrusion of glass reinforced polyester.

The extrusion of plastics is, of course, a very important subject but an aspect which is often overlooked is the need to remove volatile matter during processing: for this reason we have included a chapter on devolatilisation. Current industrial practice is towards materials modification and this attitude is reflected in the chapters on the transformation of ethylene vinyl acetate polymers and the use of wollastonite in two important thermoplastics. When assessing the performance of materials, there is a tendency to concentrate on short-term mechanical tests and ignore such topics as fatigue and longer-term testing. We are therefore very pleased to include a chapter on this subject.

Over the past few years, since the advent of cheap computing power, there has been a tremendous growth in the science of computer modelling of processes such as injection moulding and extrusion. This interest is reflected in what is the longest chapter in the book and which is devoted to computer-aided analysis of some thermoplastic processing operations. Alternative methods of polymer processing are always being sought in an effort to reduce costs, save energy or to obtain unusual properties in selected components. One such technique is solid phase compaction and a review of this interesting technique is presented.

Each of the contributions was written by a specialist in that field and to them all, we, the editors, extend our heartfelt thanks, as writing a contribution for a book such as this, whilst doing a full-time job, is no easy task.

A. WHELAN
J. L. CRAFT

CONTENTS

LIST OF CONTRIBUTORS

R. Crawford

Department of Mechanical and Industrial Engineering, The Queen's University of Belfast, Ashby Institute, Stranmillis Road, Belfast, Northern Ireland BT9 5AH, UK

K. V. Gotham

Rubber and Plastics Research Association of Great Britain, Shawbury, Shrewsbury, Shropshire SY4 4NR, UK. Present address: *1 Kensington Gardens, Minsterley, Shrewsbury, Shropshire, UK*

L. Hollaway

Department of Civil Engineering, University of Surrey, Guildford, Surrey GU2 5XH, UK

Keith T. O'Brien

Celanese Engineering Resins, 86 Morris Avenue, Summit, New Jersey 07901, USA

J. F. T. Pittman

Department of Chemical Engineering, University College of Swansea, Singleton Park, Swansea SA2 8PP, UK

E. F. Vansant

Department of Chemistry, University of Antwerp (UIA), Laboratory Adsorption and Catalysis, Universiteitsplein 1, B-2610 Wilrijk, Belgium

P. J. Wright

Blue Circle Industries plc, Research Division, London Road, Greenhithe, Kent, UK

Chapter 1

PULTRUSION

L. HOLLAWAY

Department of Civil Engineering, University of Surrey, Guildford, UK

1. INTRODUCTION

The pultrusion technique is a closed mould system for the manufacture of fibre/matrix composites. It is one of the few fully automated continuous processes used in the reinforced plastics industry. The technique is finding increasing application in industries that range from the manufacture of domestic products to aircraft construction.

The technique is a process whereby constant section shapes are produced by pulling fibre filaments through a heated die, the filament having been previously impregnated with resin. The actual processes are described in Section 3.

The products are generally straight and can take most geometrical cross-sectional shapes, although recently some products which are curved in the longitudinal direction have been manufactured.

In all processes the production procedure can essentially be divided into three steps which are assigned to corresponding parts of the production line. These are:

(a) feed-in and impregnation;
(b) shaping and curving;
(c) pulling and cut-off.

The greatest technical problems appear in steps (a) and (b) above. The first processing section is an important part because the final product depends upon the care with which the reinforcement material is fed in and impregnated. For economic reasons the highest possible operating speed is desired.

2. CHARACTERISTICS OF THE CONSTITUENT COMPONENTS OF PULTRUSION

2.1. Basic Constituents

The pultrusion composite is manufactured basically from the components:

(a) the polymer;
(b) the fibre reinforcement.

A third component, the filler, may be incorporated into the composite to either impart a specific inservice property or to displace the polymer for economy reasons.

All types of filamentary reinforcements can be utilised in the pultrusion process. These reinforcements would generally be in the form of rovings, tows, mat, cloth or any combination of these fibres. The majority of the products available commercially use E-glass filaments in their various forms but high modulus fibres, such as carbon, may also be used.

The resin systems utilised include polyesters, which constitute about 85% of the annual volume of pultrusion stock, and epoxides.

2.2. The Polymerisation Process for Polyesters

The unsaturated polyesters generally used in reinforced plastics consist of an unsaturated ester dissolved in a vinyl monomer which is able to be polymerised and which provides three-dimensional chain cross-linking. Such resins are not materially different from those used in hand lay-up processes. The solution is then polymerised under the influence of heat and/or a peroxide catalyst to provide an infusible thermoset. Polymerisation is initiated as the catalyst breaks down into free radicals under the influence of the energy derived from the heat. A chain reaction then occurs, causing the molecules to react and unite; the rate of catalyst decomposition is dependent upon the temperature. Consequently, the rate of polymerisation can be controlled by regulating the temperature and the amount of catalyst used. Catalysts for pultrusion processes are commonly based on benzoyl peroxide.

Theoretically, the reaction should proceed to the situation where all unbroken double bonds of the original unsaturated mixture have been reacted by the free radicals; however, in practice only about 92% of the unsaturated polyester is converted. When 35–40% of the unsaturated polyester is converted, gelation takes place and when about 80% is converted, an incomplete cure exists.

2.3. The Polymerisation Process for Epoxies

Epoxy resins are characterised by the presence of epoxide groupings which may exist in the raw materials or may be formed during the reaction. The curing process of the resins is achieved by using either hardeners or curing agents. The polymerisation process occurs by either:

(a) a catalytic reaction in which the epoxide groups or the resin molecules interact; or
(b) a cross-linking reaction in which copolymerising molecules unite with the epoxy resin molecules, so becoming a part of the cured resin network.

During the polymerisation of epoxies the exotherm begins before gelation commences; this is contrary to the sequence for the polyesters.

Frequently the epoxy polymers exhibit higher viscosities than the polyester polymers; consequently, the former require longer filament wet out times. In addition, it is more difficult to remove the entrapped air and to dissipate volatiles in the epoxy polymers. However, systems which are solventless are available and are particularly suited to pultrusion. In order to obtain high quality epoxy composites it is desirable to apply pressure to the system at approximately the time that gelation commences.

2.4. Comparison of Epoxy and Polyester Polymers in the Pultrusion Technique

Polyester and epoxy polymers exhibit a number of different characteristics, all of which lead to different levels of complexity in the pultrusion technique. The main differences are:

(a) the variation in the polymerisation shrinkage;
(b) the interface adhesion between the pultrusion section and the adjacent mould surface.

Referring to item (a), the volume reduction of polyesters varies between about 7 and 10% whereas the equivalent value for epoxies varies between about 1 and 2%; the majority of this volume reduction occurs at the time of gelation. As the resin cures and shrinks in the die, the stock surface separates from the die surface. The resulting decrease in frictional force allows for lower pull loads and possibly faster line speeds. The high degree of shrinkage of polyester resins can cause a problem when pultruding stock with a large cross-sectional area. In many cases, low profile additions (e.g. polyethylene) are needed to control the external cracks caused by this shrinkage. A major reason for polyester stock being capable of high speed and low full loads is the high degree of shrinkage associated with the material.

The frictional forces that accumulate due to the shear of the epoxy resin/die interface are considerably higher than those for polyester impregnated stock due to the longer period of cure and the lower shrinkage at gelation. Consequently, the epoxies are more difficult to pultrude than the polyesters and those which can be pultruded have line speeds considerably below those of the polyesters. The shrinkage characteristics of the epoxy affect the ease with which the pultruded section can be detached from the inside of the die. If, in addition, a mandrel is used to form a hollow closed section, attachment of the section to the die may take place.

With regard to (b), the epoxy resins produce higher adhesion forces during the pultrusion process than the polyesters because of the assimilation of hydrogen groups from amine curing agents and because of the effects of low surface tension.

The effects of shrinkage and internal adhesion and hence the interface traction may be minimised by techniques such as using slip sheets, waxing or plating the die, applying external die releases or using internally acting ones.

2.5. The Reinforcing Materials

There are four main types of fibres used in the pultrusion technique; these are:

(a) E-glass fibre;
(b) S-glass fibre;
(c) aramid fibre (based on aromatic polyamides);
(d) graphite (carbon) fibre.

The primary type of fibre used in the system is the E-glass; the other reinforcements are mainly for the aerospace industry and for the manufacture of advanced composites for space applications. It is possible to obtain greater stiffness from the pultruded section by replacing the E-glass fibre with a hybrid reinforcement consisting of glass and carbon fibre.

2.5.1. Glass Fibre Roving Reinforcement

Glass fibre rovings may be used in the pultrusion process to provide high longitudinal strength and stiffness. In addition, they provide the tensile strength required to pull the reinforcing fibre array through the die. The amount and location of these reinforcements are determined at the design stage.

2.5.2. Glass Fibre Mat Reinforcement

The continuous strand mat provides the most economical method of obtaining a high degree of transverse physical properties. The mats are layered with rovings and this combination forms the basic reinforcement composition for most pultruded products.

2.5.3. Glass Fibre Veils

The pultrusion technique is a low pressure process and the glass fibre reinforcement tends to position itself close to the surface of the composite. This can affect the appearance, corrosion resistance and the handling of the products. Glass fibre and polymeric veils can be added to the laminate construction to depress the reinforcement from the surface, thus providing a polymer rich surface to the composite. The two most commonly used products are surface tissues of high alkali content glass (A-glass) or a chemical resistant glass (C-glass), and polymeric veils of polyester if this is the parent material.

2.5.4. Woven and Knitted Fabrics

If the mechanical properties of the pultruded composite cannot be satisfied by conventional mat/roving arrangements, then selected woven products can be utilised to meet the end-use requirements. These products can be used by themselves or in conjunction with the standard mat construction to obtain the required strength and stiffness. The woven glass fibre reinforcements are available in directions which are balanced, mainly longitudinal, mainly transverse or 45° ply construction. As these materials are more costly than those mentioned above, the composites which utilise them are more expensive than the standard construction pultrusions.

2.5.5. Other Reinforcing Fibres
Graphite (carbon) and aramid fibres are also used in the pultrusion technique in the form of roving and woven fabric. These fibres can be used separately or in conjunction with glass fibres and add considerable stiffness to the composite. As these fibre materials are very expensive the optimum cost performance must be borne in mind when designing with them.

3. THE PROCESSING TECHNIQUES

The processing techniques for producing pultrusion sections can be divided into three categories. These are:

(a) the horizontal pultrusion process;
(b) the vertical pultrusion process;
(c) the lamination process with film tapes.

Each of these processes has particular characteristics which make it more suitable for certain areas of application.

In all the above types of pultrusion processes the production procedure can be divided into three stages; these are:

(a) the insertion of the fibre and matrix and the impregnation of the component parts;
(b) the shaping and curing of the composite;
(c) the pulling of the components through the dies and the cut-off of the composite.

The first stage of manufacture is an important one as the final composite depends upon the care with which the reinforcement material is fed into the system and is impregnated. From an economic viewpoint, it is necessary to achieve the highest possible operating speed; consequently, the impregnation process must be consistent with this speed. However, pultrusion components usually have a high glass fibre content and, therefore, the penetration of the resin mixture through the glass is difficult to achieve with high operating speeds.

3.1. The Horizontal Pultrusion Process
The horizontal pultrusion equipment for production of simple rod stock existed in the early 1950s. Round, elliptical or square cross-sectional rods, reinforced with continuous fibres of high fibre content

and tubular profiles of small diameter, are generally manufactured by this method. This is because this process was developed first and it is also more convenient to manufacture long sections horizontally. The horizontal pultrusion process can be divided into three types:

(a) the resin impregnation bath process with a steel die;
(b) the resin impregnation bath process with several short dies;
(c) the injection moulding process.

3.1.1. Resin Impregnation Bath Process with Steel Dies
The resin impregnation baths are used in most production lines. In order to facilitate the impregnation of the fibres, the strands are separated by rollers or bars which the fibres pass over and under, thereby achieving a certain milling effect which promotes the impregnation process. Before entering the die, the strands are combined together again. Strands in the form of rovings are generally used and may be combined with mat, cloth or other fibre arrangements, or the latter fibres may be used without rovings but care must be taken to ensure that sufficient fibre material in the longitudinal direction (the pull direction) is available to accommodate the pulling force.

The wetting speed of the fibres depends on their pre-treatment and on the resin mixture. Consequently, the type of sizing agent used and the type of binder play an important role, as does the length of wetting time. The length of the resin impregnation bath is limited only by practical considerations and the pot life of the resin mixture.

The shaping and the curing of the composite take place primarily in the steel dies, where the lengths vary from 500–2000 mm. The dies are generally heated electrically or they may be heated indirectly with oil as the heat carrier. The gelling and the curing of the composites takes place within the die.

The distance, the maximum temperature and the length of the gelling zone from the entrance to the die are dependent upon:

(a) the reactivity and the amount of resin;
(b) the peroxides used;
(c) the pulling speed;
(d) the thickness of the composite profile walls.

Figure 1 is a photograph of a horizontal pultrusion machine manufactured by Pultrex Ltd, Clacton, Essex, UK. Figure 2 is a typical diagrammatic view of a Pultrex machine which is a free

FIG. 1. Photograph of horizontal pultrusion machine. (Reproduced by kind permission of Pultrex Ltd, Clacton, Essex, UK.)

standing unit made up of seven units. These are the material supply unit, the infeed and resin unit and five modular sections. The five sections are made of heavy duty weldments and are joined together to give the rigidity and accuracy of alignment required in the pultrusion process.

The infeed and die are mounted in the first section. The bed has

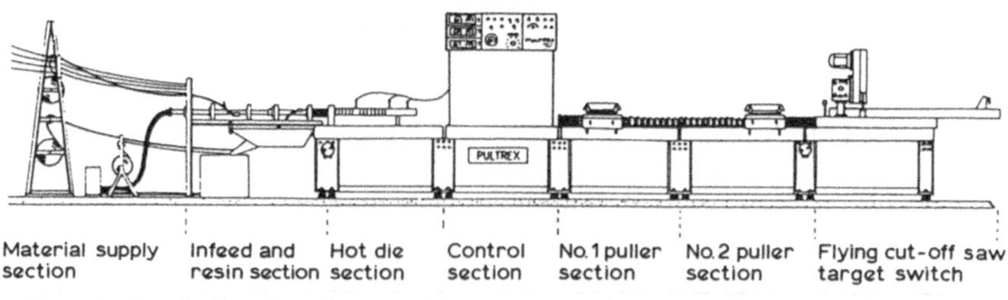

FIG. 2. Typical pultrusion machine section. (Reproduced by kind permission of Pultrex Ltd, Clacton, Essex, UK.)

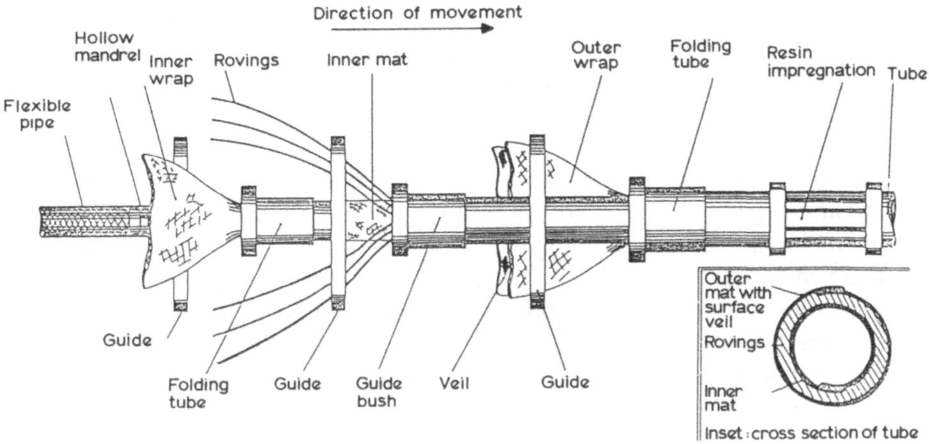

FIG. 3. Infeed of typical pultrusion machine. (Reproduced by kind permission of Pultrex Ltd, Clacton, Essex, UK.)

accurately machined ways so that tooling can be interchanged and positioning and alignment can easily be repeated. The second bed section is similar in construction and provides the rigid reaction link from the puller to the die. The control panel is mounted above and behind this section. This machine is fully described in Ref. 1.

The insert to Fig. 3 shows a typical cross-section of a pultruded tube. The main reinforcements shown are rovings with a randomly orientated fibre mat positioned on the outside and inside surfaces of the tube. This latter reinforcement is overlapped on itself at one side of the tube section on the outer surface and on the other side of the tube on the inner surface. This overlap arrangement of mat reinforcement gives a better result for the hoop strength than the method using overlaps on one side of the tube only.

Figure 3 shows diagrammatically the infeed system of a Pultrex machine: the inner mat is formed first, then the rovings and finally the outer mat and veil. The resin impregnation in this system is undertaken by pumping the component through the mandrel and out through the reinforcements. The excess flows back to a holding tank for re-circulating.

3.1.2. Resin Impregnation Bath Process with Several Short Dies

In this process the resin impregnation technique is the same as that described for the resin impregnation bath process with steel dies, but in place of one steel die, several short, heated dies arranged one behind the other are used for shaping and curing. The composite profile passes between the dies through heated channels.

3.1.3. The Resin Injection Moulding Process

The resin impregnation of the fibre takes place under pressure within the die segment. There are limitations to this procedure in that the fibres in the die are very tightly compressed and penetration of the resin in large wall thickness composites requires high pressures and long time periods. In addition, experience in the design and sizing of the injection channels is highly desirable. A further problem is that the tightly packed fibres can act as a filter and may partially filter out additives in the resin.

3.1.4. The Lamination Process with Film Tapes

In this process fibres, generally mat or cut rovings, are drawn into the required amount of resin which is regulated by means of a doctor blade. The laminate is sandwiched between the two layers of film tape which are sealed at their edges. The tape is the instrument of transportation through the machine. Either the steel die is used or several shaping dies made from a heat resistant wood or laminate plastic arranged one behind the other in a heated air channel.

3.2. The Vertical Pultrusion Process

In this process the pultrusion machine is arranged vertically, and the impregnation procedure can be divided into the three types as described for the horizontal pultrusion process. In the impregnation bath procedures, the baths are located above the entrance end of the die.

3.3. New Processes in Pultrusion

New pultrusion techniques will be discussed in this section. The industrial emphasis is being directed towards technical innovation and the use of high technology materials which have the potential for opening up markets previously unavailable to the pultrusion process.

Four subject areas will be discussed:

(a) the variable cross-section pultrusion die;
(b) the pre-form/post-form technique;
(c) the thermoplastic matrix pultrusion section;
(d) the large and powerful pultrusion machine.

3.3.1. Variable Cross-section Pultrusion Dies

It is possible to produce variable cross-section pultrusion composites by using variable cross-section dies. The production of these sections is dependent upon an articulating die. Richard, Plumer and Driscoll[2] have identified the key technical areas to which attention must be addressed as:

(a) the specific kinematic configuration of the die (i.e. the way the shape changes with time);
(b) the control of the die articulation;
(c) the restriction on the die surface (i.e. continuous or discontinuous).

The technology for producing variable cross-section pultrusion composites is currently in the development stage. The production machines mainly relate to one-off special systems.

3.3.2. The Pre-form/Post-form Technique

The pre-form/post-form techniques make it possible to produce products exhibiting combinations of curve and twist, previously considered unattainable. The initial applications are in the aerospace field but the potential is industry wide.

The technique involves a combination of prepreg roving and cloth, together with dry roving or tow, which is fed through two short heated dies to compact the laminate. The tackiness of the prepreg bonds the dry roving to it and produces a composite which can be handled. The resin, however, is not actually cured at this stage.

The resulting 'pultruded pre-form' can be put into cold storage for final moulding at a later date. When it is needed it can then be brought to room temperature and placed in position in the mould. If it is to be moulded to another part, this latter is also placed in the mould. A normal autoclave moulding procedure follows in which the pre-form and part are cured.

3.3.3. The Thermoplastic Matrix Pultrusion Section

The thermoplastic matrix pultrusion uses high technology matrices which have high softening temperatures and high physical properties with relatively low melt viscosities. Compared with thermosetting polymers, thermoplastics matrices will greatly improve the toughness of the composite. Much of the initial work has been undertaken using standard prepregs such as the solvent applied polysulphone, and poly(ether sulphone) polymers in combination with glass, Kevlar or carbon roving tow, tape or cloth reinforcements. These component parts are fed through short heat and chill dies to be pultruded into profiles.

It is becoming difficult for manufacturers to utilise the solvent application methods because of growing government legislation against work and environment exposure to these chemicals. Adequate solvent has to be retained in the material to enable it to be sufficiently pliable for machine and manual handling but the excess has to be drawn off and this operation considerably increases the processing time.

3.3.4. The Large and Powerful Pultrusion Machines

Recently the design of large profile pultrusion equipment has been required in certain industrial areas, particularly by Japanese firms. These requirements are primarily in large underground and office/industrial buildings; also for electrical wire conduits and cable trays. To enable these sizes of components to be made, very large cross-sectional capacity, and therefore powerful, pultrusion machines are required.

The available pulling envelope on some of the largest machines are of the order of 810 mm wide × 180 mm high. These machines would require a pulling capacity in the range of 140 kN and a flying cut-off saw of diameter 625 mm to produce a maximum cutting depth of 250 mm.

4. RESIN CURING TECHNIQUES

Pultrusion machines use one of the following curing methods:

(a) thermal curing with conductive heat in metallic dies;
(b) dielectric curing with radio frequency (RF) energy in RF transparent dies.

Steel dies for thermal curing can be manufactured with the use of conventional machine tools even for complex shapes. Service life of steel dies, if properly chromed, is almost unlimited. The disadvantage of the steel die is that it is not transparent to dielectric energy and it very effectively shields the fibre/matrix material from any heating effect.

Dielectric curing is fast but the choice of materials for the manufacture of the die is very limited. Steel, as stated above, cannot be used. Glass filled Teflon is the material generally used but it is difficult to machine complex shapes from it and it tends to wear out more rapidly than steel, thus making the tolerance control difficult to achieve.

The dielectric heating in RF transparent dies with radio frequency energy is instantaneous throughout the fibre/matrix material regardless of the cross-sectional size. Consequently, the curing speed is much faster than conventional methods.

On the other hand, thermal curing with conductive heating of the steel die using hot oil or electric strip heaters is quite slow because the material is a relatively poor conductor and consequently, the rate of heat injection to provide uniform cure is necessarily limited. All the thermal energy which is required to bring the fibre/matrix composite to the curing temperature must be applied by the die to the outside of the profile. The progression of heat into the centre of the composite, a necessary addition for resin polymerisation, is quite slow compared with the dielectric heating. It is necessary to heat a general purpose polyester resin to about 80°C to initiate gelation, which occurs inside the steel die.

5. MECHANICAL PROPERTIES

As pultrusion composites are composed of several different component materials, they can be designed with properties to meet specific requirements. Varying amounts of fibre reinforcement can be positioned in the composite to produce a wide range of strengths and stiffness properties.

5.1. Fibre Reinforcements

Typical GRP pultruded composites will contain between 45 and 55% of glass by weight. Although it is possible to increase this percentage, the required pull to draw the material through the die increases significantly above a fibre content of about 60%. Generally the fibre array will contain unidirectionally aligned rovings and continuous strand mat and the percentage of these two elements will depend upon the stiffness and strength in the longitudinal and hoop directions.

Pultrusion composites consisting of hybrid reinforcement of glass and carbon can be produced. The higher modulus carbon component

TABLE 1
TYPICAL PROPERTIES OF THE MAJOR FIBRES USED IN THE PULTRUSION PROCESS

Property	*E-Glass*	*S-Glass*	*Kevlar*	*Carbon*
Specific weight	2·60	2·49	1·47	1·72
Tensile strength ($MN\,m^{-2}$)	3447·5	4585·2	2964·9	1896·1
Tensile modulus ($GN\,m^{-2}$)	72·4	86·9	131·0	379·2
Elongation to break (%)	4·8	5·4	2·3	0·5

would normally be positioned in the longitudinal direction of the composite, but a unidirectional carbon arrangement in conjunction with the glass fibre could be used to provide increased stiffness in the hoop direction.

The properties of the fibres that are generally utilised in the pultrusion process (viz. E-glass, S-glass, Aramid and carbon), are given in Table 1. These fibres are also available in a variety of woven and knitted goods which add further to the range of strength and stiffness properties of the final composite.

5.2. Resin Matrices

5.2.1. Polyester Resin

Polyesters are the primary resins used in the pultrusion technique; the reasons for this have been given in Section 2.4. The polyesters cover a wide range, all with distinctive characteristics but the capabilities and functions of these various resins can only be determined by end-use applications. Typical mechanical properties are given in Table 2.

5.2.2. Vinylester

The vinylester resins are primarily used to improve chemical resistance. Because of their chemistry they have a better coupling effect than the polyesters, and exhibit a 15–20% improvement in the strength of the composite. Typical mechanical properties are given in Table 2.

TABLE 2
TYPICAL MECHANICAL PROPERTIES FOR POLYMERS USED IN THE PULTRUSION TECHNIQUE

Property	*Polyester*	*Vinylester*	*Epoxy*
Specific gravity	1·13	1·12	1·28
Tensile strength ($MN\,m^{-2}$)	77·2	81·4	75·8
Tensile modulus ($GN\,m^{-2}$)	3·3	3·4	3·3
% Elongation	4·2	4·5	6·3
Flexural strength ($MN\,m^{-2}$)	122·0	134·0	115·0
Flexural modulus ($GN\,m^{-2}$)	3·2	3·1	3·3
Heat distortion (°C)	77	99	166
Barcol Hardness	40	35	50

5.2.3. Epoxy Resins

Epoxy resins will enhance the mechanical properties of a composite compared with those of polyester; this enhancement can be of the order of 25%. However, a comparison of the mechanical properties for the two polymers (Table 2) shows that there is little difference between them, with the exception of the heat distortion values which vary by 100%.

5.3. Fillers

Various fillers can be used in the pultrusion process. Aluminium silicate (kaolin clay) is used to improve chemical resistance, opacity, surface finish and insulation properties. Calcium carbonates offer improved surface whiteness, opacity and general lowering of costs. Alumina trihydrate and antimony trioxide are used for fire retardancy (Section 6.3). The latter filler can also be used in conjunction with clays for improved insulation properties. Pigments can be used to provide a required colouring to the pultrusion composite (Section 6.4).

5.4. Advances in the Range of Resins

Other resins which could be used in the future for the pultrusion technique are the recently developed range of urethane methacrylate resins introduced by ICI Mond Division, UK. The company claims that the resins have very high reactivity compared with unsaturated polyester resins, low viscosity and, in combination with alumina trihydrate, they give excellent low smoke fire retardance. The chemical resistance also is good, particularly to alkalis and hydrocarbons. Currently, however, the costs are some 20–30% higher than the unsaturated polyester resins.

6. INSERVICE PROPERTIES

Pultruded composites may be used in a variety of applications (see Section 7). The following sections discuss inservice properties which may be required in selected situations.

6.1. Temperature

Glass fibre/polyester matrix pultruded composites undergo some loss of strength from continuous exposure to high temperatures. If a high strength retention is required after a continuous exposure to high temperature, it is possible to select a high performance resin system designed specifically for elevated temperatures. It is advisable always to ask a resin manufacturer for advice on the correct resin system to be used for any particular application.

6.2. The Weathering of GRP Pultrusion

Glass fibre/polyester matrix composites undergo some deterioration and change in appearance during exposure to weathering. In order to minimise this effect with pultruded profiles a number of options are available.

6.2.1. UV Stabilisers

Sunlight, and particularly its ultraviolet component, is the most significant single factor responsible for the breakdown of polymer composites and for producing changes in colour. Other factors influencing this degradation are temperature, humidity and the permeability of the exposed surface; the latter controls the penetration of oxygen and the escape of the products of degradation.

UV stabilisers are usually added to the resin system at the time of manufacture of the composite. They will retard the effect of weathering but eventually the surface resin will deteriorate and a condition called 'fibre blooming' will occur on the surface of the composite with subsequent slight deterioration in physical properties.

6.2.2. Surface Veils
To enhance the weathering characteristics, the UV resistance and the corrosion resistance of the pultruded composite, a polymeric surface veil can be applied to the section during processing. The veil creates a resin-rich surface about 0·035 mm thick without detracting from the section's performance. The surface veil coats the material with the base resin on-line and eliminates the need for secondary coating operations on areas other than the end cut or fabricated holes.

6.2.3. Urethane Coatings
If the pultruded section is to be exposed to the weather for long periods and in addition is required to have a long life, urethane coatings can be applied. This process is performed as a secondary operation with a typical coverage of 0·0025–0·003 mm.

These coatings protect the composites from UV degradation and are generally recommended for parts which will be exposed to a variety of weathering conditions. They are also employed where uniformity of colour is essential or a variety of colour is required in a low volume application.

6.3. Fire Behaviour
One of the chief concerns of the designer using polymers in the construction industry is the problem associated with fire. In the context of pultrusions it is most important for the reader to remember that glass fibre reinforced plastic (GRP) is a composite made from two materials, one of which is organic, the other inorganic. The glass fibre, which has the greater volume fraction, does not burn, whereas the polymer is combustible. Although the polyester is combustible compared with a number of other plastics, its fire behaviour is good. GRP is not easy to ignite but in certain grades it has a fairly high rate of surface spread of flame and the burning is generally accompanied by considerable smoke. The rate of burning does not accelerate greatly during the progress of the fire and at its completion a blackened mass of glass fibre reinforcement remains.

To classify the combustibility of materials, small scale simulated tests are generally undertaken; currently the standard fire tests concentrate on:

(i) combustibility;
(ii) surface spread of flame;
(iii) fire penetration;
(iv) fire propagation.

In the United Kingdom, items (ii), (iii) and (iv) are covered by BS 476.[3] The tests are severe, requiring specialised equipment and skilled handling and are usually undertaken in testing houses specialising in such work.

The fire penetration tests in the United Kingdom (BS 476 Part 3) apply specifically to roofing.

6.4. Pigments and Their Effects on Pultrusion Sections

To enable colouring of the pultruded section to be undertaken, a pigment can be added to the polymers at the manufacturing stage, but this additive is not an inert material. Pigments can react with the polymer to reduce or alter the composite's properties. Little research work has been done to date on the effects of adding pigments to pultrusions. A possible reason for this is the large amount of work and time involved in trying various mixes, especially when cross-contamination between them must be eliminated. Anderson and Riddell[4] have investigated the addition of titanium dioxide to pultrusions (to provide the section with a white colouring) and have shown that this additive adversely affects properties of pultrusions, especially the lengthwise and crosswise notched Izod impact strength. On the other hand, the addition of zinc sulphate causes little or no reduction in the properties of the composite.

7. THE END USE OF PULTRUSION SECTIONS

During the last two decades great technological advances have been made in the manufacturing process and the uses of polymer composites which have led to their acceptance as high performance structural materials. The high strength-to-weight characteristics of the composites was one of the main factors that attracted the aerospace industry to them and thus in turn encouraged significant research and development to enable a wide range of highly efficient composites to be produced.[5] The construction industry has benefited from the development work undertaken in the aerospace industry although acceptance of the material by the former industry has been much slower than by the latter. Consequently the field of composite materials in civil/structural applications is in its infancy and only now has progress commenced.

7.1. Semi-structural Applications for Pultrusions

The following sections provide a short review of the possible markets for pultruded composites.

7.1.1. Consumer/Recreation Market

One of the largest current openings for pultruded profiles is the consumer/recreation one, and the principal product that started the market penetration of fibre/matrix pultrusions was the fishing rod. The market, however, is not a sophisticated one in that most of the products are the result of the need for strength as well as some flexibility. The flexural modulus of the fibre/matrix pultrusions makes them ideal for use in many applications, particularly as they do not fatigue as readily as metals.

Some applications are as follows:

(i) fishing rods,
(ii) radar antennae,
(iii) archery bows and arrows,
(iv) sail battens,
(v) tent poles,
(vi) fence posts,
(vii) ski poles,
(viii) playground equipment,
(ix) hockey sticks.

Undoubtedly the outlet will expand but probably will lack the stability of other markets; to make progress in this field several high volume applications are needed.

7.1.2. Electrical Equipment Market

The electrical market can be broken down into three areas;

(i) electrical OEM;
(ii) electrical and communication utility;
(iii) electrical maintenance where the desired properties are strength and insulation.

The OEM market uses the fibre/matrix pultrusion sections for transformer spacing sticks, motor winding pins and motor topsticks.

Significant new developments have occurred in the concept of the manufacture of fibre/matrix printed circuit boards. It is now possible to pultrude a sheet profile continuously, with copper added, prior to the raw materials entering the die; this reduces significantly the cost per square metre of the composite.

Fibre/matrix pultrusions are extensively used in the electrical and maintenance market where the composite is the dielectric between the electrical conductor and the person; this is the largest sector of the electrical market.

The fibre/matrix composite maintains the safety factor in the following applications:

(i) pultruded fibre/matrix ladders;
(ii) tool handles;
(iii) third rail coverboard.

A third rail coverboard is a protective component of the conductor rail for mass-transit railways in the USA and other countries (not the UK).

7.1.3. Corrosion Resistant Composite Material

The non-corrosive property of fibre reinforced polyester composites is utilised in many markets where corrosion resistance is the primary characteristic required. The major markets are;

(i) chemical processing;
(ii) water and waste water treatment;
(iii) agricultural products.

Whilst corrosion resistance is the important property required for the composite, the pultruded fibre reinforced polyester profiles are frequently used because of their structural integrity. For this reason, many of the pultruded composite products could be considered for the construction market but are classified only as corrosion resistant materials.

In the part of the construction market concerned with the effects of corrosion, the fibre reinforced polymer composites in general, and the pultruded composites in particular, are used in applications such as:

(i) gratings,
(ii) stairs,
(iii) handrails,
(iv) ladder cages,
(v) bridges and platforms,
(vi) piping supports,
(vii) electrical cable trays,
(viii) structural supports.

The paper industry as well as the water and sewage disposal authorities are beginning to replace metal and redwood in applications such as:

(i) weirs,
(ii) baffles,
(iii) gutters,
(iv) grate guides,
(v) sludge paddles.

7.1.4. Transportation Markets

The transportation market includes applications that fall into the automotive, material handling and containerised freight areas.

Fuel economy in the automotive market has created the impetus to search for a lightweight, structural material that will enable car manufacturers to meet legislated mileage ratings. This has led manufacturers to consider pultruded fibre reinforced polyester as a material that is lighter in weight than the conventional one. It has been suggested that some potential automotive applications lend themselves to pultruded carbon fibre composites. These are:

(i) bumper beams,
(ii) frames and cross members,
(iii) drive shafts,
(iv) leaf springs.

Considerable weight savings over the traditional materials are possible with carbon fibres. Assuming that the premiums for these composites can be paid for by the consumer, carbon fibres in pultruded profiles could be the solution to many of the fuel economy problems of the motor manufacturers.

7.1.5. Construction Market

As pultruded composites are mainly unfamiliar to the structural engineer, their strength and rigidity properties are not realised and thus many structural applications that would be possible are never considered. However, this market may soon come into its own owing to the need for more energy efficient structures. The thermal conductivity of GRP is much lower than that for steel and aluminium. Consequently this will make composites a desirable material for many structural applications in the building industry.

7.2. Structural Applications for Pultrusions

Although the pultrusion technique for producing GRP and CFRP (carbon fibre reinforced plastic) has been in existence for more than two decades, it is only now that the product has reached the point where its importance as a structural engineering material is becoming more widely realised.

7.2.1. Sectional Structural Components

It is clear that one area in which the product has potential is that of sectional structural components which could be in the form of plane and double layer skeletal structural systems. In the latter systems the members are mainly under axial forces although they could also be subjected to bending moments.

Legislation relating to the compulsory installation of the side under-run protection equipment on most new vehicles was introduced into the United Kingdom in May 1984. The Sidesafe System was developed to satisfy this legislation and is shown in Fig. 4. It utilises pultruded sections in the form of a frame which is attached to the body of the vehicle. In comparison with steel systems, Sidesafe guards

FIG. 4. Photograph of Sidesafe System under test. (Reproduced by kind permission of Maunsell Structural Plastics Ltd, London, UK.)

FIG. 5. Photograph of footbridge manufactured from pultruded tube sections. (Reproduced by kind permission of BTR-Permali Ltd, Gloucester, UK.)

combine exceptional strength with very low weight, they have enhanced impact absorption, do not corrode and are virtually maintenance free.

Figure 5 shows a footbridge manufactured from pultruded sections. The plane frame sides are 25 mm diameter, 2 mm wall thickness pultruded tubes with pultruded channel sections for the walkway. The bridge is 4·3 m long, 1·2 m wide and deep and can carry a distribution load of about 500 kg on its 0·25 m wide walkway. The all-up weight is 34·5 kg and it can be lifted readily by one person and carried around easily by two. All structural fittings to the rods and tubes are crimp-bonded. (See Section 7.2.2.1.)

7.2.2. Jointing of Sections

The main difficulty in the manufacture of space structures is in the jointing of members; this difficulty is probably the principal reason why skeletal systems in fibre/matrix composites have made little impact in the structural engineering fielo. There are three options open to designers when considering joints. These are to form:

(a) a bolted joint;
(b) a bonded joint;
(c) a combination of a bolted and bonded joint.

Steel[6] has shown that bonded joints are generally structurally more efficient on a weight basis than are mechanical joints when constructing prototype space frames using pultruded CFRP tubes. Hollaway and Ishakian[7] reached a similar conclusion when jointing CFRP rods. However, both these references have indicated that the fabrication of the skeletal structure presents some difficulty.

7.2.2.1. The crimped and bonded aluminium method. A method of jointing is to utilise crimped and bonded aluminium end fittings.[8,9] Such a joint has been proposed by Green and Phillips and has been successfully applied to a variety of fibre/matrix composites in tubular form. This technique involves crimping an oversize aluminium sleeve on to the composite tubular section and, in so doing, forming a uniform thin, epoxide glue line. The sleeve may then be threaded to allow the connection of members via a nodal joint or alternatively the region outside the crimped overlap may be flattened to accept a bolted type of connector. This technique provides a lightweight fitting that requires only stock material and avoids the need for costly precision machinery.

A tool to crimp the aluminium sleeve to the pultruded section can be manufactured from mild steel, and a typical one is shown in Fig. 6. By way of illustration in this and subsequent sections a pultruded tube, of fibre/matrix percentage ratio by weight of 60/40 and having an external diameter of 25 mm and wall thickness of 2 mm, will be used. The radius of the tool for crimping this size of pultruded section, including an aluminium sleeve of wall thickness of 3·175 mm and allowing for a glue line of 0·15 mm thickness, would be 32·05 mm. The tool is designed such that the gap between the two halves is 6·35 mm (i.e. twice the aluminium wall thickness) when the two arcs of the die coincide on the same circle.

Prior to the crimping operation, the aluminium sleeves are solution treated at 530°C for one hour and then cold water quenched. This softens the aluminium and prevents cracking in the regions of high deformation during the crimping process. The aluminium is then exposed to a chromic–sulphuric acid surface etch treatment prior to bonding.

FIG. 6(a). Crimping tool to crimp aluminium sleeve on to pultruded section.

FIG. 6(b). Aluminium sleeve/pultruded tube in position in crimping tool.

Bonding pretreatment for the pultrusion is confined to wet abrasion with silicon carbide papers and wiping with acetone soaked paper wipers until all abrasion debris is removed.

Prior to assembling the crimping tool, adhesive is applied to the bonding surfaces of the pultruded tube and the aluminium sleeve. In order to allow excessive adhesive to be squeezed out, the vertical load is applied slowly. This load is increased to a maximum value (typically a load of 75 kN for an overlap length of 50 mm) when the pultruded tube is just gripped firmly by the crimped sleeve.

When using a viscous adhesive considerable hydrostatic pressure may be generated by rapid loading, and to prevent crushing of the composite section a temporary internal supporting tube is recommended. The completed end grip is then removed from the tool, excess adhesive being wiped off, and the joint is then given the required curing treatment. If the pressure applied during the crimping operation is controlled, the following may be achieved:

(a) reasonably uniform thin bond line around most of the crimped joint circumference;
(b) the transverse crushing of unsupported tubular pultrusions.

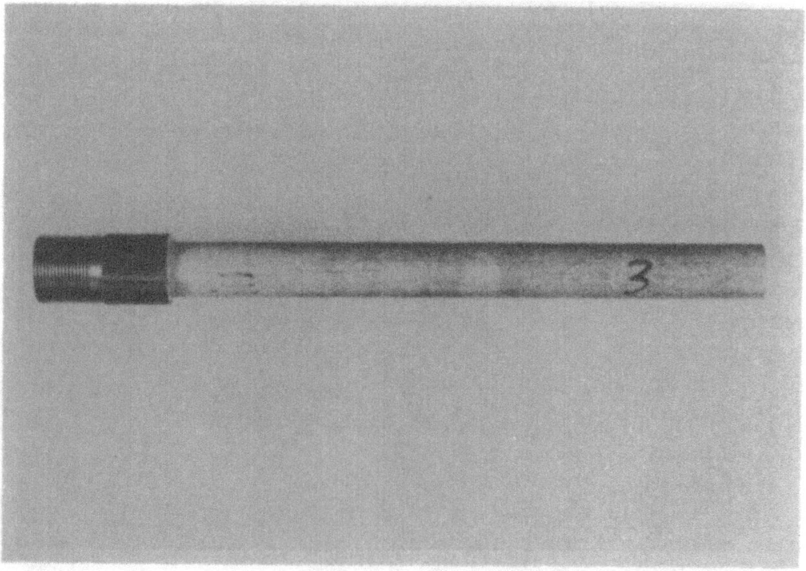

FIG. 7. Aluminium sleeve crimped and bonded on to a pultruded tube.

FIG. 8. Nodal joint to receive eight pultruded members.

The crimped and bonded joints are efficient in terms of the ultimate magnitude of the load that the joint can take and also in terms of weight savings. They are also suited to a skeletal plane frame and are relevant to a double layer frame where the crimped aluminium sleeve would be threaded and screwed into the appropriate nodal joint of the space frame or where the aluminium outside the crimped and bonded region would be flattened and a pin joint connector through the flattened aluminium would be secured to the nodal joint by bolting. Figure 7 shows an aluminium sleeve crimped and bonded onto a pultruded tube; the end of the aluminium sleeve is threaded. Figure 8 shows eight pultruded members which have been screwed into a nodal joint. The latter is made of aluminium block machined to size and shape and provided with the threaded holes in the right location with the required orientation to accommodate the members.

7.2.2.2. The bonded end-cap method. A second method of forming nodal joints to skeletal structural systems manufactured from fibre/matrix pultruded composite members has been developed at the University of Surrey.[10] It is an economic method for forming nodal

FIG. 9. End-cap bonded to a pultruded tube.

joints and readily enables unskilled labour to erect a fibre/matrix skeletal structure. It consists of a GRP 'bonded' end-cap to the pultruded tube and connecting plates. The former is shown in Fig. 9. The end-cap material can be either glass filled nylon or a conventional glass/polyester composite; the end-cap is manufactured by an injection moulded process and the connecting plate is also made from an injection moulded process or from a hot pressed dough moulding compound (DMC).

The end-cap consists of an inner core and an outer sleeve which are joined at the base. The outer surface of the inner core and the inner surface of the outer sleeve have a thread moulded into them. Into the outside surface of the outer sleeve five pairs of tapered ribs are moulded which will mate with slots on the connecting plate when the joint is assembled. The end cap is made in a single injection moulding operation. A cross-section of the cap is shown in Fig. 10(a), and Fig. 10(b) shows the section with a pultruded tube in position.

To form the bond between the tube and end-cap a measured quantity of epoxy adhesive is injected into the space between the inner core and outer sleeve, known as the annulus of the cap; the tube is pushed into the space until it reaches the cap's base. This forces the

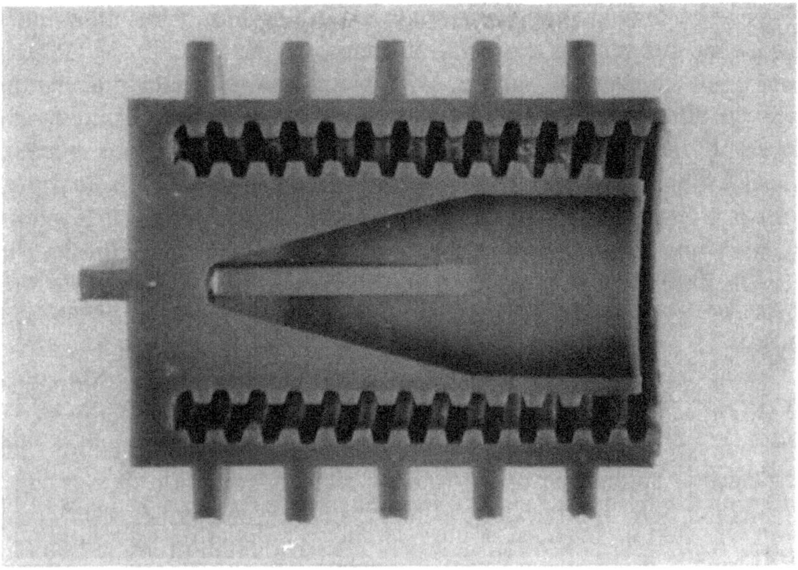

FIG. 10(a). Cross-section of fibre/matrix end-cap.

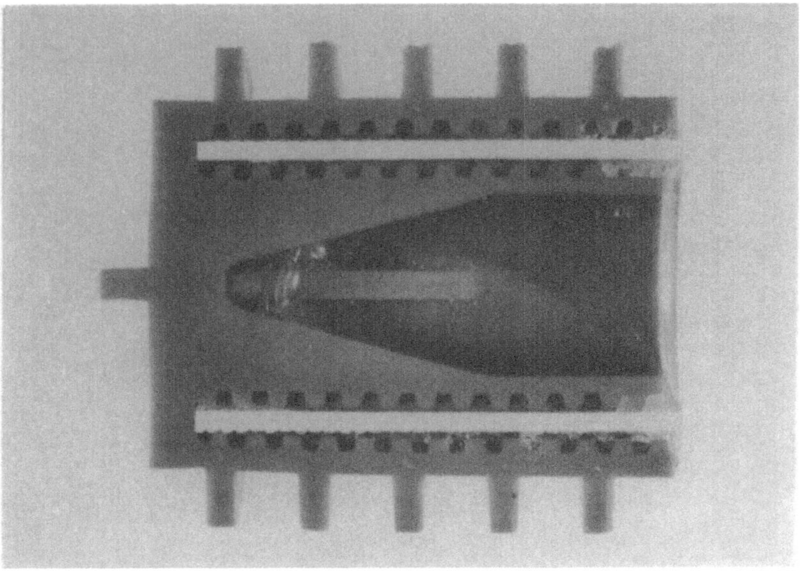

FIG. 10(b). Cross-section of fibre/matrix end-cap with pultruded tube in position.

adhesive to flow around the threads on either side of the tube and to surface at the top of the cap; it then polymerises. If the adhesive cannot be seen at the surface of the annulus it would indicate that either insufficient adhesive has been used or that the pultruded tube, for some reason, has not been inserted into the end-cap correctly. These faults, however, should never occur in practice. Clearly the threads are an essential part of the design of the cap as they control the glue line thickness and ensure quality control. The adhesive runs into the threads and forms a chemical bond with the pultruded GRP but forms a mechanical bond with the glass filled nylon (if this material is used) or a chemical bond if a GRP composite end-cap is used.

The node connecting plates can be injection moulded from glass filled nylon 66 for internal structures not exposed to the elements. For external systems it may be preferable to use other fibre filled polymeric systems because of the possible degradation as a result of ultra violet light and because of the adverse creep characteristics of the nylon 66 when under load. Figure 11 shows a detail of a possible node connecting plate. The thickness of the connecting plates will vary so as

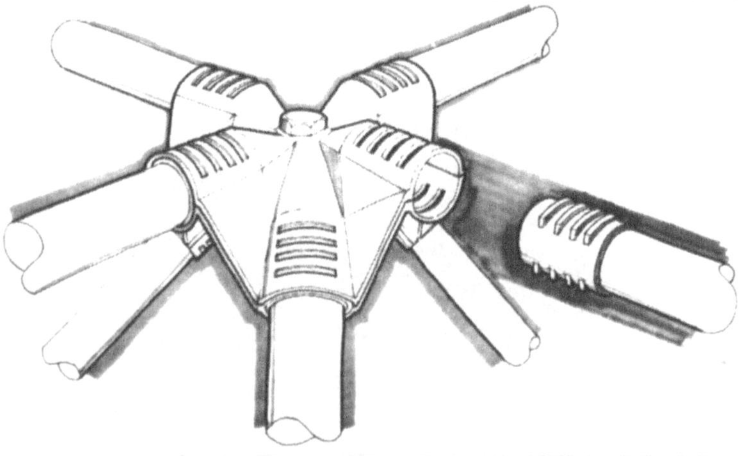

FIG. 11. Detail of a possible node connecting plate for a pultruded fibre/matrix composite structure.

to take account of the differing stress levels within the composite; the greatest thickness of the plates will be located at the slots which receive the tapered ribs of the end-caps and at the bolt holes through which the bolts pass to secure the two plates together. Other areas of the plates will be folded to stiffen them against bending moments.

7.3. Research Being Undertaken into the Use of Pultruded Fibre/Matrix Composites in Structural Skeletal Systems

It would be helpful, at this stage, to discuss some research topics which are currently being undertaken to promote a greater understanding of the structural properties of pultruded sections. The crimped and bonded joint and the GRP end-cap with cover plate have opened up a whole new field in skeletal structural systems which can be manufactured from fibre/matrix composites.

A theoretical linear solution would be required to analyse a fibre/matrix skeletal structural system in the low strain region but a non-linear analysis would be performed in order to study its characteristics to failure as a change in geometry takes place. There will be no discussion here of the necessary steps for the derivation of the two methods which have been covered in Ref. 11.

7.3.1. System Imperfection

In conventional steel space structures the effects of the nodal joints in the stiffness of members is not considered and the actual length of the members is taken as the distance between the centres of the nodes to which the member is joined. If the members are manufactured from a relatively low modulus material the joints and fixings do have an effect on their stiffness and this increase in stiffness should be taken into account when analysing skeletal systems.

It is difficult to produce pultruded sections which are completely straight with the result that if these sections are used as compressive members they will behave as initially curved struts, buckling at a lower load than if they were perfectly straight. A structure manufactured from these members will deform and fail under a smaller applied load than will a perfect structure.

A further imperfection in the structure could result from an initial lack of fit in the system.

The fibre/matrix ratio of the pultruded tube used in skeletal systems is such that the mechanical property characteristics of the composite follow closely those of the fibre, which has a linear stress–strain relationship to failure. Consequently, if the compressive member is placed under a severe load which causes it to deform and which is then removed, the member will return to its initial undeformed state without any apparent damage to the material. This is a considerable advantage over a steel skeletal system where permanent deformation would have resulted from buckling.

7.3.2. Effects of Imperfections

To illustrate the consequences of the above imperfections, four double-layer skeletal structures analysed by numerical and experimental techniques will be considered. The first model was manufactured from pultruded GRP tubes with a fibre/matrix percentage ratio by weight of 60/40. The outside diameter of the tubes was 25 mm and the wall thickness was 2 mm. The node points were made by crimping and bonding aluminium tube on to the GRP tube.

The second model was geometrically similar to the first and was manufactured from GRP pultruded tubes except for the four external corner diagonal members which were made from hybrid composites of glass/carbon fibres and polyester resin with a percentage weight ratio of 30/30/40, respectively; the external diameter and wall thickness of these composite members were the same as those for the GRP ones. The span of models 1 and 2 was 2 m; Fig. 12 shows a plan view of them. Model 3 was manufactured wholly from GRP pultruded tubes to the same specification as above and model 4 was also manufactured from GRP tubes except for the top two centre longitudinal members (members 79–132 and 81–137), which were made from the hybrid. The span of models 3 and 4 was 4 m; Fig. 13 shows a plan view of them. Figure 14 shows a photograph of the experimental model 3 which is typical for all models considered.

To investigate the effects of the imperfections, two node points and two members will be considered in each structure. The top four nodes of models 1 and 2 and the four top centre nodes of models 3 and 4 were loaded. Figures 15 and 16 show the relationship between the load factor, which is a function of the load carried, and the deflection for the analytical and experimental solutions to models 1 and 2 for node points 5 and 55, respectively. The graphs have been drawn for different degrees of imperfections. The two which have been considered are initial curvature compatible with the first buckling mode and given to the members of the structure with variations in maximum deflection of 0·1–1·0 mm, and an eccentricity of 2 mm at the top node

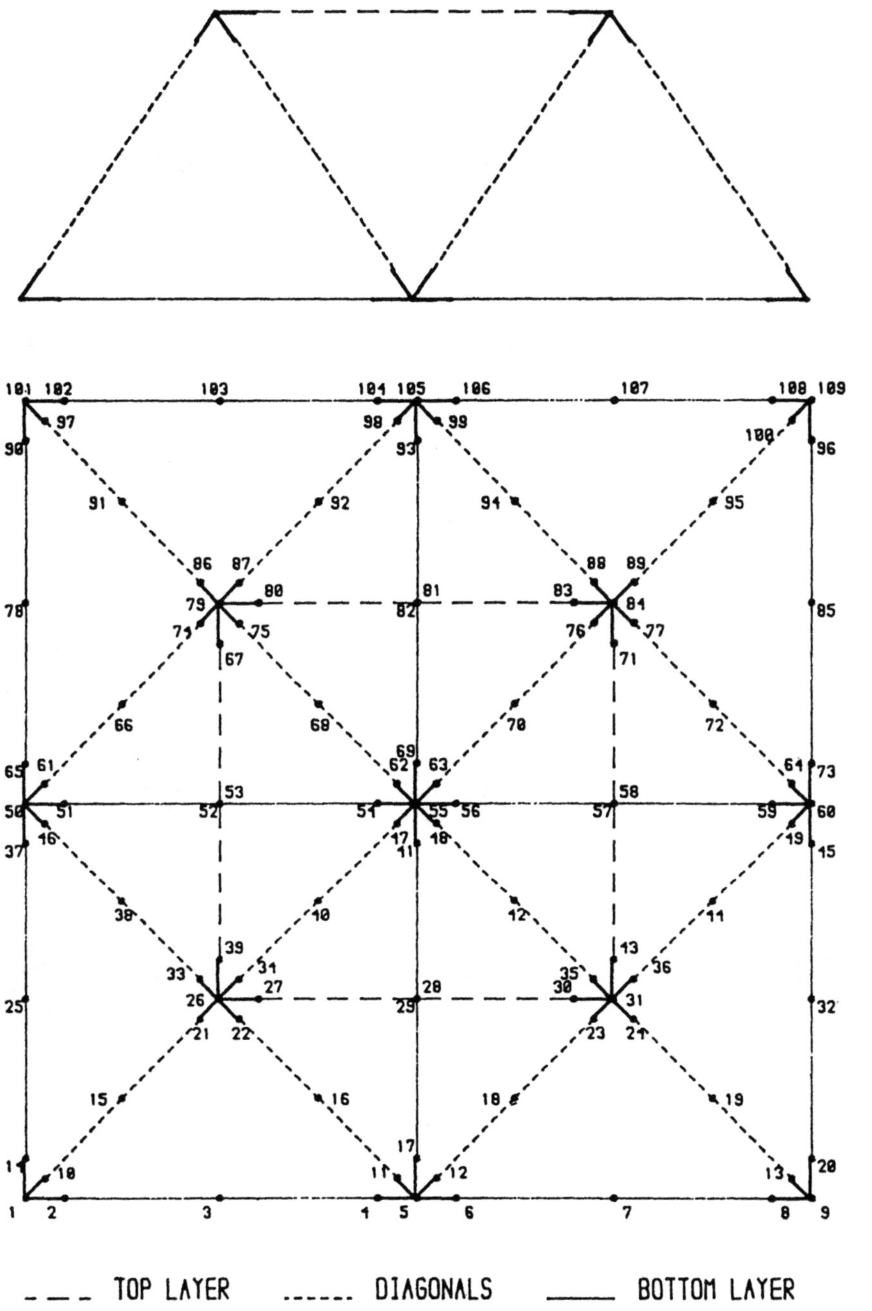

FIG. 12. Plan view, configuration and joint numbering system for models 1 and 2.

Loading
Positions
79, 81, 132, 137.

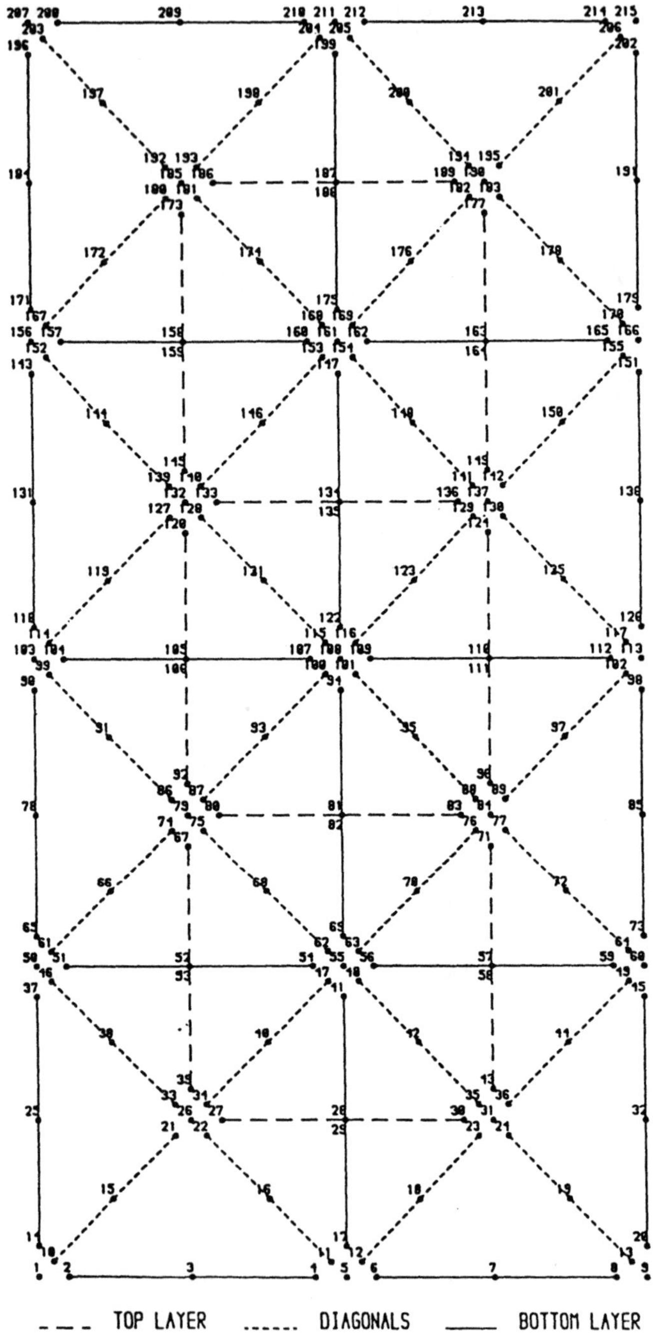

FIG. 13. Plan view, configuration and joint numbering system for models 3 and 4.

FIG. 14. Photograph of experimental model 3.

of the system. It may be seen that the latter imperfections produce slightly different behaviour in the linear analysis. Figure 17 shows a similar relationship to the above for models 3 and 4.

In skeletal structural systems it is well known that, provided external loads are applied through the node points and not directly on to the members, axial forces are the predominant member loadings and buckling moments are generally small. Consequently only axial forces are drawn here. Figures 18 and 19 show relationships between load factor and force in members 9–31 and 26–79, two of the most highly stressed compressive members in model 1, and Figs 20 and 21 show a similar relation for model 2, each plot containing the results of the linear and non-linear analysis as well as the experimental solution.

Relationships between load factor and axial force in members 9–31 and 26–79 for models 3 and 4 are given in Figs 22 and 23; in addition, the same relationships are given for members 55–77, 60–84 and 84–137.

For all cases considered, it may be seen that there is very little difference between the linear and non-linear analytical analysis under the applied loads below 80% of ultimate, and, indeed, above this ultimate value the difference is small. Also, imperfections had little effect on the external forces of the highly stressed members.

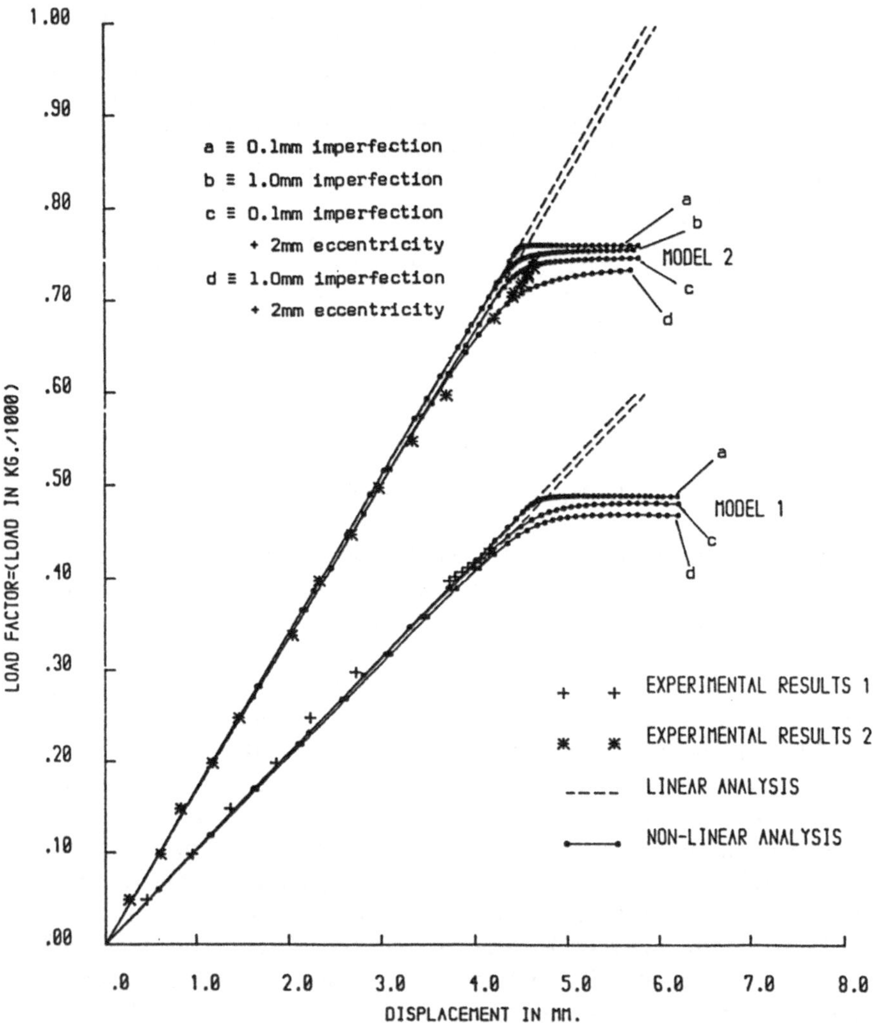

FIG. 15. Load factor–z displacement of joint no. 5 for analytical and experimental models 1 and 2.

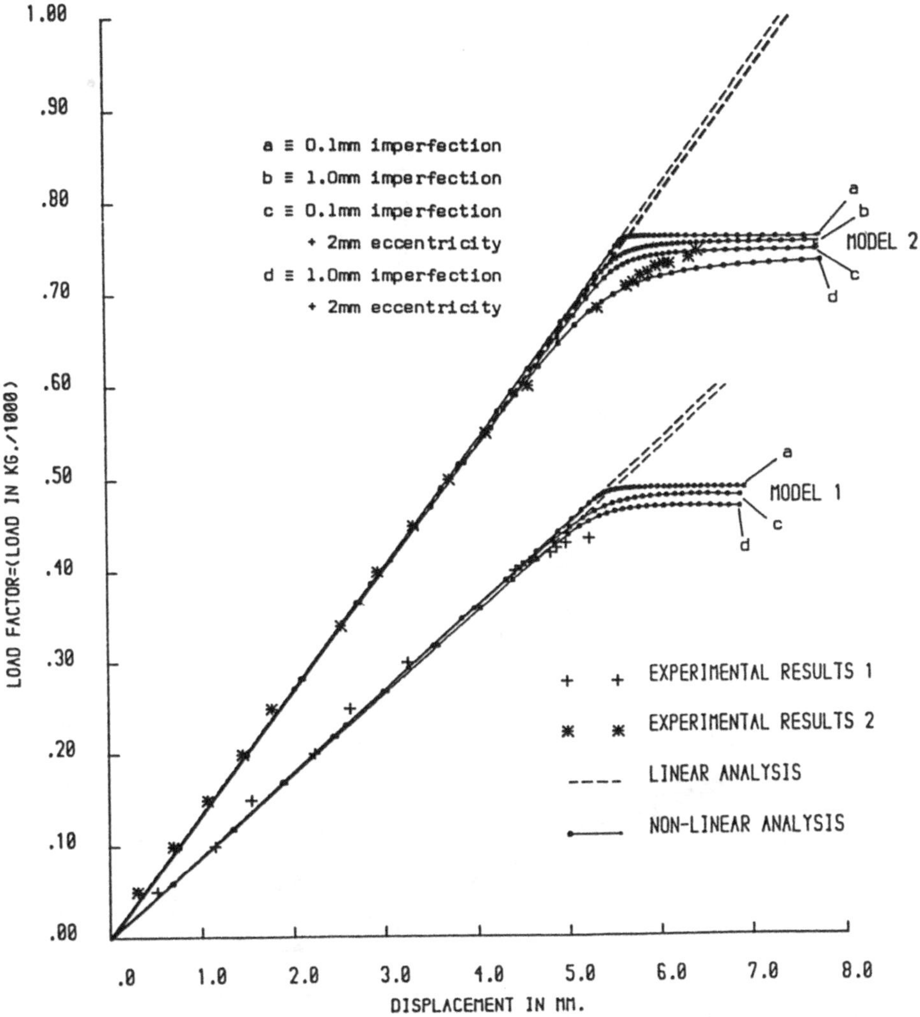

FIG. 16. Load factor–z displacement of joint no. 55 for analytical and experimental models 1 and 2.

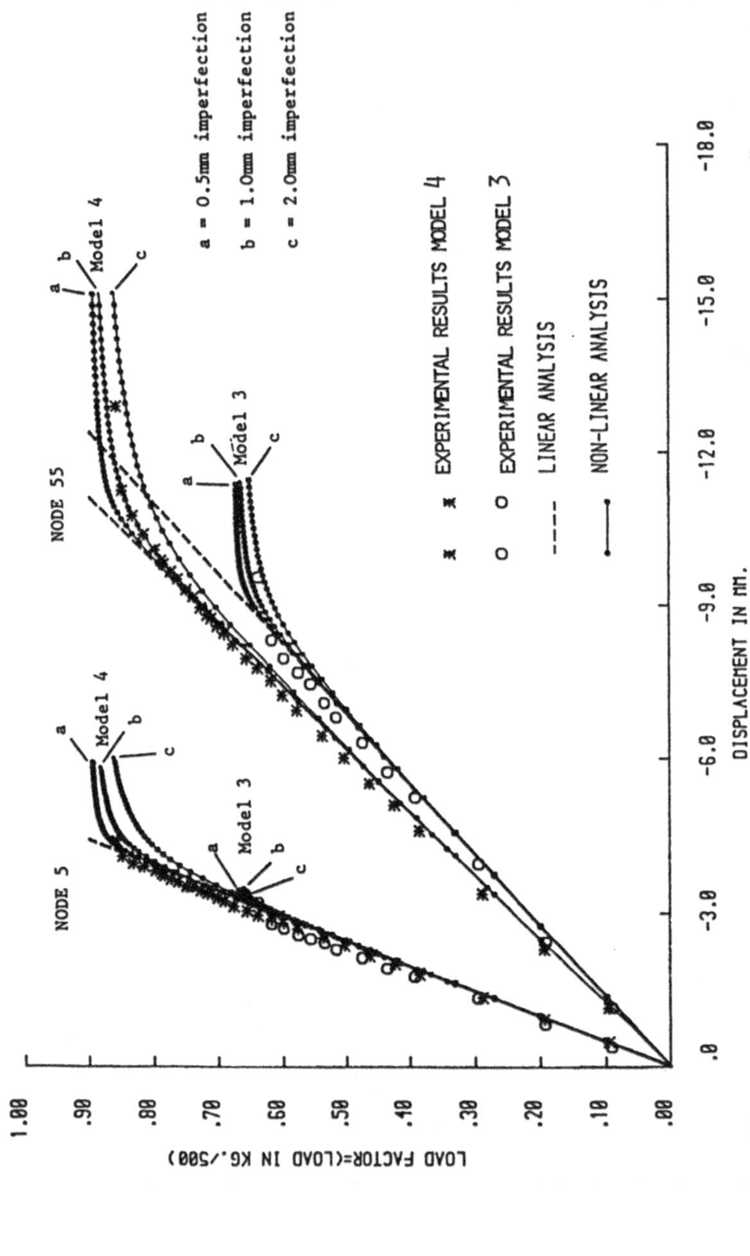

FIG. 17. Load factor–z displacement of joint nos. 5 and 55 for analytical and experimental models 3 and 4.

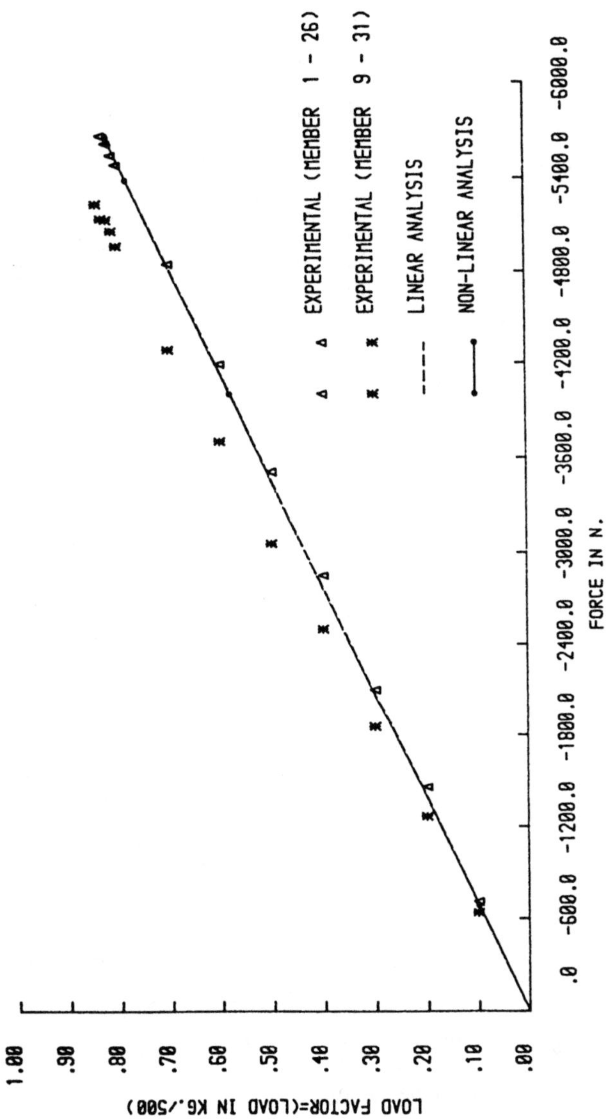

FIG. 18. Load factor–axial force for analytical and experimental model 1: members 9–31 and 1–26.

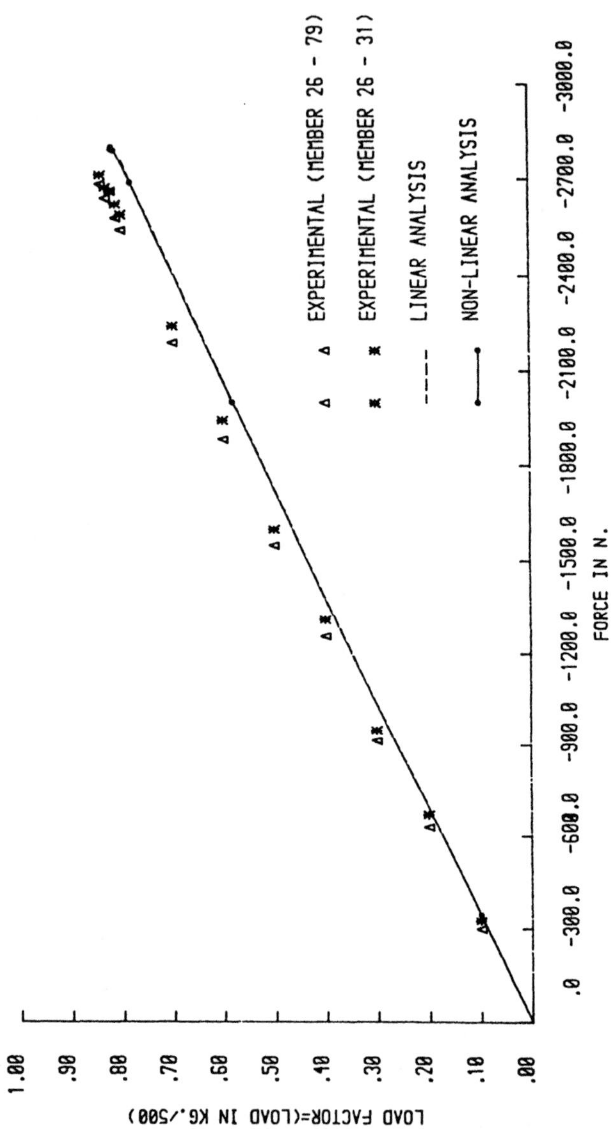

FIG. 19. Load factor–axial force for analytical and experimental model 1: members 26–31 and 26–79.

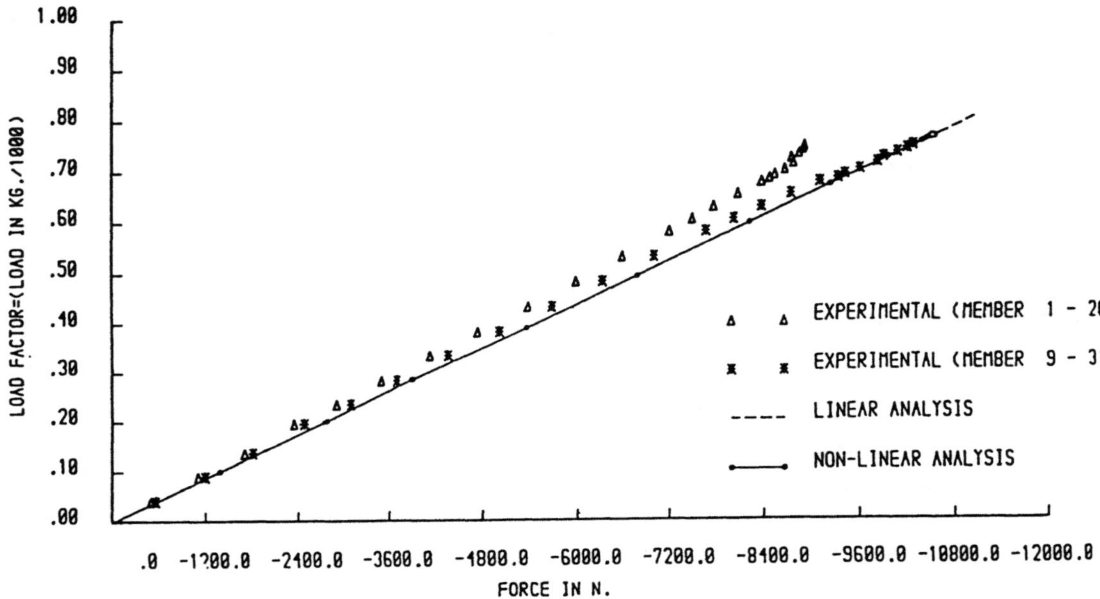

FIG. 20. Load factor–axial force for analytical and experimental model 2: members 9–31 and 1–26.

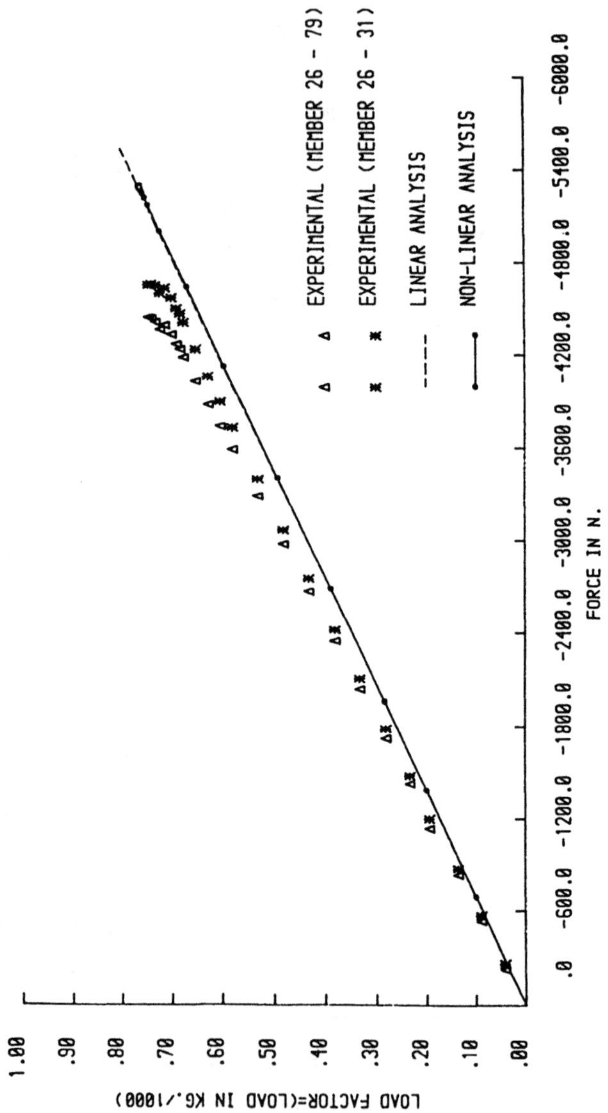

FIG. 21. Load factor–axial force for analytical and experimental model 2: member 26–79.

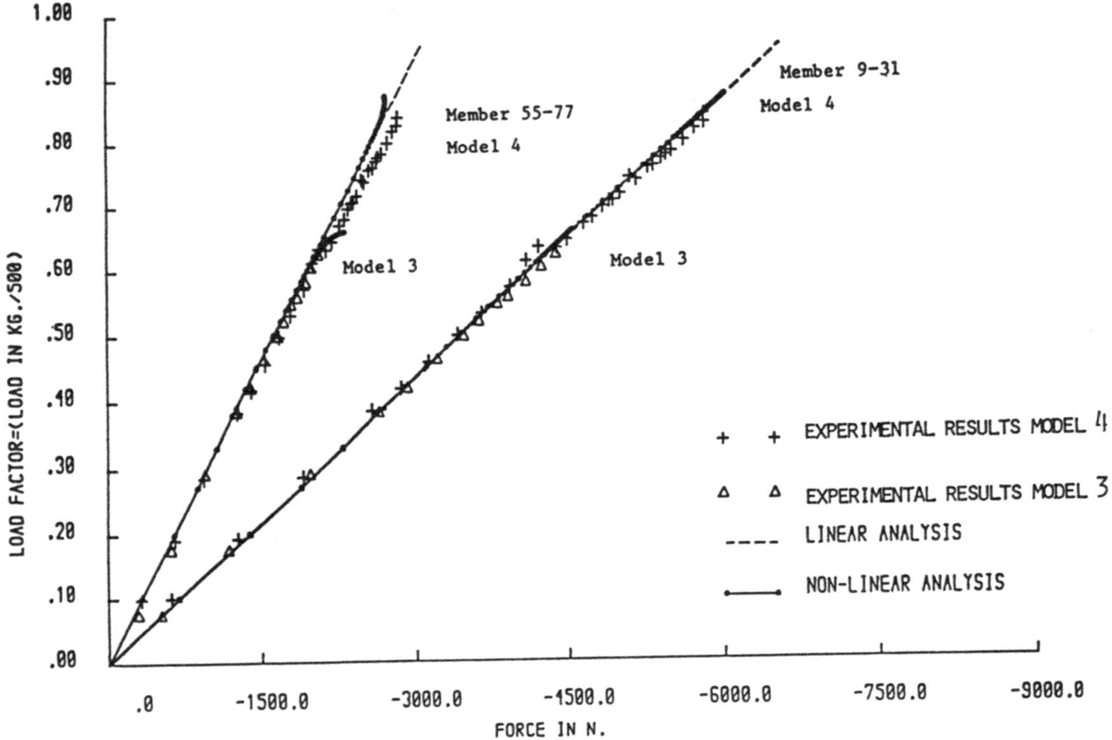

FIG. 22. Load factor–axial force for analytical and experimental models 3 and 4: members 9–31 and 55–77.

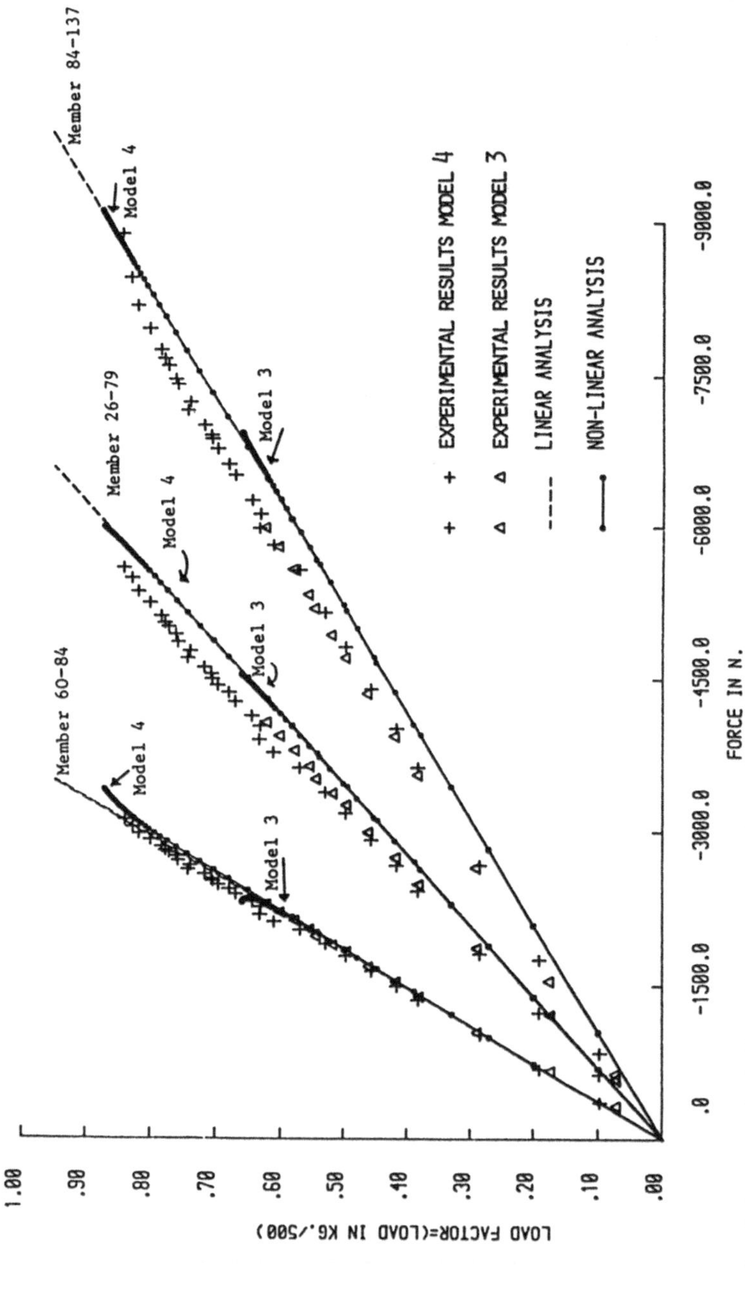

FIG. 23. Load factor–axial force for analytical and experimental models 3 and 4: members 26–29, 60–84 and 84–137.

8. CONCLUSIONS

It is clear that double-layer skeletal systems manufactured by the pultrusion technique using essentially linear fibre/matrix composites are a viable structural form for the construction industry. In addition, the behaviour under load of these systems can be predicted with considerable accuracy from a non-linear structural analysis, and, indeed, on the low strain region from a linear analysis provided that certain imperfections in the material and structural systems are taken into account.

Clearly systems manufactured from glass fibre polyester resin pultruded tubes will have a place in the structural fields of civil engineering in the medium to small span roof systems (up to 12 m × 12 m span). Considerable improvement in the overall structural behaviour can be obtained by replacing critical compressive members with a stiffer hybrid member also manufactured by the pultrusion technique.

The time has now come when all forward looking engineers are beginning to take a serious interest in the structural use of this exciting fibre/matrix composite material.

REFERENCES

1. Shaw Stewart, D., 'Production of FRP tubing by the pultrusion process', *35th Annual Technical Conf. Reinf. Plastics/Composites Institute,* New Orleans, LA, USA, Paper 4-D, February 1980.
2. Richard, R., Plumer, J. and Driscoll, S., 'Development of a variable cross-section pultrusion (VSP) die', *38th Annual Conf. Reinforced Plastics/Composites Institute,* Houston, TX, USA, Paper 6-E, February 1983.
3. 'External fire exposure roof test' BS 476, Part 3, 1975; 'Non-combustibility test for materials', BS 476, Part 4, 1970; 'Ignitability test for materials', BS 476, Part 5, 1968; 'Surface spread of flame tests for materials', BS 476, Part 7, 1971; 'Test methods and criteria for the fire resistance of elements of building construction', BS 476, Part 8, 1972; British Standards Institute, London.
4. Anderson, R. and Riddell, R., 'Effects of pigments on pultrusion physical properties and performance', *38th Annual Conf. Reinf. Plastics/Composites Institute,* Houston, TX, USA, Paper 6-H, February 1983.
5. Zweben, C., 'Advanced composites for the aerospace applications, a review of current status and future prospects'. *Composites* **12** (4), October 1981, 235–40.

6. Steel, D. J., 'The use of carbon fibre composites for axially loaded tension/compression components', *Conf. on Designing with Fibre Reinforced Materials,* Paper C234/77, Institution of Mechanical Engineers, London, September 1977.
7. Hollaway, L. and Ishakian, V. G., 'Analysis of a pultruded carbon fibre/epoxy skeletal structure', *Conf. on Designing with Fibre Reinforced Materials,* Paper C235/77, Institution of Mechanical Engineers, London, September 1977.
8. Green, A. K. and Phillips, L. N., 'The development of metal end fittings for axially loaded composite components', *Proc. 3rd Int. Conf. on Composite Materials,* Pergamon Press, Oxford, August 1980, p. 1247.
9. Green, A. K. and Phillips, L. N., 'Improvements in or relating to connectors', UK Patent Application GB 2052 766A, February 1981.
10. Hollaway, L. and Baker, S., 'The development of nodal joints suitable for double layer skeletal systems made from fibre/matrix composites', *Proceedings of the 3rd International Conference on Space Structures,* University of Surrey, 11–14 September 1984, Elsevier Applied Science Publishers, London, pp. 908–14.
11. Rustum, A. and Hollaway, L., 'The structural behaviour of a double-layer skeletal system manufactured from a pultruded fibre/matrix composite', Paper 39 presented at *The Reinforced Plastics Congress,* 1982, Brighton, Sussex, 8–11 November 1982.

Chapter 2

DEVOLATILISATION

Keith T. O'Brien

Celanese Engineering Resins, Summit, New Jersey, USA

1. INTRODUCTION

Devolatilisation is not a new unit process, but it is only recently that it has been accorded the attention it warrants. There are now a number of theoretical treatments of this topic and a variety of equipment items which have been adapted to, or developed for, the devolatilisation process. In its broadest sense devolatilisation is the removal of volatiles, including moisture, from a substance. In this dialogue the only substances allowed will be plastics and, in the main, attention will only be paid to devolatilisation from a molten mass. The volatiles may be extracted as a gas or vapour (condensable gas).

A summary of the possible states of the host plastic and extractable volatiles is given in Table 1. The theory behind the unit process for each of these combinations of states will be given in Section 3 and the equipment used, in Section 5.

By treating both the fundamentals and the available hardware of devolatilisation for the situations outlined in Table 1 the text should provide the reader with all the essentials of the devolatilisation of plastics as it is understood and practised today. But before this, a broad overview of the polymer production and conversion operations is required to acquaint the reader with the scenario under which devolatilisation becomes a necessary and essential unit operation.

TABLE 1
POSSIBLE COMBINATIONS OF STATE IN THE DEVOLATILISATION PROCESS

Host	*Volatile*
Type 1	
Solid	Gas/vapour
Type 2	
Liquid	Gas/vapour

2. OVERVIEW OF POLYMER PRODUCTION AND CONVERSION

2.1. Production

In the production of polymers, solvents, unreacted monomers and reaction by-products often remain in the polymers. These residuals usually have low boiling points in relation to the polymer and have come to be called 'volatiles'. It is not always essential to remove them from the polymer since they may become innocuous and reside harmlessly in the polymer matrix. However, it becomes necessary to remove them when their presence would result in an unacceptable reduction in polymer properties, or create health dangers, odour or the like. Another statement of this same fact would be that quality improvements can be achieved by extracting the volatiles. Of course the costs of extraction have to be weighted against the quality improvements in a commercial operation.

Polymers are usually solid at the reactor exit, but may also be liquid. They may be ready for shipment to customers after the reaction process, but more usually require subsequent stabilisation and/or finishing steps. For solution polymerisations the solvent must be removed, for bulk polymerisations the residual monomer must be removed and in both instances reaction by-products must be removed.

2.2. Conversion and Compounding

In the stabilisation process, chemicals (such as antioxidants and heat stabilisers) are added, which do not radically change the polymer, but substantially increase its thermal stability so that it may be used several times without significant reduction in properties. During this process of stabilisation there is another opportunity for volatiles to be produced, and possibly another reason to devolatilise the polymer.

To further enhance the properties of the now stable polymer other types of additives are frequently compounded into it. Lubricants, pigments, flame retardants, drip suppressants, nucleating agents, UV stabilisers, minerals, elastomers and glass fibres are examples of such additives. When the additives are blended with the molten polymer there is a third opportunity for volatiles to be produced, usually from side reactions, and moisture content from the additives. During the compounding operation it is also frequently possible to remove residual amounts of solvents, monomer and water remaining from previous operations.

In general, as the polymer progresses sequentially through these three processes the amount of volatiles reduces, and the volatiles themselves are more complex and difficult to define. These factors influence the choice of equipment for the extractions of volatiles, as do the inherent properties of the polymer.

Now that the needs have been established and the scenario set, a discussion of the fundamental mechanisms involved and the underlying theory is appropriate.

3. FUNDAMENTAL MECHANISMS AND THEORY

In Table 1 an outline was given of the combinations of states possible between the host polymer and the volatiles. Consideration will now be given to the way in which the devolatilisation takes place.

3.1. Theoretical Considerations

Consider the extraction of a gaseous volatile from a solid polymer slab, the volatile being distributed within the solid in an unknown manner. To illustrate the mechanism involved, it will be assumed that there is a constant concentration, c_0, at a distance **x** from the surface. This is illustrated in Fig. 1. If the concentration at the surface is also c_0 nothing will happen, but if the concentration is less, say $c^{\#}$, then there is a concentration gradient pulling the volatile to the surface. This may be expressed simply as

$$\frac{\partial c}{\partial t} = \frac{D\partial^2 c}{\partial x^2} \tag{1}$$

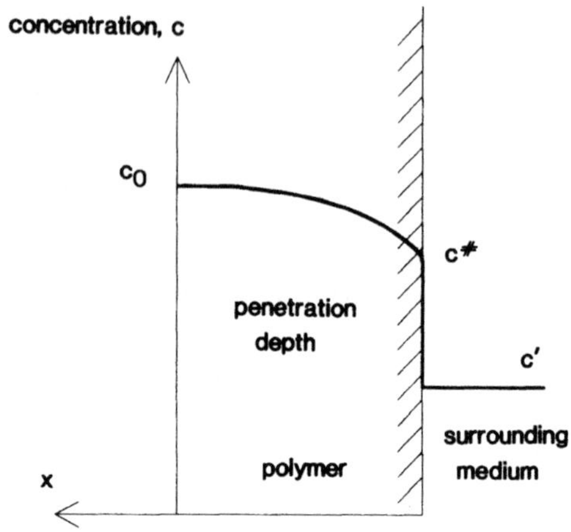

FIG. 1. Concentration gradient at a polymeric free surface.

where c = concentration, D = diffusivity, t = time and x = coordinate in diffusion direction.

Of course, the assumptions made in arriving at this equation are not necessarily valid, so the exactness of eqn (1) is in doubt. Often, the polymer is particulate and cannot be assumed to be an infinite solid. However, it does illustrate the mechanism of diffusion to the surface which must certainly occur during devolatilisation.

Now let us examine what is happening at the surface. The concentration at the surface is $c^{\#}$, and it is assumed that it is $c^{\#}$ at every point on the surface. It is further assumed that the concentration in the surrounding medium (usually air or vacuum) is c' where c' is less than $c^{\#}$. The difference in concentration causes volatiles to convect from the polymer to the surrounding medium. This mechanism may be expressed mathematically as

$$W = k(c^{\#} - c') \qquad (2)$$

where W = diffusion rate/unit area and k = mass transfer coefficient.

If this surrounding medium is stagnant, the concentration of volatiles in it will gradually rise to $c^{\#}$ and the volatile migration will terminate. However, if the concentration is held at c' by constantly changing the medium then the extraction will continue.

The principles which apply to the extraction of a gaseous volatile from a solid also apply to the extraction of a gaseous volatile from a liquid (or melt). However, with a polymer melt it is possible to constantly renew the mass interface through mechanical agitation. Volatiles are transported to the surface continually and there is only minimal difference in volatile concentration throughout the bulk polymer. Thus volatile removal from an agitated liquid is more readily achieved than from a solid.

3.2. Practical Implications

Having established the mechanisms of devolatilisation for very simplistic situations, it is already possible to state three tenets for the design of devolatilisation equipment. Firstly, the volatiles concentration at the surface should be maintained as close as possible to that in the bulk. For liquids, some agitation mechanism is preferable, and for solids, small particle sizes and agitated beds are optimal. Secondly, large surface areas per unit mass will allow more rapid devolatilisation since the rate of extraction is proportional to the surface area and the total extractable volatiles is related to the mass. Thirdly, the distance for the volatiles to diffuse through the polymer should be as small as possible. This ensures steep gradients of concentration and therefore high diffusion rates. It is worthwhile noting that the first and third tenets are contradictory in terms of their geometrical requirements. However, they are not mutually exclusive and both mechanisms occurring at once would be ideal, but either works alone.

As well as diffusion occurring, there is also the possibility that bubbles may form in the liquid. This occurs when the polymer pressure is reduced to a level below the equilibrium partial pressure of the dissolved volatile by an amount sufficient to overcome surface tension effects created by the bubble. The liquid (or melt) is then supersaturated and bubbling begins. The bubbles then convect the volatiles to the surface, in enhanced quantities, where the bubbles must be ruptured to allow the volatiles to escape.

Let us now consider the state of equilibrium at the interface between the polymer and the volatile. The equilibrium curve for styrene in polystyrene is shown in Fig. 2. It is very clear that a reduction of pressure in polystyrene will cause a decrease in the equilibrium concentration, and thus enhance the devolatilisation process. So here is a fourth tenet to be remembered in equipment design. Pressures in the devolatilisation region should be maintained low. Increasing temperature also decreases the equilibrium concentration and so high temperatures are also desired, a sixth requirement.

Another way to improve devolatilisation is by the use of carrier

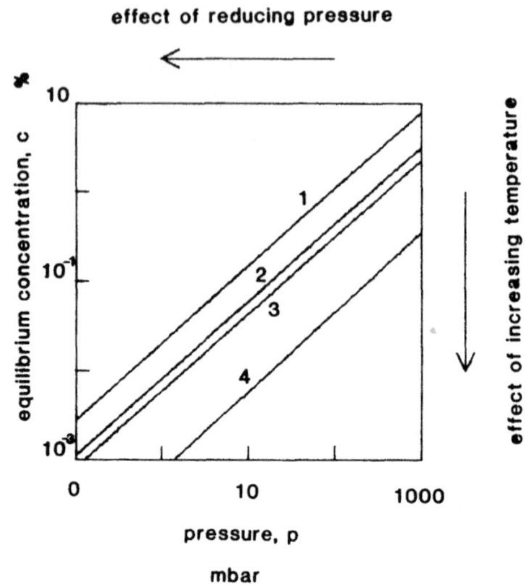

FIG. 2. The influence of pressure and temperature on equilibrium concentrations. (1) Polystyrene–styrene—Flory–Huggins theory; (2) PMMA–MMA—Flory–Huggins theory; (3) HDPE–cyclohexane—Flory–Huggins theory; (4) LDPE–ethylene—experimental.

substances. Such carriers are mixed into the melt and during the extraction phase increase the partial pressure of the volatiles at equilibrium, making extraction easier.

In looking at the fundamental mechanisms underlying the devolatilisation (DV) process, six requirements have been established to ensure high extraction levels. There are three more which are very fundamental and which should not be overlooked.

Since the process is transient the time of the process becomes important and it is very clear that long residence times are desirable to allow the process to proceed. This is especially true with high viscosity melts where bubble transport is severely hindered. The last two requirements are that the extraction path is adequately sized, and that the extraction rate is adequate.

The cross-section of the extraction path should be large enough and the volume flow rate sufficient that the volatiles are not restricted in this region. These requirements for ideal devolatilisation are summarised in Table 2.

TABLE 2
IDEAL DEVOLATILISATION

Property	*Desired value*
Surface area	Large
Diffusion path	Small
Rate of surface renewal	High
Temperature	High
Residence time	Long
Pressure	Low
Equilibrium concentration	Low
Extraction path	Large
Cross section rate of extraction	Large

4. VACUUM SYSTEMS

Vacuum systems for devolatilisation usually consist of a vacuum pump, flanges and pipelines, measuring and control instrumentation and if necessary, a vacuum tank, separators, baffles and traps. On occasions multi-stage steam jets may be utilised to provide the vacuum instead of a vacuum pump.

4.1. Pump Types

The choice of the system depends upon the application, which is usually classified according to Table 3. In some instances a vacuum

TABLE 3
VACUUM RANGES OF COMMERCIAL VACUUM PUMPS

Description	*Pressure range p (mbar)*
Low vacuum	10^3–1
Medium vacuum	1–10^{-3}
High vacuum	10^{-3}–10^{-7}
Ultra-high vacuum	$<10^{-7}$

TABLE 4

CLASSIFICATION OF VACUUM PUMPS

Gas transfer	Displacement	Reciprocating	Diaphragm
			Piston
		Rotary	Liquid ring
			Rotary valve
			Rotary piston
			Oval gear
	Kinetic	Mechanical	Gas ring
			Axial turbo
			Radial turbo
			Molecular
			Turbomolecular
		Fuel diffusion	Liquid jet
			Gas jet
			Vapour jet
			Diffusion
			Diffusion ejector
		Ion transfer	
Gas binding	Adsorption		
	Getter	Solid	
		Sublimation	
		Ionisation	
	Cryogenic		

may not be necessary due to the low level of volatiles and their ease of release, but when a vacuum is required a vacuum pump is essential. These units may be classified according to their mode of operation, as presented in Table 4.

In the gas transfer type the volatiles are pumped to atmosphere, whereas in the gas adsorption type they are adsorbed inside the pump. Usually gas transfer pumps are employed. The displacement type transport the gas, whereas the kinetic type accelerate the gas. The remaining details of pumps should be self evident from Table 4.

4.2. Operating Performance

Consideration may now be given to the operating performance of the different pump types so that they may be matched against specific applications. There are two key operating performance variables which concern devolatilisation, the pressure and the extraction rate. The advantanges of low pressure were discussed in Section 3, as was the need for adequate extraction rates and cross-sectional areas of extraction paths. Figures 3 and 4 indicate the operating pressures and

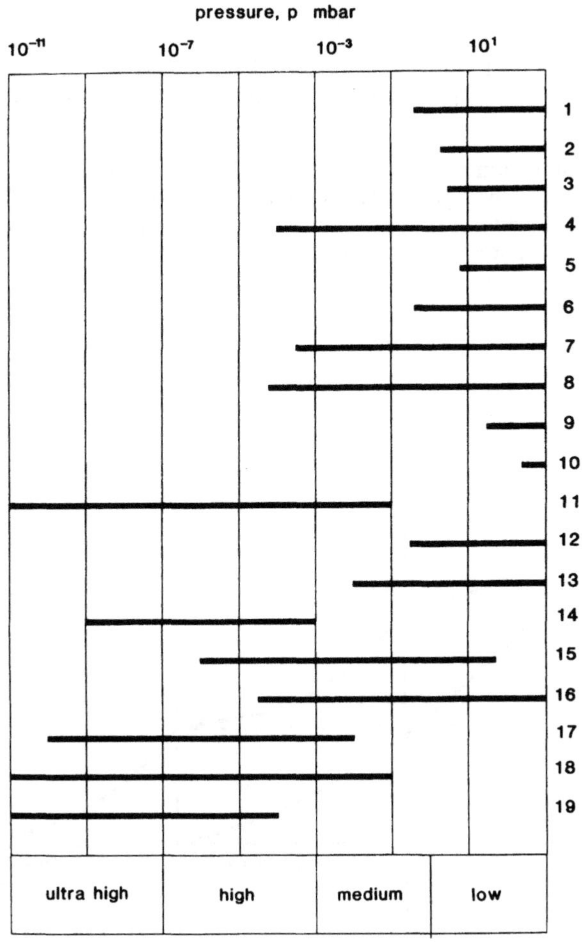

FIG. 3. Classification of extraction pumps by operating pressure. (1) Reciprocating piston; (2) diaphragm; (3) liquid ring; (4) rotary valve; (5) vane; (6) rotary piston; (7) oval gear; (8) gas ring; (9) turbo; (10) turbomolecular; (11) liquid jet; (12) vapour jet; (13) diffusion; (14) diffusion ejector; (15) absorption; (16) sublimation; (17) ionisation.

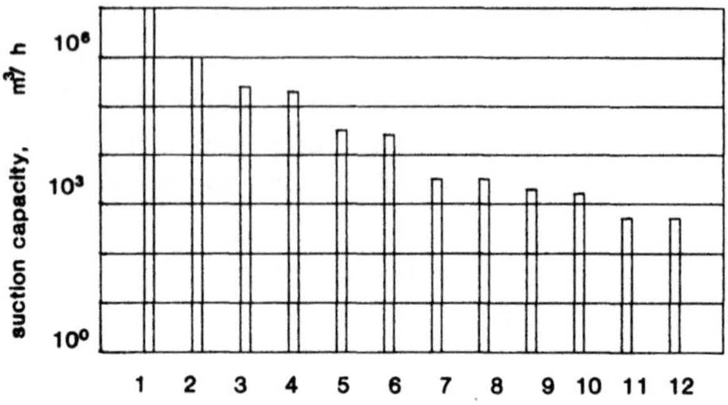

FIG. 4. Classification of extraction pumps by extraction rate. (1) Vapour jet (multi-stage); (2) oval gear; (3) cryogenic; (4) diffusion (oil); (5) liquid ring; (6) piston; (7) vane; (8) turbomolecular; (9) diffusion (mercury); (10) ion getters; (11) rotary valve; (12) rotary piston.

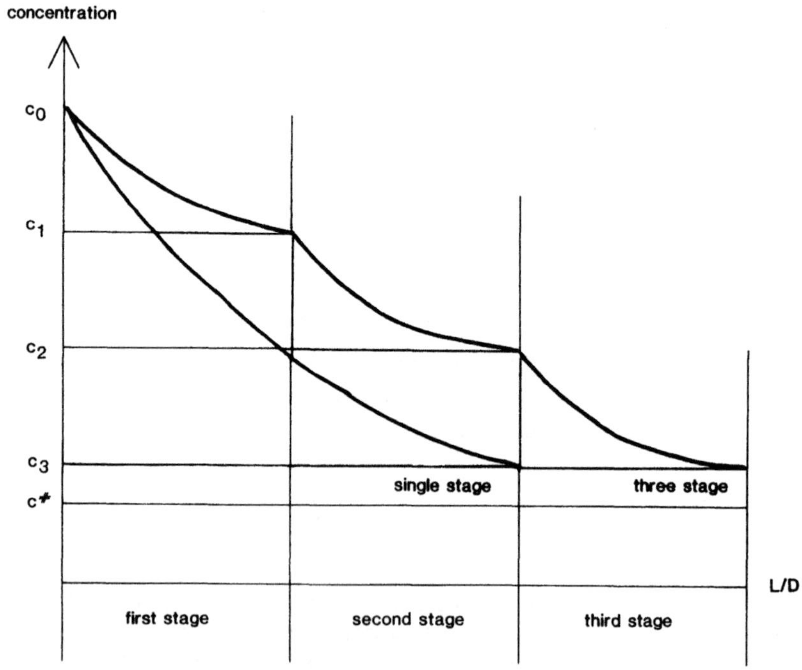

FIG. 5. Effect of staged reaction on devolatilisation length.

extraction rates available with some of the pump types presented in Table 4. It is a relatively easy task to use these figures for equipment design purposes. This is especially true when the quantity of volatiles to be removed is high. For example, if one is required to remove 100 m^3/h at a vacuum of 10 mbar, one could enter each graph in turn at these values and determine which pumps satisfy both conditions. There would be seven possibilities:

reciprocating piston	vane	vapour jet
liquid ring	rotary piston	
rotary valve	oval gear	

Cost would then determine the choice.

One item not immediately apparent when using Figs 3 and 4 is that the pump selected may be excessively large. In such instances, it is frequently possible to use several smaller pumps by staging the extraction process. This is illustrated in Fig. 5. The same end concentration of volatiles in the polymer is achieved using the three smaller pumps rather than the larger pump. However, a longer devolatilisation length, consisting of three devolatilisation sections, is now required.

5. EQUIPMENT

As stated earlier, in Section 1, the occurrence of devolatilisation may be in a post reactor, or in a stabilisation or a compounding operation. In general, a post reactor situation would often involve large amounts of volatiles in a low viscosity solution. A stabilisation operation would involve medium amounts of volatiles in a medium viscosity prepolymer and a compounding operation would involve small amounts of volatiles in a high viscosity polymer. The exceptions to these generalisations are many, and this has served to confuse the area even more, but it remains a fact that devolatilisation occurs from materials with a wide range of viscosities.

A simplistic diagram to illustrate this generalisation is shown in Fig. 6. Proceeding from this point it is possible to consider viscosity as the dominant variable which dictates equipment configuration, since equipment may then be sized appropriately. This is illustrated in Table 5 where the viscosity limits of major devolatilisation equipment items

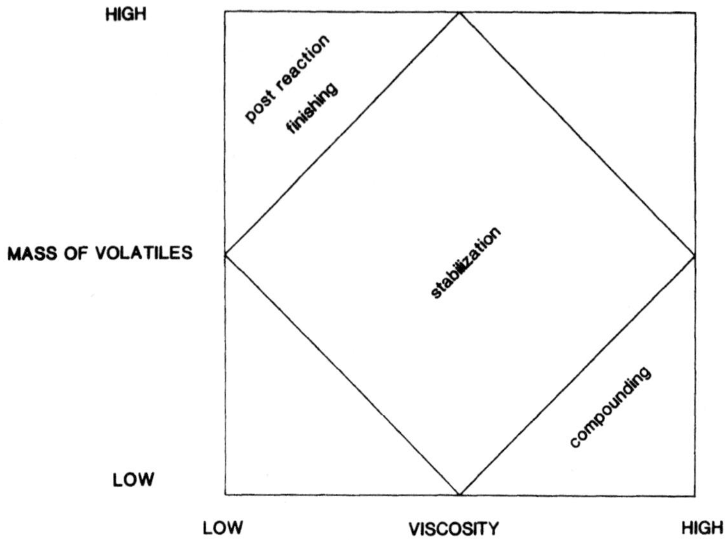

FIG. 6. Scenario of devolatilisation situations.

are tabulated. The limits on throughputs of polymer and the permissible levels of extractables are also given in this table.

This is not the only criterion for selecting DV equipment since, for a given viscosity, more than one machine may usually be used. In such circumstances trials may be necessary and attention must be paid to the power requirements and the separator efficiency. Both these

TABLE 5

OPERATING WINDOWS FOR COMMERCIAL DEVOLATILISERS

Equipment	*Viscosity range* (Ns/m^2)	*Throughput range* (kg/s)	*Volatiles level* (%) *maximum*
1. Flash evaporator	10^{-3}–10^{-2}	$6{\cdot}3 \times 10^{-4}$–$1{\cdot}26 \times 10^{-1}$	10
2. Flowing film	10^{-3}–5×10^{-2}	$6{\cdot}3 \times 10^{-4}$–$1{\cdot}26 \times 10^{-1}$	5
3. Falling film or strand evaporator	10^{-2}–10^{-1}	$6{\cdot}3 \times 10^{-4}$–$1{\cdot}26 \times 10^{-1}$	5
4. Thin film evaporator	1–10	$1{\cdot}26 \times 10^{-1}$–$1{\cdot}26$	5
5. Filmtruder	10–10^{3}	$1{\cdot}26 \times 10^{-1}$–$1{\cdot}26$	5
6. Single-screw extruder	10^{2}–10^{5}	$6{\cdot}3 \times 10^{-3}$–$6{\cdot}3 \times 10^{-1}$	3
7. Diskpack	10^{2}–10^{5}	$6{\cdot}3 \times 10^{-3}$–$1{\cdot}26 \times 10^{-1}$	3
8. Twin-screw extruder	10^{2}–10^{5}	$1{\cdot}26 \times 10^{-3}$–$10{\cdot}04$	3
9. Multi-stage extruder	10^{2}–10^{5}	$8{\cdot}82 \times 10^{-3}$–$10{\cdot}04$	3

variables depend in turn on the separation efficiency, the surface area, the surface renewal (back mixing), the exposure time, the residence time (mean and distribution) and whether or not staging is used.

Now the available commercial equipment has been succinctly classified, let us proceed to examine the relationships between the devolatilisation mechanisms of Section 3 and the equipment listed in Table 5. It should be noted that only the devolatilisation process in the equipment is discussed here. The cleaning, maintenance, reliability and overall performance are not treated.

5.1. Flash Evaporator

In this unit the polymerisation products are passed through tubes, usually spiral coils, which are heated to a temperature above the boiling point of the volatiles. This is shown schematically in Fig. 7.

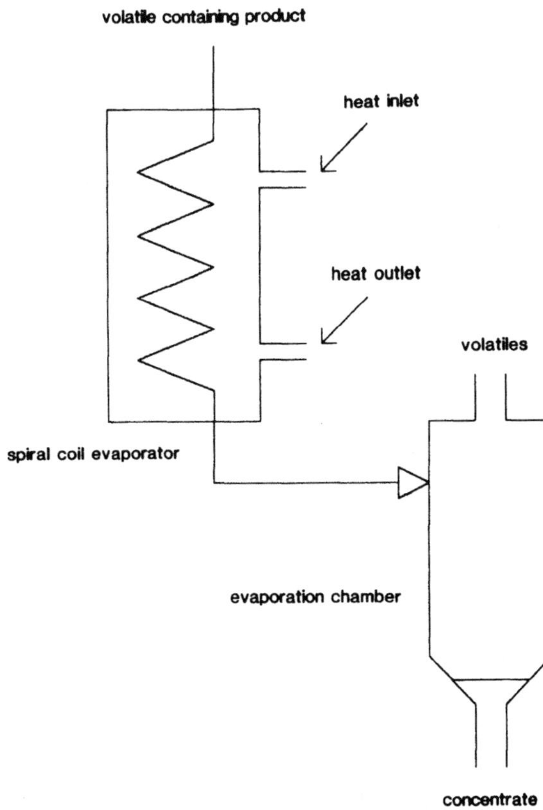

FIG. 7. Flash evaporation.

This operation is conducted at enhanced pressures so as to ensure no evaporation. Then the heated mass is discharged to an evaporation chamber, where the pressure is used to control the flash evaporation. The volatiles are extracted using a vacuum pump and the concentrated polymer, which collects at the bottom of the chamber under gravity, is pumped away.

5.2. Falling Film and Strand Evaporators

These units have commonality in that gravity is employed to convey the polymer, after it has been arranged into a film or a strand of thickness or diameter similar to the penetration depth, which is defined in Fig. 1 as the distance between the surface and a point in the interior where changes at the surface have negligible effect. Lower

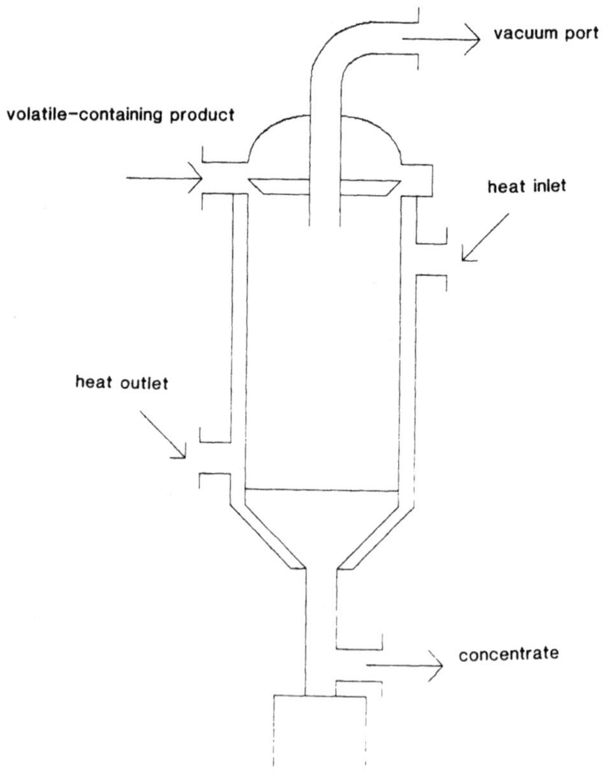

FIG. 8. Falling film evaporation.

viscosity polymer solutions can be devolatilised in the flowing film vaporiser. Higher viscosity polymer melts can be devolatilised in the falling film or falling strand evaporator.

The falling film vaporiser allows the polymer solution to flow down the walls of a chamber, which is heated. DV takes place only from one side of the film. The concentrated polymer solution then collects at the base of the chamber and is pumped away. This process is illustrated in Fig. 8. In falling film evaporators a polymer film is extruded into the chamber. DV takes place from both sides of the film. The film collapses into a pool at the base of the chamber and is pumped away. In the falling strand evaporator, shown in Fig. 9, polymer strands are

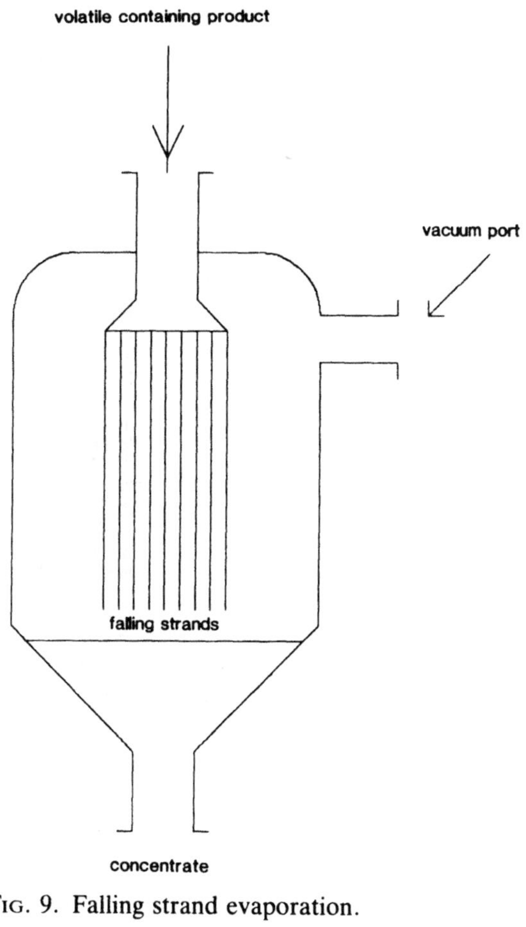

FIG. 9. Falling strand evaporation.

extruded into the chamber. DV takes place from the outer surface, radially. The strands then fall into a pool at the base of the chamber and are pumped away.

5.3. Thin Film Vaporiser

This is not a single unit but really a family of machines which operate on the same principle.

5.3.1. Basic Construction

In addition to the geometry of the flowing film evaporator (see Section 5.2), a central shaft has been added. This shaft has paddles mounted

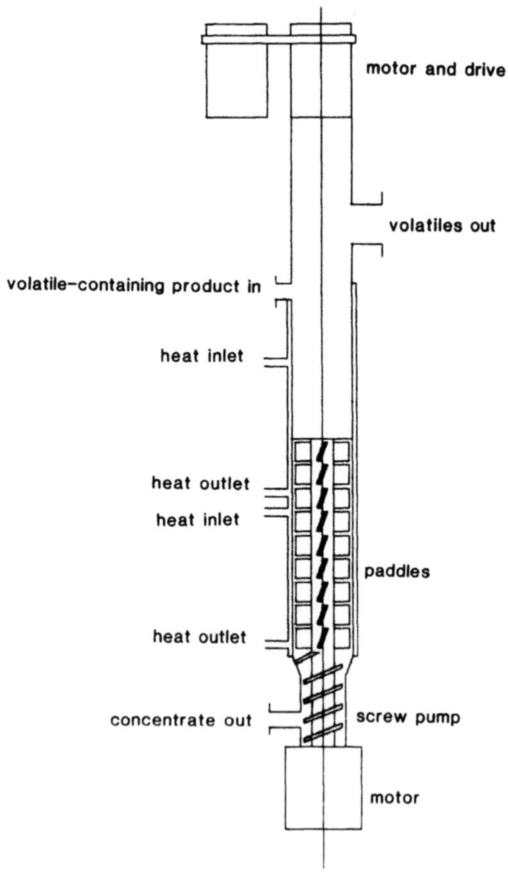

FIG. 10. Thin film vaporiser.

around its periphery and the paddles extend to the film at the outer wall. The film created between the wall and the paddle, as the paddle passes over the wall, creates an annular region where devolatilisation can readily proceed. A schematic drawing is shown in Fig. 10.

The paddles, inclined to the vertical, push the polymer downward toward the exit, while simultaneously renewing the mass transfer surface and back mixing to avoid extraneous heat transfer from the heated wall. With the addition of this drag force mechanism the allowable viscosity of the polymers which can be devolatilised is greatly increased.

The Filmtruder® is essentially a more robust version of the other units in this family, and may be used for viscosities up to 10 kP; it therefore competes with the single screw extruder, the Diskpack®, the kneaders, the twin screw extruders and the multi-stage extruders. This is shown clearly in Table 5.

The thin film vaporisers are readily attached to a vacuum system which removes the volatiles and also reduces the internal pressure to promote a mass transfer. Lower bulk temperatures are maintained due to the extremely low thickness of the polymer layer and availability of a large heat transfer surface. Residence time may be controlled from the length of the unit, the inclination of the pumping blades, the shaft speed and polymer film thickness.

5.3.2. Flow Configurations

The thin film vaporiser is very flexible in its configuration which may be changed by using a variety of wiper blades, as shown in Fig. 11. The changes in wiper blades cause changes in the agitation pattern, the

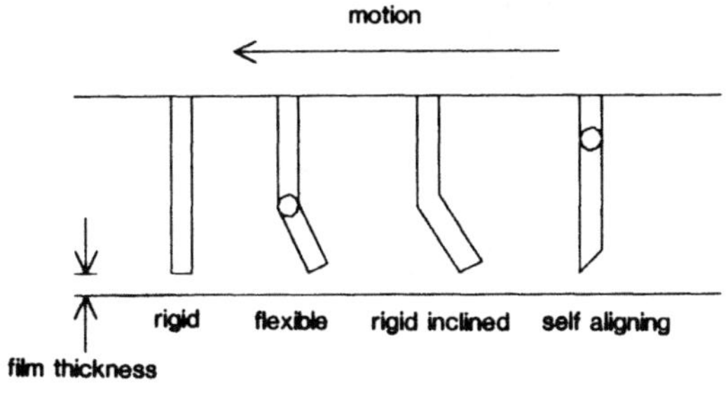

FIG. 11. Thin film vaporiser blades.

Filmtruder® is the registered trademark of the Luwa Corporation.
Diskpack® is the registered trademark of the Farrel Corporation.

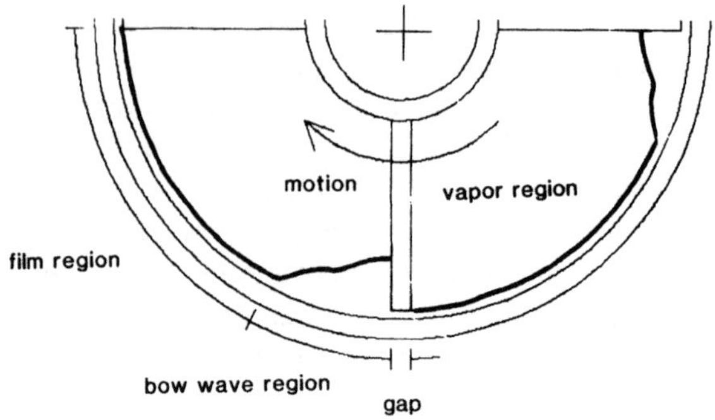

FIG. 12. Cross-section of the thin film formation in a thin film vaporiser.

so called bow wave, which is a recirculating flow in front of the wiper blade. A typical bow wave is shown in Fig. 12. The size and shape of the bow wave, in turn, influences the surface renewal, the mixing and the residence time. If the unit is viewed in the axial plane, as shown in Fig. 13, it may be seen that the wiper blades are not continuous but discrete. This allows mixing of the polymer streams as the shaft rotates, as well as the axial conveyance.

The thin film vaporiser may also use counterflow, unidirectional flow or a combined flow. These possible flow configurations are shown in

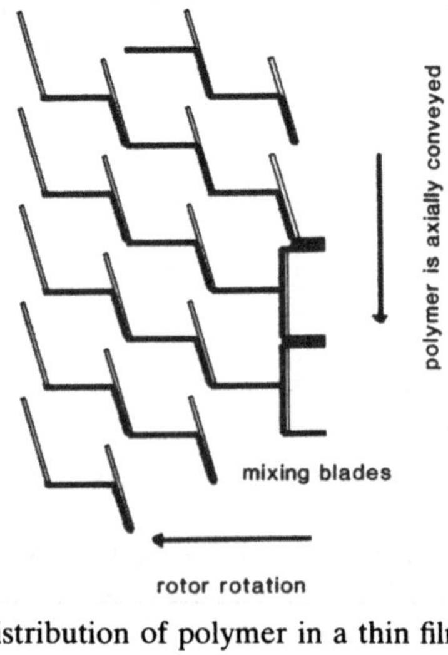

FIG. 13. Distribution of polymer in a thin film vaporiser.

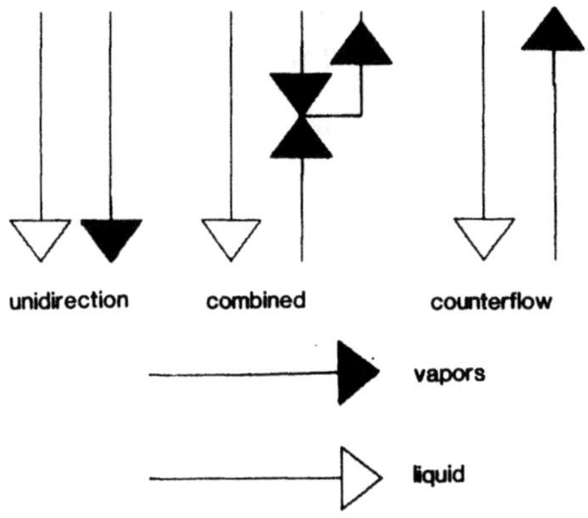

FIG. 14. Configurational options in thin film vaporisation—line diagrams.

Fig. 14 schematically, and in more detail in Fig. 15 where the additional changes to the basic unit are included.

For the removal of volatiles in trace amounts the counterflow arrangement is best used, whereas if large amounts of a volatile must be removed the unidirection arrangement is best suited. This is true

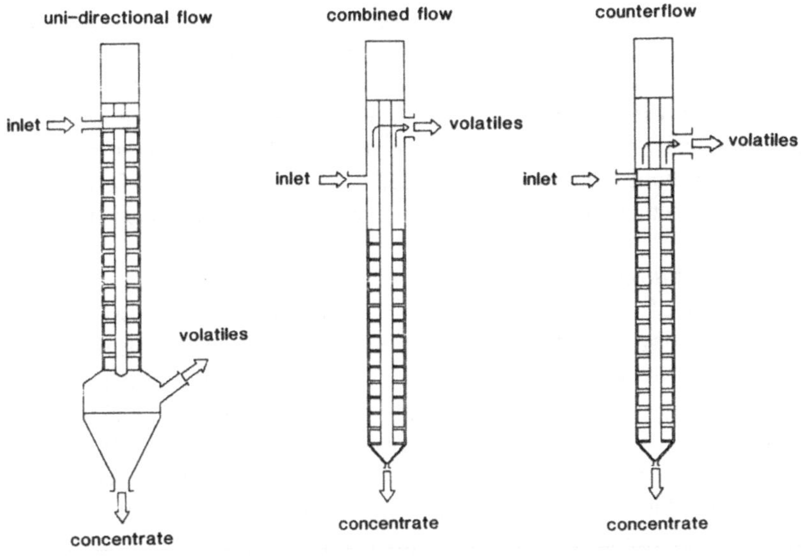

FIG. 15. Configurational options in thin film vaporisation—schematic.

TABLE 6

CONFIGURATIONS OF THE THIN FILM VAPORISER AND ITS DERIVATIVES

	Counterflow	*Uni-directional flow*	*Counter/uni-directional flow*
Commercial uses	Concentration Removal of monomer Removal of residual volatiles reaction	Concentration Solvent recovery	Concentration Solvent recovery Polycondensation
Vapour velocity	1–8 m/s	5–40 m/s	5–40 m/s
Evaporation ratio	Medium	Very high	High
Separating efficiency	Good	Very good	Very good
Combination with flash vaporisation	Limited	Excellent	Very good
Foaming products	Poor	Good	Good
Viscosity limits	Up to 10 000 Pa s	Max. 500 Pa s	Up to 10 000 Pa s
Final concentration	Ppm	0·5–5%	Ppm

because, as well as diffusion controlled or bubble induced devolatilisation, flash evaporation is also probably occurring. With the counterflow arrangement bubble induced devolatilisation at the bottom causes operating difficulties and the separating efficiency is reduced. If the viscosity of the polymer is high, however, then the downward transport to the reservoir is limited. So the counter/unidirection hybrid must be used. Now the vapours are drawn off at the centre of the unit and the polymer may be discharged as with the counterflow version. These differences due to flow configuration are summarised in Table 6.

5.3.3. Operating Characteristics

The basics of thin film vaporisers having now been considered, the operating characteristics will now be treated. The film thickness has been shown to be a key influencing variable in devolatilisation. For the thin film vaporiser and the Filmtruder the possible film thicknesses are shown as a function of the polymer viscosity in Fig. 16. From a knowledge of the process kinetics it is a relatively easy task to match the unit to the process. In a somewhat undefined process the operating film thickness must be experimentally determined.

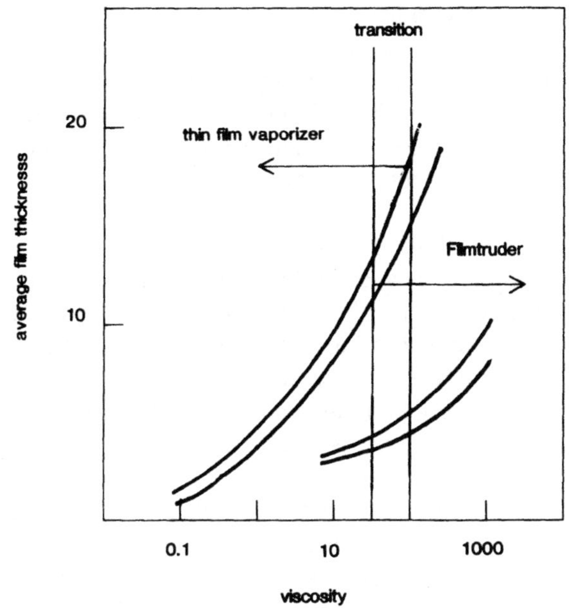

FIG. 16. Operating regions of the thin film vaporiser, and Filmtruder.

It was stated earlier that the thin film vaporiser and the Filmtruder are much the same, except that the latter is more sturdily built for use with higher viscosity polymer melts. This is true, but the operating characteristics do change somewhat. Since the Filmtruder is processing higher viscosity feedstocks the heat created by viscous dissipation is now no longer insignificant. This in turn means that less heat needs to be supplied through the outer walls, and that the heat is more evenly distributed within the polymer mass. The heat is thus more insulated from the free surface by the polymer film and volatile removal is more efficiently achieved. The viscous dissipation increases with the rotor speed, so the conduction of heat from the walls ceases to be the rate controlling feature. The viscous dissipation also increases with viscosity so, as the polymer becomes more concentrated and more viscous, the mass transfer is hardly impeded.

The wide variety of configurations possible with the thin film vaporiser family make the units suitable in more situations than any of its competitors. Some examples of its use include the concentration of epoxy resins, the removal of residual styrene from polystyrene and solvent recovery from atactic polypropylene.

5.4. Single-screw Extruder

When the single-screw extruder is being used for devolatilisation it is invariably also being used to convey solids, to melt, to mix and/or to pressurise the polymer. This limits the options on devolatilisation, since a decrease in speed may cause inadequate pressurisation while attaining an improved level of devolatilisation.

5.4.1. The Devolatilisation Zone

In a typical single-screw extruder the channel of the screw in the devolatilisation zone runs about one third full of polymer melt. Most of this material is in a rotating pool in front of the leading edge of the flight. It is contained on three sides by the barrel, screw flight and screw root respectively and has one free surface where devolatilisation takes place. This geometry is shown schematically in Fig. 17 in two dimensions.

The rotating melt pool, in fact, is a helical ribbon which follows the screw flight. At each end of the helix the free surface disappears due to changes in the channel geometry at the relief and transition regions. This means that the vapour region consists of a helical ribbon also, but closed at the ends. This region is connected to the atmosphere via a hole (or vent) in the barrel, and this vent allows the extraction of volatiles.

There is also a region between the flight tip and the barrel where the polymer is not wiped and so a polymer film may reside in this area. It has little thickness and its surface is constantly renewed as the flights sweep across it. Since it has a free surface, devolatilisation must take place from it.

If the transport phenomena exhibited in the single screw extruder

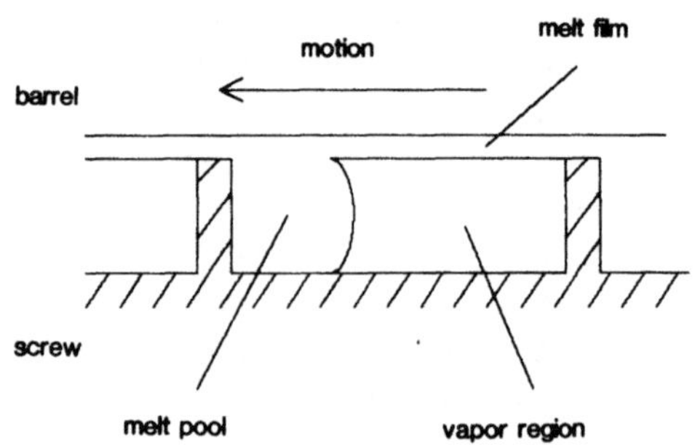

FIG. 17. Cross-section of the devolatilisation zone of a single-screw extruder.

TABLE 7
EFFECT OF DESIGN AND OPERATING VARIABLES ON THE DEVOLATILISATION EFFECTIVENESS OF EXTRUDERS

	Design		*Operating*			
	Channel depth	*Zone length*	*Screw speed*	*Zone temp*	*Vacuum level*	*Vent X section*
Surface area	I[a]	I	—	—	—	—
Diffusion path	R[b]	—	—	—	—	—
Rate of surface renewal	I	—	I	—	—	—
Temperature	—	—	I	I	—	—
Residence time	—	I	R	—	—	—
Vapour pressure	—	—	—	—	I	—
Solvent equilibrium point	—	—	—	—	—	—
Extraction path cross-section	—	—	—	—	—	I

[a] I = increases.
[b] R = reduces.

are now compared to the requirements for ideal devolatilisation classified in Table 2 it may be seen how to design a devolatilisation region in such a unit.

The surface area available for mass transfer and the diffusion path and the rate of surface renewal may be optimised by filling the channel as little as possible with melt. But this reduces throughput or requires a channel so deep that the mechanical strength of the screw is exceeded. Obviously trade-offs must be made. The surface area may also be increased, as may residence time, by increasing the length of the devolatilisation section. Neglecting secondary effects, the effect of operation and design variations on the devolatilisation performance of a single-screw extruder are depicted in Table 7.

5.4.2. A DV Model

Latinen developed a simplistic model for devolatilisation in extruders. Although not strictly accurate, it is helpful in giving an understanding of what is transpiring. Assuming the melt is distributed as in Fig. 17, and that there are no variations in concentration of the volatile within the melt except at the interface, that the melt surface has a concentration equal to the equilibrium concentration in the surrounding space and that the diffusion coefficient is constant, then the system can be described mathematically as

$$\frac{d^2c}{dx^2} - (Pe)\frac{dc}{dx} - (Pe)(E)(c - c^{\#}) = 0 \tag{3}$$

where c = concentration, x = dimensionless/axial coordinate, $c^{\#}$ = equilibrium concentration, Pe = Péclet number, E = devolatilisation index and L = length of the devolatilisation zone.

The solution of this equation takes the form

$$\frac{c_f - c^{\#}}{c_0 - c^{\#}} = f(Pe, E) \tag{4}$$

where c_f = final concentration and c_0 = initial concentration.

The solution is shown in Fig. 18.

To ensure that the final concentration of volatiles is low, it is clear from this graphical solution that the devolatilisation index should be as

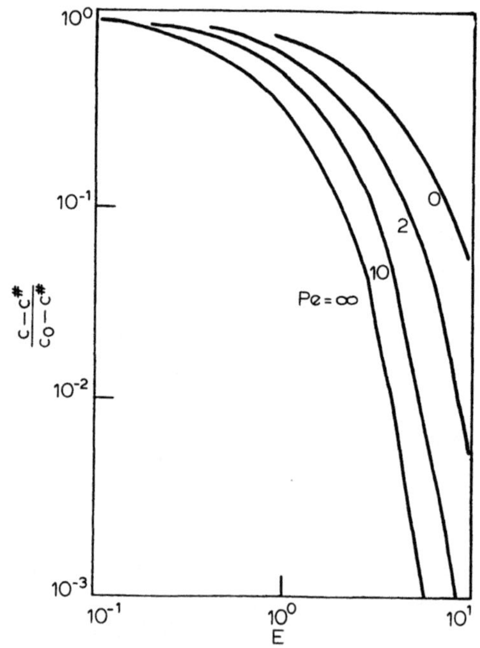

FIG. 18. Performance curves for a single-screw extruder, showing the effect of Péclet number and devolatilisation index on concentration.

high as possible. The devolatilisation index is defined by Latinen as

$$E = \frac{(A)k}{(Q)} \tag{5}$$

where k = mass transfer coefficient, A = exposed area and Q = volume throughput.

Since k is a material system function, then it becomes apparent that (A/Q) should be maximised. In other words the exposed area per unit of throughput should be designed to be as great as possible, within economic and mechanical constraints.

From Fig. 18 it is also clear that the Péclet number should be as large as possible, which implies that pure drag flow is desirable. In the devolatilisation zone of a single-screw extruder there is more constancy of pressure, compared to the gradients experienced in the pumping regions, and so axial mixing is minimal. But axial mixing does occur as a result of the motion of the melt pool, and the transfers between the melt film and the melt pool. Although necessary to ensure surface renewal, the mixing causes the residence time distribution to be wider.

The solution of eqn (3) may also be recast in terms of the devolatilisation efficiency (c_f/c_0). The result is shown in Fig. 19.

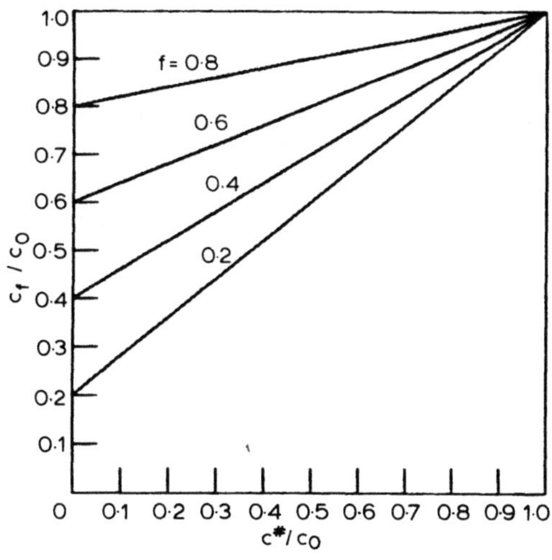

FIG. 19. Performance curves for a single-screw extruder, showing the effect of equilibrium concentration on devolatilisation efficiency.

If $c^{\#}$ and c_0 are known, then the final concentration (c_f) depends only upon the $f(Pe, E)$.

5.4.3. Point of Devolatilisation

The point at which devolatilisation may be performed in a single screw extruder may also be varied to meet the requirements of the process. Forward venting is used predominantly when solids are being fed to the extruder, and is illustrated in Fig. 20. If a melt is being fed to the extruder, for example, post reaction compounding of LDPE, then back venting, as shown in Fig. 21, is desirable. The choice between the two modes depends upon where the volatiles are occurring. In the case of PVC dry blends the volatiles are already present in the hopper, and may be removed by using a vacuum hopper. If volatile levels are very high, multiple venting may be necessary, utilising forward and/or back venting, as shown in Fig. 22. The remaining influencing variable, the equilibrium concentration, may be controlled by the techniques discussed in Section 2.

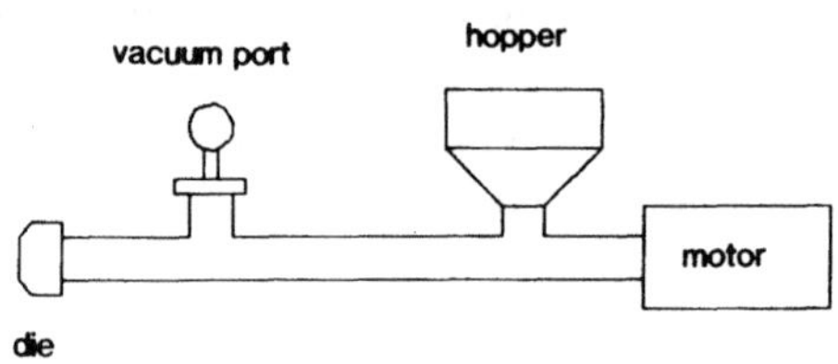

FIG. 20. Forward extraction.

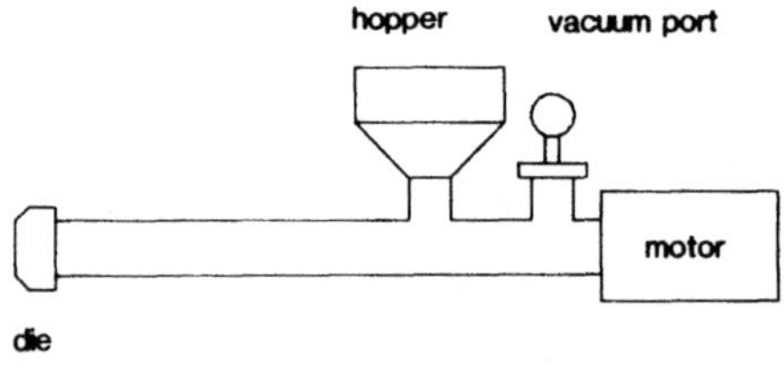

FIG. 21. Rear extraction.

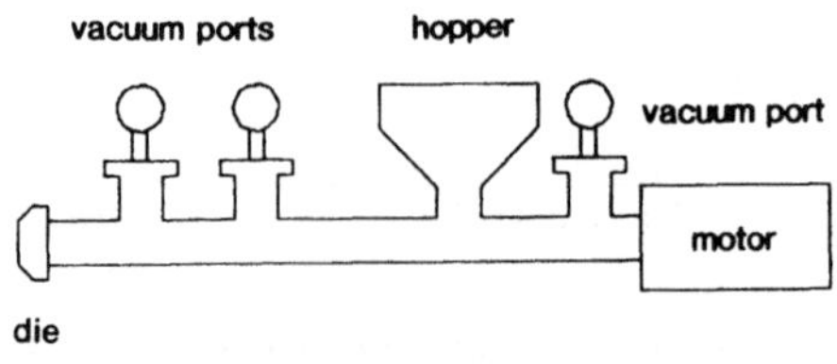

FIG. 22. Combined extraction.

5.5. Diskpack

This unit was introduced to the marketplace only a few years ago, and is only just making inroads into industrial practice. Unlike other polymer processing units it uses two moving surfaces to convey. The surfaces are made up from disks which form an annular channel between them. The disks are mounted on a shaft which rotates. The transit path for the polymer is then extended by transferring the polymer from one pair of disks to another pair until the process is complete.

This is readily achieved by mounting a battery of disks on the same shaft. The annular channels are then terminated by channel blocks which force the polymer into the next annular space. This is shown in Fig. 23, in three different views.

The function of each channel may vary as the polymer is conveyed as a solid, melted, mixed, devolatilised and pressurised. The channels may also be connected in parallel as well as in series.

Devolatilisation may be performed in two distinct Diskpack units, the Compounder and the Devolatiliser. The Compounder takes polymer pellets and additives, including melt fed additives, and melts, mixes and pressurises them ready for pelletisation. This unit is arranged as shown in Fig. 24. Note that the DV is carried out at one end of the unit and then the melt discharged upstream. This arrangement ensures no vent flows.

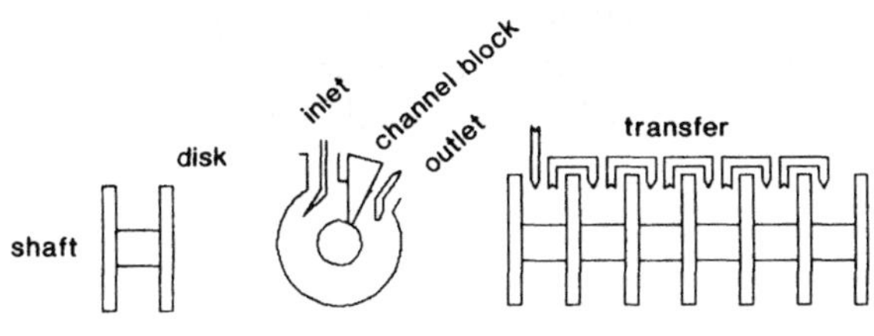

FIG. 23. Schematic of the Diskpack.

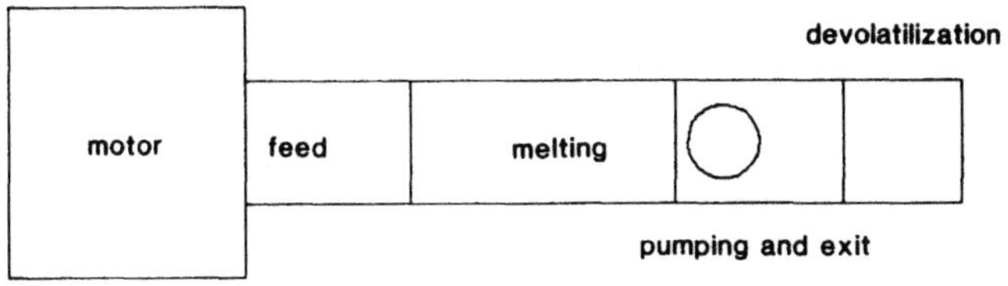

FIG. 24. Functional zones of the Diskpack.

The second unit, the Diskpack Devolatiliser, is a post-reaction unit which takes the melt feed, spreads it over the disks to ensure thin films and then pumps it to the succeeding stage. The residence time is short inside a Diskpack and little surface renewal takes place.

5.6. Kneaders

These units consist of a screw which can oscillate axially as well as rotate in the conventional way. The flight on the screw is interrupted, and the axial and rotational speed synchronised so that protrusions from the barrel wall, called kneading pins, mesh with the flight interruptions. A schematic of a section of the unit is shown in Fig. 25.

Due to the complexity of the combined motions in cylindrical coordinates it is simpler to consider the screw as a flat surface with the flight segments as protrusions. This transformation is shown in Fig. 26. The kneading pins, shown in this figure, mesh with the flight segments. By correctly synchronising the axial and rotational speeds, and by correctly designing the geometry of the segments and pins, the meshing can be closely controlled. To emphasise this point the process will be followed through one cycle, in Fig. 26. Consider that the screw flights are stationary and that the barrel is moving down the page. The kneading pins first wipe the upper left face of the screw flight, then the upper right face, then the lower left face and finally the lower right face. Now four conjunctions have been made between the flight segment and the pin, making the unit self wiping. By controlling the thickness of the kneading gaps and the shear intensity, the thickness of the exiting melt films may be controlled. The bubble rupture process is

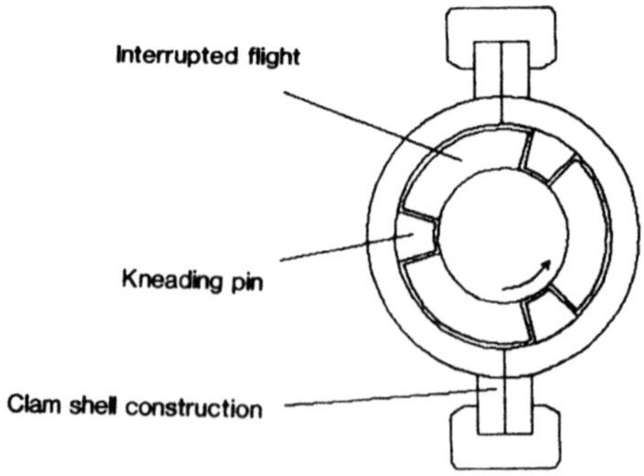

FIG. 25. Cross-section of a kneading extruder—radial.

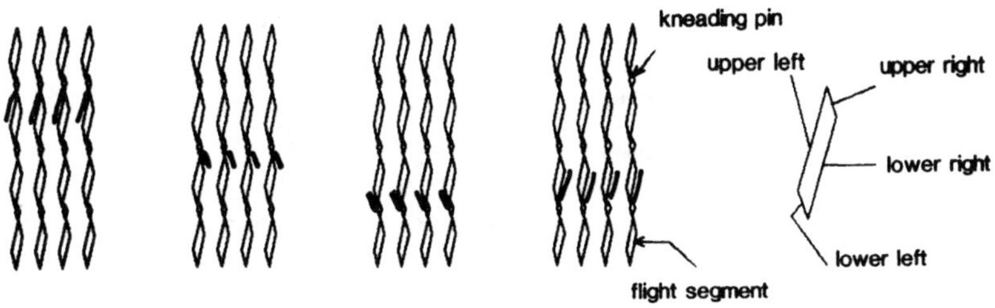

FIG. 26. Kneading actions in the kneading extruder.

controlled by the intensity of shear in the conjunction. The mixing intensity of the unit ensures good backmixing.

The channels are usually only partly filled, and a large free space for volatile escape is available. However, the DV process may be further controlled by judicious use of restriction rings, as indicated in Fig. 27. The restriction ring enables the formation of a melt seal between the screw and the barrel by partially damming the flow. Vacuum may now be pulled in the DV zone without any risk of extracting fine additives which are not yet fully mixed.

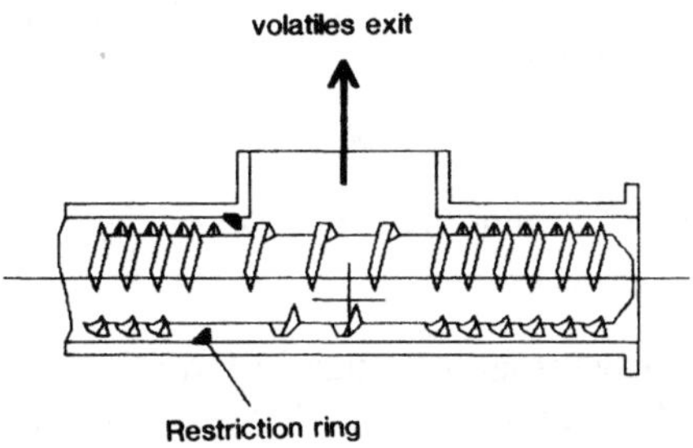

FIG. 27. Cross-section of a kneading extruder—longitudinal.

5.7. Multi-screw Extruders

In most instances an extruder with multiple screws is a twin-screw extruder, but since units with four screws are sometimes used, they will not be excluded. Twin-screw extruders can be configured in many ways—co-rotating or counter-rotating, non-intermeshing, partially intermeshing and fully intermeshing. The three arrangements of practical use in the devolatilisation of plastic melts are shown in Fig. 28.

Each of these configurations can be arranged for multiple vent

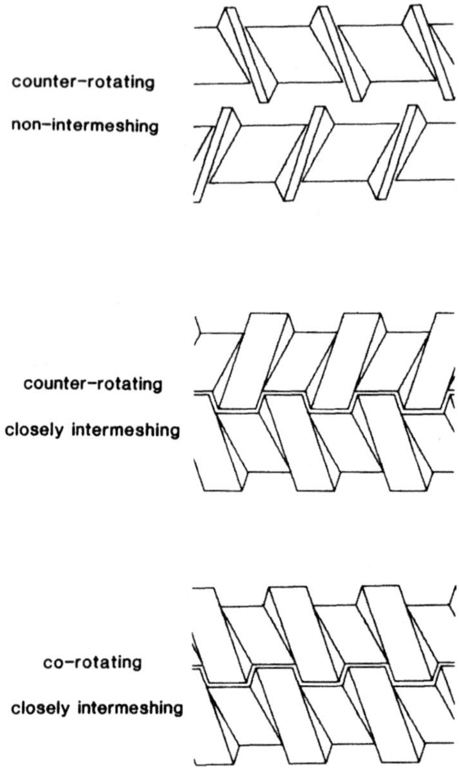

FIG. 28. Twin-screw extruder configurations.

ports, and so take advantage of staged extraction where large amounts of volatiles exist. A brief summary of the characteristics of these units is given in Table 8.

Some of the uses of these devolatilisers are also included in this Table.

The analysis that Latinen performed for single-screw devolatilisation is applicable with modification to all but the closely intermeshing counter-rotating units. If the devolatilization index is modified as follows:

$$\mathbf{X} = \frac{D}{kN(L/D)} \tag{6}$$

where D is the screw diameter, and the other symbols are as defined

TABLE 8

PROCESSING FEATURES OF TWIN-SCREW POLYMER DEVOLATILISERS

Screw system	*Transport*	*Screw flights axially*	*Exposed melt surface*	*Self-Cleaning*	*Multi-stage venting*
Counter-rotating closely intermeshing	Positive displacement	Closed	Near the vent only	No	Possible
Counter-rotating partially intermeshing	Combined drag and pressure	Open	Entire, partly filled zone	No	Possible
Counter-rotating non-intermeshing	Combined drag and pressure	Open	Entire, partly filled zone	No	Possible
Co-rotating closely intermeshing	Combined drag and pressure	Open	Entire, partly filled zone	Yes	Possible

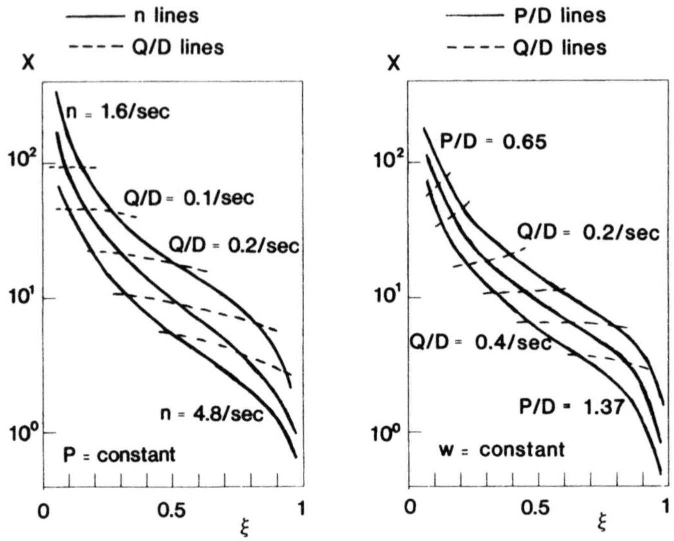

FIG. 29. Performance curves for two lobe co-rotating twin-screw extruders, showing the effect of degree of filling on the devolatilisation index.

previously, then

$$\mathbf{X} = \mathbf{F}(P, w, N, \xi)$$

where P = pitch, w = screw speed, N = number of flights, Q = volume throughput, Q_D = volume throughput under open discharge and $\xi = Q/Q_D$, degree of flight filling.

This functional relationship is presented in Fig. 29 for the case where $N = 2$.

It is clear that for constant ξ, **X** increases as the screw speed decreases and as the pitch decreases. Also for a constant throughput, **X** increases as the screw speed increases, but stays nearly constant (slight decrease) as the pitch increases.

It is also an instructive exercise to calculate the normalised throughput (actual throughput/open discharge throughput), G, as a function of the initial concentration, c_0, and the final concentration, c_f. This is performed using Figs 18 and 29 and taking realistic values for the remaining quantities. The result is shown in Fig. 30.

The conclusion which may now be drawn is that at a constant throughput, the ratio of final concentration to initial concentration

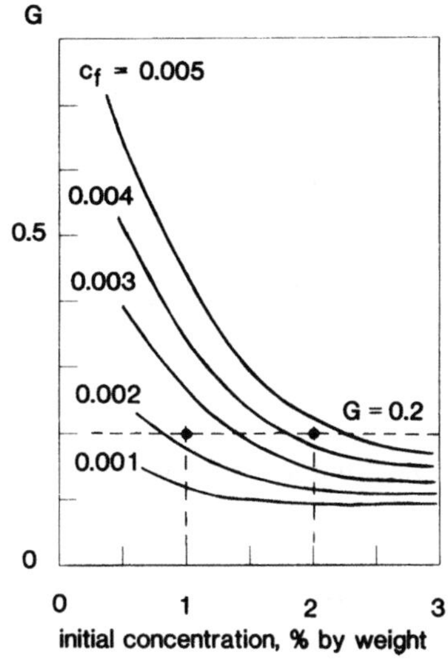

FIG. 30. Performance curves for two lobe co-rotating twin-screw extruders, showing the effect of initial concentration on throughput.

(c_f/c_0) is nearly constant. (For $G = 0{\cdot}2$, $c_f/c_0 = 0{\cdot}225$.) This fact is usually confirmed in practice.

5.8. Multi-stage Units

Multi-stage units are available with twin-screw units feeding single-screw units, with kneaders feeding single-screw units and with any combination derived on a custom built basis.

For PVC devolatilisation a planetary gear extruder feeding a single screw extruder has been used very successfully. This arrangement is shown in Fig. 31. With multi-stage units it is usual to perform the extraction at the interface of the individual units as shown. Multiple venting is readily performed by using vent ports in the individual units in addition.

The MP mixer, a twin-screw to single-screw unit, has been very successful as a devolatilising extruder, e.g. compounding, solvents from polypropylene, and hexane from ethylene–propylene rubber. The devolatilisation index of such units may be calculated from the region where extraction is performed, and integrated into the machine description as a multi-stage unit.

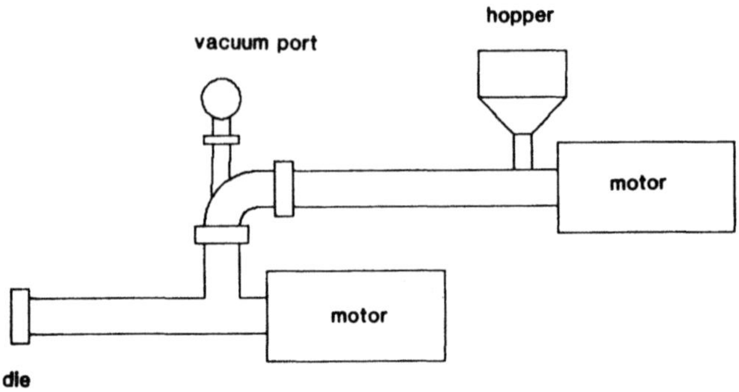

FIG. 31. Multi extruder devolatilisation.

5.9. Injection Moulding Machines

DV is conducted on injection moulding machines to remove the moisture content of hygroscopic polymers. It is preferred over pre-drying since the energy requirements are not so great. DV used during the moulding process helps eliminate moisture-related quality defects in the mouldings. However, the moulding machine is necessarily longer to accommodate the relief, DV and second transition zones. The intermittent axial movements of the plasticating screw as it retracts and injects dictates that the DV zone is at least as long as the stroke plus three diameters. It also means that the extraction is less effective since the port is not always central in the zone, and so the escape paths for the volatiles are longer. As the screw stroke increases this issue becomes even more critical. Since only water is being extracted a vacuum pump is not required and venting to atmosphere is used predominantly.

6. VENT PORT DESIGN

In Section 3 the need for adequate vent port cross section area was discussed. In this section vent port design will be discussed taking this factor and many others into account. The key parameters which must be considered are listed in Table 9. The amount of vapour to be removed must be determined, and correlated with the size of the vacuum system. It is possible that polymer particles, from bubble rupture, may be entrained if the mean velocity of the melt stream is

TABLE 9

FACTORS AFFECTING VENT PORT DESIGN

1. Amount of vapour to be removed
2. Vapour flow rate
3. Propensity of the melt to foam
4. Polymer/wall attachment force
5. Polymer melt rheology
6. Screw speed operating range
7. Melt temperature
8. Throughput rates of melt

excessive. Empirical results indicate that

$$v = 0{\cdot}21(p_v/p_p) \tag{7}$$

where v is the maximum allowable average velocity, p_v is the density of the volatile and p_p is the density of the polymer.

For single-screw extruders, multi-screw extruders, kneaders, the Diskpack and multi-stage machines where polymer may enter the port, the design is most critical. If foaming is excessive and would in general cause closure of the vent port then some device with a counteraction must be inserted into the vent port. Self cleaning stuffing screws are held in high regard for this purpose and are shown schematically in Fig. 32 for a twin-screw co-rotating extruder. As a guideline, if v exceeds 10 ft/s, a counteraction device should be installed in the vent. For low velocity flows there is a wide choice of vent ports available ranging from the circular vent of Fig. 33(c) through the circular vents with inserts in Figs 33(a) and 33(b) to the longitudinal slit vents shown in Figs 33(d) through 33(f). The use of inserts in the vents is commonplace to help overcome the polymer/wall attachment force

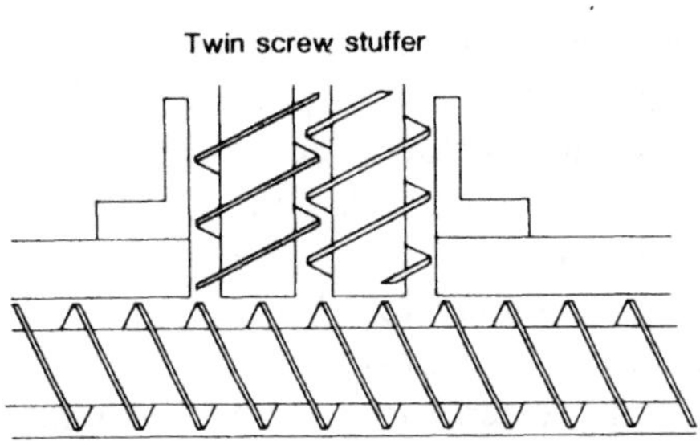

FIG. 32. The twin-screw stuffer to prevent vent flows.

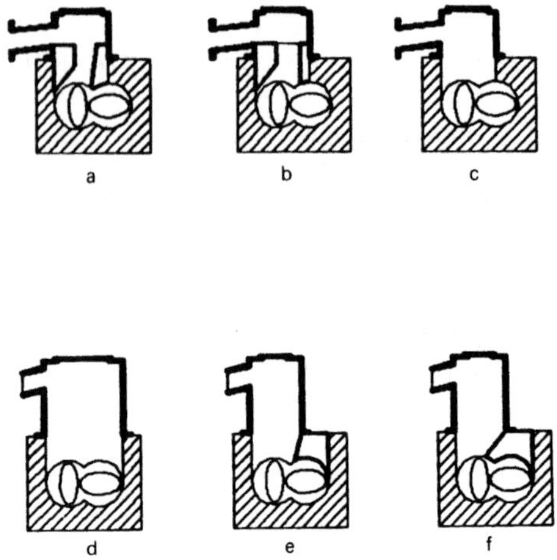

FIG. 33. Vent port types. (a) Circular vent, intermeshing zone covered and feed pocket; (b) circular vent, with horizontal outlet and feed pocket; (c) circular vent, with vertical outlet; (d) longitudinal slit vent, with vertical outlet; (e) longitudinal slit vent, with horizontal outlet; (f) longitudinal slit vent, intermeshing zone covered.

which aids a polymer in blocking the vent. The design of the vent inserts should ensure that the polymer is diverted back into the screw channels whilst the volatiles escape. The inserts can be supplied in standard shapes, but sometimes it is necessary to empirically determine the shape required. The longitudinal slit vents offer greater flow areas for the volatiles and provides less risk of polymer blocking the vent.

7. CLOSING REMARKS

Devolatilisation, as practised today, is a very necessary unit operation in the production and conversion of polymers. There is a great deal of equipment available to perform devolatilisation, and the original equipment manufacturers have collected a plethora of information on the subject. Some has found its way into the technical literature (see Bibliography), or has been disclosed in patents. Much more is held as trade secrets. The theory of devolatilisation has been developed for simple geometries and tested successfully. For many equipment items this has been translated into useful design equations and performance curves, even in the absence of an exact theoretical model. By judiciously using the available information, most of which is included in this text, it is invariably possible to gauge the equipment required for a new task from knowledge of viscosities and flow rates. However, well chosen experimental evaluations are still necessary to determine which is the optimum choice for a particular situation.

BIBLIOGRAPHY

Fundamentals

BIESENBERGER, J. A., 'Polymer devolatilization: theory of equipment', in *Proceedings of the 37th Annual Technical Conference of the Society of Plastics Engineers,* New Orleans, LA, 1979, pp. 971–7.

BIESENBERGER, J. A., 'Polymer devolatilization: theory of equipment', *Poly. Eng. and Sci.,* **20,** 1980, 1015.

BIESENBERGER, J. A., 'Two routes to effective resin devolatilization', *Plastics Engineering,* June 1984, 43–8.

BIESENBERGER, J. A. and TODD, D., 'Fundamentals', in *Devolatilization of Polymers,* J. A. Biesenberger, Ed., Hanser, Munich, 1983, pp. 3–31.

FLORY, P. J., *Principles of Polymer Chemistry,* Cornell University Press, Ithaca, NY, 1953.

LATINEN, G. A., 'Devolatilization of viscous polymer systems', *ACS Advances in Chemistry Series,* **34,** 1962, 235.

NEWMAN, R. E. and SIMON, R. H. M., 'A mathematical model of devolatilization promoted by bubble formation', *Seventy-third Annual A.I. Chemical Engineering Conference,* Chicago, IL, 1980.

PADBERG, G., 'Fundamental principles of polymer devolatilisation', in *Devolatilisation of Plastics,* transl. by S. Welling, VDI-Verlag, Düsseldorf, 1980, pp. 1–23.

TODD, D. B., 'Polymer devolatilization', in *Proceedings of the 32nd Annual Technical Conference of the Society of Plastics Engineers,* San Francisco, CA, 1974, pp. 472–5.

Vacuum Pumps and Systems

PAHL, M. H., 'Vacuum systems for the devolatilisation of polymers', in *Devolatilisation of Plastics,* transl. by S. Welling, VDI-Verlag, Düsseldorf, 1980, pp. 155–79.

Flash Evaporators

NEWMAN, R. E. and SIMON, R. H. M., 'A mathematical model of devolatilization promoted by bubble formation', *Seventy-third Annual A.I.Ch.E. Conference,* Chicago, IL, 1980.

PAHL, M. H., 'Devolatilisation through flash evaporation', in *Devolatilisation*

of Plastics, translated by S. Welling, VDI-Verlag, Düsseldorf, 1980, pp. 49–68.
TODD, D., SIMON, R. H. M., LANE, J. and BIESENBERGER, J. A., 'Equipment', in *Devolatilization of Polymers,* J. A. Biesenberger, Ed., Hanser, Munich, 1983, pp. 35–85.

Falling Film/Strand Evaporators

GRACE, J. R., *Trans. Inst. Chemical Engineers,* **51,** 1973, 116–20.
NEWMAN, R. E. and SIMON, R. H. M., 'A mathematical model of devolatilization promoted by bubble formation', *Seventy-third Annual A.I.Ch.E. Conference,* Chicago, IL, 1980.
TODD, D., SIMON, R. H. M., LANE, J. and BIESENBERGER, J. A., 'Equipment', in *Devolatilization of Polymers,* J. A. Biesenberger, Ed., Hanser, Munich, 1983, pp. 35–85.

Thin Film Vaporisers

HEIMGANENER, E., 'Devolatilisation in the thin-film vaporiser', in *Devolatilisation of Plastics,* translated by S. Welling, VDI-Verlag, Düsseldorf, 1980, pp. 69–97.
MCKELVEY, J. M. and SHAMPS, G. V., 'Fluid transport in thin film polymer processors', *Poly. Eng. & Sci.,* **19**(10), 1979, 651–9.
TODD, D., SIMON, R. H. M., LANE, J. and BIESENBERGER, J. A., 'Equipment', in *Devolatilization of Polymers,* J. A. Biesenberger, Ed., Hanser, Munich, 1983, pp. 35–85.

Single Screw Extruders

ANDERS, D., 'The use of single screw extruders for devolatilising thermoplastics', in *Devolatilisation of Plastics,* translated by S. Welling, VDI-Verlag, Düsseldorf, 1980, pp. 133–54.
BIESENBERGER, J. A., 'Polymer devolatilization: theory of equipment', *Poly. Eng. & Sci.,* **20,** 1980, 1015.
BIESENBERGER, J. A. and KESSIDIS, G., 'Devolatilization of polymer melts in single-screw extruders', *Poly. Eng. & Sci.,* **22**(13), 1982, 832–35.
COUGHLIN, R. W. and CANEVASI, G. P., *A.I.Ch.E. J.,* **15,** 1969, 560–4.
FRANZKOCH, B., 'Devolatilisation during extrusion', in *Devolatilisation of Plastics,* translated by S. Welling, VDI-Verlag, Düsseldorf, 1980, pp. 199–225.
LATINEN, G. A., 'Devolatilization of viscous polymer systems', *ACS Advances in Chemistry Series,* **34,** 1962, 235–46.
ROBERTS, G. W., *A.I.Ch.E. J.,* **16,** 1970, 878–82.
TODD, D., SIMON, R. H. M., LANE, J. and BIESENBERGER, J. A., 'Equipment', in *Devolatilization of Polymers,* J. A. Biesenberger, Ed., Hanser, Munich, 1983, pp. 35–85.

Diskpack

HOLD, P., TADMOR, Z. and VALSANNIS, L., 'Applications and design of a novel polymer processing machine', in *Proceedings of the 37th Annual*

Technical Conference of the Society of Plastics Engineers, New Orleans, LA, 1979, pp. 205–11.

Kneaders

FRANZ, P., 'Devolatilisation in plastics compounding' in *Devolatilisation of Plastics,* translated by S. Welling, VDI-Verlag, Düsseldorf, 1980, pp. 181–98.

Twin Screw Extruders

COLLINS, E. P., DENSON, C. D. and ASTARIA, G., 'The length of a transfer unit (LTV) for polymer devolatilization processes in screw extruders', in *Proceedings of the 41st Annual Technical Conference of the Society of Plastics Engineers,* Chicago, IL, 1983, pp. 138–41.
FRANZKOCH, B., 'Devolatilisation during extrusion', in *Devolatilisation of Plastics,* translated by S. Welling, VDI-Verlag, Düsseldorf, 1980, pp. 199–226.
GRAS, E. and EISE, K., 'The use of multi screw extruders for devolatilizing low solids polymer solutions', in *Proceedings of the 33rd Annual Technical Conference of the Society of Plastics Engineers,* Atlanta, GA, 1975, pp. 386–9.
HESS, M. and EISE, K., 'Devolatilization of residual styrene monomers in twin-screw extruders', in *Proceedings of the 35th Annual Technical Conference of the Society of Plastics Engineers,* Montreal, 1977, pp. 460–3.
TODD, D., SIMON, R. H. M., LANE, J. and BIESENBERGER, J. A., 'Equipment', in *Devolatilization of Polymers,* J. A. Biesenberger, Ed., Hanser, Munich, 1983, pp. 35–85.
WERNER, H., 'Devolatilisation of polymers in multi-screw devolatilisation', in *Devolatilisation of Plastics,* translated by S. Welling, VDI-Verlag, Düsseldorf, 1980.

Multi-State Extruders Devolatilisation

TODD, D., 'Polymer devolatilization', in *Proceedings of the 32nd Annual Technical Conference of the Society of Plastics Engineers,* San Francisco, CA, 1974.

Vent-Port Designs

WERNER, H., 'Devolatilisation of polymers in multi-screw devolatilisers', in *Devolatilisation of Plastics,* translated by S. Welling, VDI-Verlag, Düsseldorf, 1980.

Chapter 3

ETHYLENE VINYL ACETATE AND ITS ACETOXY–HYDROXIDE TRANSFORMATION

E. F. VANSANT

Department of Chemistry, University of Antwerp, Wilrijk, Belgium

1. INTRODUCTION

Ethylene vinyl acetate (EVA) copolymers belong to the family of polyolefin resins. They are thermoplastic materials which are produced by polymerising ethylene gas with vinyl acetate. They can be synthesised to contain only a few per cent by weight of the vinyl acetate comonomer or they can be produced with amounts of vinyl acetate well in excess of 50%. Those polymers with a low vinyl acetate content have properties closely related to those of a low density ethylene homopolymer (LPDE) while those with a high vinyl acetate content are similar in appearance to gum rubber.

The effect of introducing vinyl acetate during the polymerisation process is twofold:

(a) The crystallinity of the copolymer (as compared to LPDE) is progressively decreased with increasing VA content, and
(b) polarity is increased.

These properties result in a lowering in rigidity, softening point and surface hardness and increased clarity, permeability, solubility, environmental stress crack resistance, low temperature toughness, compatibility with other polymers and resins, acceptance of fillers and coefficient of friction. Also important are the absence of plasticisers, resistance to weathering and ozone attack and the relative ease with which the materials can be crosslinked using chemical or irradiation techniques.

As mouldable/formable thermoplastics, EVA copolymers are finding applications across a very broad front. EVA copolymer films can be produced on conventional tubular film equipment. Also, EVA copolymers can be extruded into tube, piping or any other profile with extruders and dies normally used for processing LDPE or PVC. EVA resins are easily processed on any type of injection or blow moulding machine. Again, conventional extrusion coating equipment can be used with EVA copolymers. The use of EVA as a component in hot melt coating is also an important and growing industrial application.

In order to improve the processing and performance properties of EVA copolymers, acetoxy–hydroxide transformation processes can be carried out to obtain a new range of thermoplastic co- and terpolymers. A critical evaluation of the comprehensive patent literature on the acetoxy–hydroxide transformation of EVA was given by Koopmans *et al.*[1] This chapter will therefore discuss the transformation processes and look at the effects of such transformations on the physical properties of the resultant materials.

2. EVA PRODUCTION

Ethylene vinyl acetate copolymers are produced from the monomers ethylene and vinyl acetate (Fig. 1). The chemical coupling of both monomers can be achieved by a high-pressure, a solution or an emulsion process. A variety of EVA copolymers, having different vinyl acetate incorporation and molecular weight, can be prepared

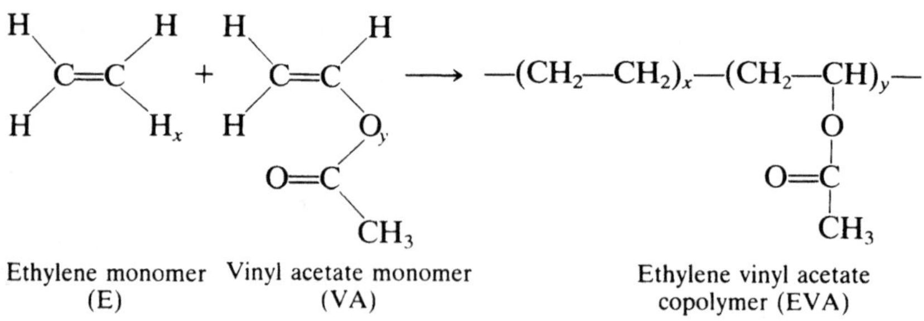

FIG. 1. The preparation of ethylene vinyl acetate (EVA) copolymers. (The subscripts x and y indicate the mol % of each monomer present in the copolymer.)

using the copolymerisation equation:[2]

$$\frac{d(M_1)}{d(M_2)} = \frac{(M_1)}{(M_2)} \frac{r_1(M_1) + (M_2)}{(M_1) + r_2(M_2)}$$

where (M_1) and (M_2) are the concentrations of the monomers 1 and 2 respectively: r_1 and r_2 are rate constants.[3,4]

3. ACETOXY-HYDROXIDE TRANSFORMATION PROCESSES IN EVA COPOLYMERS

EVA copolymers contain reactive acetoxy groups which can be transformed into OH-groups (Fig. 2). This kind of acetoxy–hydroxide exchange is known under different names: hydrolysis, saponification, alcoholysis and transesterification. This diversity of nomenclature can lead to erroneous interpretations. The various reaction names were given to indicate specific carboxylic acid derivative reactions.

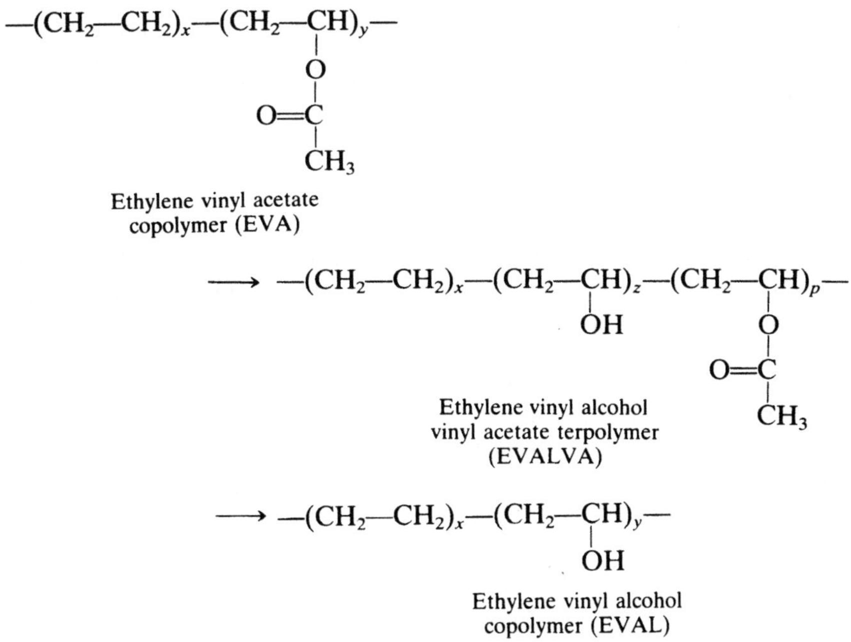

FIG. 2. The acetoxy–hydroxide transformation of an ethylene vinyl acetate copolymer. (The subscripts x, y, z, p, refer to the mol % of each monomer in the macromolecule.)

3.1. Hydrolysis and Alcoholysis

Hydrolysis implies a reaction of an ester with water, a generally slow reaction usually strongly catalysed by either acid or base (Fig. 3). Acid catalysed ester hydrolysis is an equilibrium process driven to completion by using a large excess of water. Base catalysed hydrolysis, however, is not an equilibrium reaction but always proceeds to completion. Here the base, a hydroxide ion, is a true reagent instead of a catalyst. Of the two nucleophiles present, H_2O and OH^-, the latter is the strongest, and adds more rapidly to the carbonyl of the ester than the weak nucleophile, H_2O. The base catalysed hydrolysis of an ester is better known as a saponification.

Alcoholysis is, analogous to hydrolysis for water, a general term to indicate reaction of carboxyl acid derivatives with alcohols. The alcoholysis of esters, however, is given the name 'transesterification', for alcohols reacting with esters to give new esters and alcohols. The reaction is catalysed either by acid or base, and the mechanisms for the transesterification (Fig. 4) involve steps identical to those for acid and base catalysed hydrolysis, with one significant exception. In the base catalysed transesterification, the acid–base reaction never occurs because the free carboxylic acid is never formed. Thus, base catalysis is subjected to the same equilibrium conditions that apply to the acid catalysed reaction. The base catalyst is always an alkoxide ion, usually corresponding to the primary or secondary alcohol used, and not a hydroxide ion. The reaction can be driven to completion by distilling off the newly formed ester.

Because of the different mechanisms, it is more appropriate to name the new polymers according to their structure, i.e. ethylene vinyl alcohol (EVAL) co- and ethylene vinyl alcohol vinyl acetate (EVALVA) terpolymers.

3.2. Transformation Techniques

Each of the reaction mechanisms specified in Figs 3 and 4 is applicable to EVA copolymers. In general, two techniques can be used for the polymer transformation reactions: these are homogeneous transformation and heterogeneous transformation.

3.2.1. Homogeneous Transformation

The EVA copolymers are processed in an homogeneous reaction medium, i.e. a one phase system. To achieve such a medium, one can dissolve the EVA copolymer, add the required reagent and finally

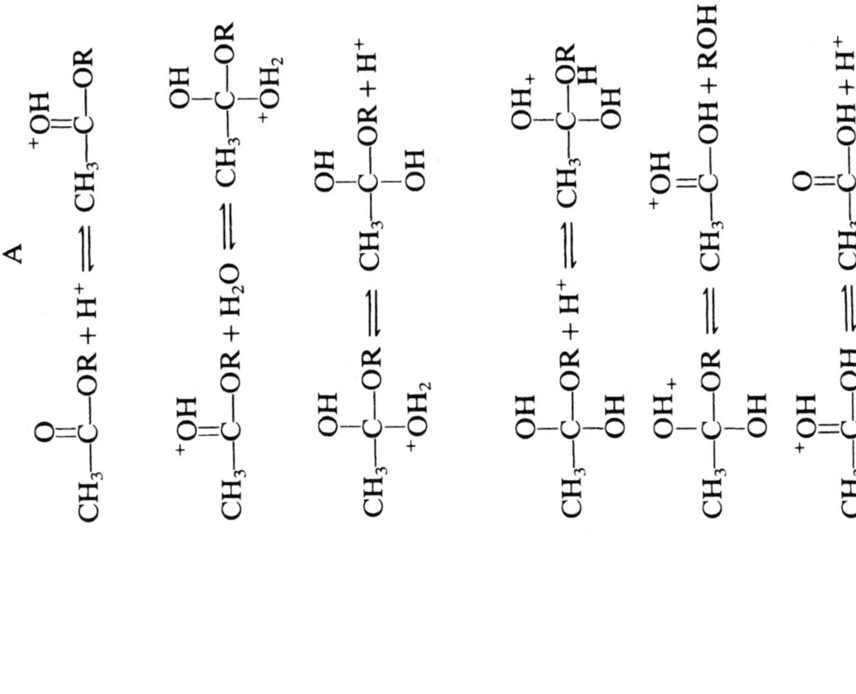

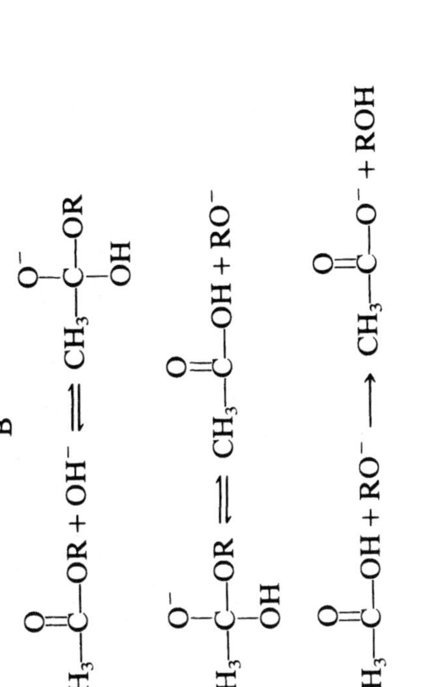

FIG. 3. The reaction mechanism of an acid (A) and a base (B) catalysed hydrolysis of an acetoxy ester group. (R represents a secondary carbon atom as part of the hydrocarbon backbone of the ethylene vinyl acetate copolymer.)

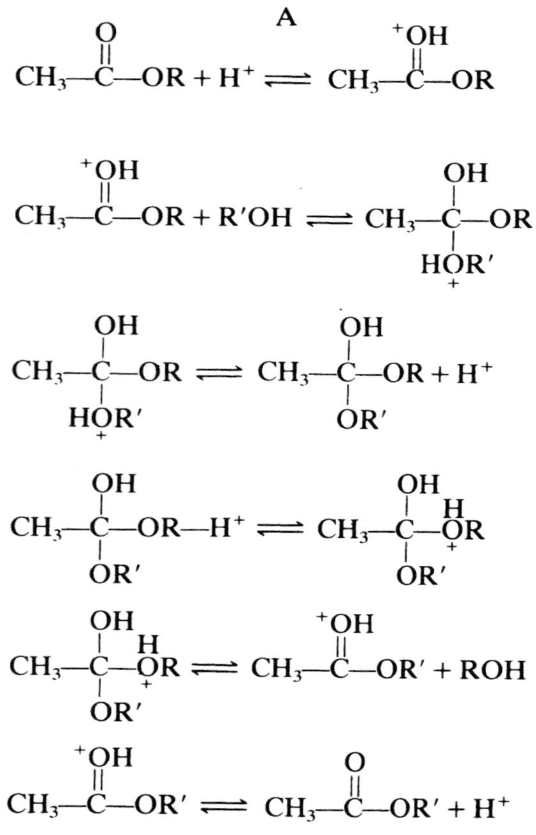

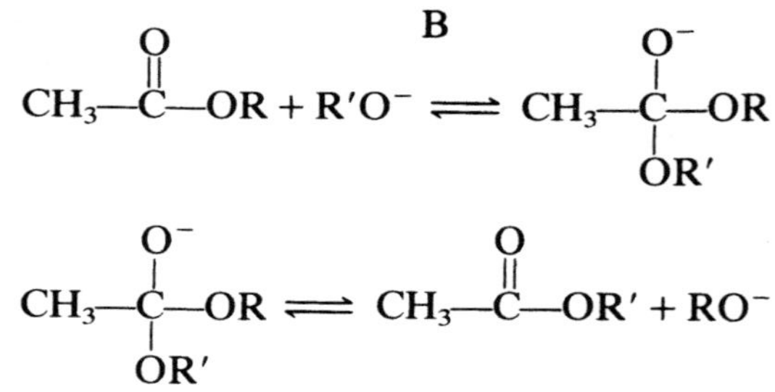

FIG. 4. The reaction mechanism of an acid (A) and a base (B) catalysed transesterification of an acetoxy ester group. (R represents a secondary carbon atom as part of the hydrocarbon backbone of the ethylene vinyl acetate copolymer. R′ represents the hydrocarbon part of the usually primary or secondary alcohol used as reagent. For the base catalysed reaction (B) the alkoxide anion normally corresponds to the alcohol used.)

recover pure EVAL or EVALVA and the solvents. This approach has several disadvantages: (1) large amounts of solvent which are difficult to recover, (2) basic media are preferred instead of acidic ones (too corrosive), as end-product discoloration—yellow brown—tends to occur and (3) if alkali metal hydroxides are used in excess, the formed acetate salts are hard to remove.

3.2.2. Heterogeneous Transformation

In this case we have a two phase system, composed of the copolymer and the reaction medium. The EVA copolymer can either be present in pellet, or powder form in a non solvent, in which case there is a risk of a suspension in which reaction occurs only at the surface (this is undesirable for EVAL production), or as drops of molten EVA in a non-solvent. This last approach has received a lot of attention in recent patents. The method consists basically of: production of a melt; the formation of a melt dispersion in water, containing an emulsifier and reagent; and finally the cooling of the entire system. The main advantage of this process is the possibility of controlling the particle size of the prepared EVAL or EVALVA polymers. Grinding techniques can be avoided if the polymers are recovered as crumbs which can be used for powder coating.

Detailed information about the transformation processes is summarised by R. J. Koopmans *et al.*[5]

4. THE CHARACTERISATION OF EVA AND ITS DERIVATIVES

The properties of EVA are mainly determined by the molecular weight distribution and the vinyl acetate content. Also, for the EVALVA and EVAL derivatives, the amount of vinyl alcohol incorporated is important.

4.1. Molecular Weight Distribution (MWD)

The molecular weight distribution is very dependent on the synthesis process and the reaction conditions. Unfortunately, quantitative MWD measurements of co- and terpolymers are extremely difficult to make.[6,7] A rough indication of the MWD can be found using the melt index (MI) (ASTMD 1238-73). Compared to the parent EVA, the

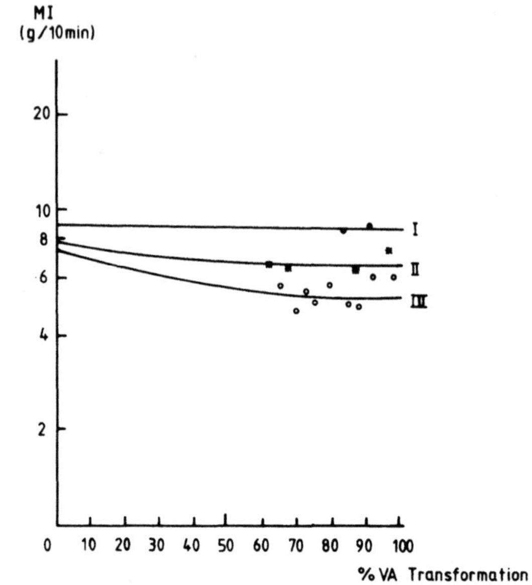

FIG. 5. Melt index as a function of the % VA transformation for three EVA copolymer types, containing originally 3·5 (I), 5·0 (II) and 12·0 (III) mol % VA.

TABLE 1

HIGH PERFORMANCE SIZE EXCLUSION CHROMATOGRAPHIC RESULTS OF SEVERAL EVA AND EVAL/EVALVA POLYMERS

The number average ($\bar{A}_n$) and weight average ($\bar{A}_w$) molecular size were calculated using a polystyrene calibration curve.

Type		($\bar{A}_n$)	($\bar{A}_w$)	$\bar{A}_w/\bar{A}_n$
EVA (I)	3·5% VA	19·154	187·940	9·8
(II)	12% VA	14·856	127·679	8·6
(III)	18·5% VA	12·031	105·873	8·8
EVAL/EVALVA[a]	$I_{91\cdot3}$	25·085	170·118	6·8
	$II_{98\cdot4}$	9·582	48·188	5·0
	$II_{66\cdot0}$	26·370	109·810	4·2

[a] The EVAL/EVALVA samples refer to the parent EVA (I, II, III), while the subscript indicates the VA % transformation.

EVAL/EVALVA polymers have lower MIs (Fig. 5). The MI decreasing with increasing % VA transformation was also observed by Schillgalies *et al.*[8] and they attributed this to the increased chemical bonding forces between hydroxyl groups.

A qualitative idea of the molecular weight distribution can be obtained with high performance size exclusion chromatography (HPSEC). The obtained distribution gives the molecular size distribution (MSD). Table 1 shows the number average ($\bar{A}_n$) and weight average ($\bar{A}_w$) molecular size for EVA and their derivatives. The $\bar{A}_w/\bar{A}_n$ ratio indicates the width of the MSD. These values showed for the EVA copolymers a size reduction with increasing mol % VA incorporation. This feature may be due to slightly different reaction conditions used in the EVA synthesis. Furthermore, the MI seems to reflect the $\bar{A}_n$ and $\bar{A}_w$ (MI $\sim 1/\bar{A}_n$ or $\bar{A}_w$) values well. This is in agreement with the results of Barlow and Campbell.[9] A comparison of the EVAL and EVA MSD curves (Fig. 6) reveals: (1) a shift to lower molecular size with increasing mol % VA and (2) a reduction of the low molecular size fractions for the EVAL copolymers.

4.2. Quantitative Vinyl Acetate Determination

The methods available to determine the VA content in polymers can be divided into chemical and instrumental methods.

4.2.1. Chemical Methods

On considering the EVA copolymer structure, it is readily seen that these macromolecules contain highly reactive acetoxy ester groups. Therefore, a quantitative reaction of any chemical compound with the VA units and the possibility of detecting formed or disappeared reaction products provides a suitable method of determining the EVA composition.

4.2.1.1. Acetoxy–hydroxyl transformation. The reaction of a base (sodium or potassium hydroxide) with EVA copolymers, usually referred to as a saponification, is (in principle) a very simple chemical method of determining the VA content.

Depending on the solubility of the EVA copolymer, a sample of about 1–2 g is dissolved in an ethanol/potassium hydroxide (KOH) solution, with or without the aid of an aromatic solvent (e.g. toluene).[10,11]

It is, however, preferable to have one general procedure applicable

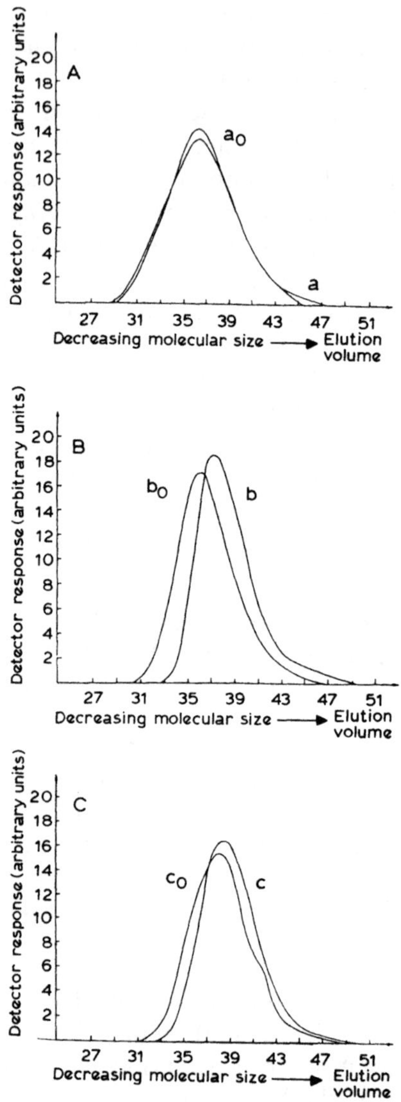

FIG. 6. Comparison of the molecular size distributions for the parent EVA (a_0, b_0, c_0) and the prepared EVAL copolymers (a, b, c). (A) Parent EVA with 3·5 mol % VA; (B) parent EVA with 12 mol % VA; (C) parent EVA with 18·5 mol % VA.

to materials containing 0% up to and including 100 wt % VA. Such a method[12] uses a 50/50 (by volume) ethanol/toluene reaction medium. This procedure seems to give only a 0·5 wt % VA reproducibility for reactions with EVA copolymers having a 15 wt % or higher VA content.

As an excess of KOH is used, the unreacted base can be titrated with a hydrochloric acid solution, using phenolphthalein as indicator, after the reaction in eqn (1) is completed.

$$R—OOCCH_3 + KOH \rightarrow R—OH + KOOCCH_3 \qquad (1)$$

where R = hydrocarbon polymer chain.

After an additional blank reaction, the wt % VA can be calculated by

$$\text{wt \% VA} = (V_2 - V_1) \,.\, N \,.\, 8{\cdot}609/m$$

where V_2 = ml acid sample titration, V_1 = ml acid blank, N = normality of acid solution and m = weight sample (g).

At temperatures of 80°C or less, the reaction time ranges from 3 to 10 h.[10–12]

Consequently, this method is time-consuming. Moreover, the calculated wt % VA values should be considered with care, because:

—A saponification in an alcoholic medium automatically produces a base-catalysed transesterification, which (in turn) produces a new ester also capable of reacting with KOH.[13]

—Due to incomplete dissolution of some EVA copolymers (wt % VA < 10) and/or the newly formed macromolecules, not all VA groups can be reached by the base.[14]

—Absorption and diffusion processes on gel particles might reduce the reaction rate and influence the KOH concentration.

Still, a saponification remains an absolute method that needs no calibration. According to Mayor *et al.*,[11] reproducible results for EVA copolymers with a VA content of up to 35 wt % of about 2 wt % VA can be obtained.

4.2.1.2. Formation of acetic acid in acid medium. Two different methods can be used, a distillation method[15] and a melt procedure.

The distillation method has a very complicated experimental set-up and procedure. About 1 g EVA sample is brought into contact with chromic acid. After a 3-h reaction period, the acetic acid formed is distilled off with water. The experimental set-up and reaction conditions are carefully chosen in order not to contaminate the acetic acid solution with other acids during distillation.

The newly formed acetic acid is titrated with a sodium hydroxide (NaOH) solution, using phenolphthalein as indicator. From the amount of NaOH, the wt % VA is calculated. It should be noted that this DIN method does not give reproducible results to within 0·1 wt % VA when, additional to the VA groups, other esters are present in the macromolecules or in the EVA blend. Apart from this restriction, the method is applicable to any EVA copolymer as a pure resin or as a composite material.

The melt procedure has the advantage that no contamination of the acetic acid solution during a distillation is possible. An EVA sample of 100–500 mg is reacted with 1–2 g molten *p*-toluenesulphonic acid (*p*Ts) for 2–2·5 h at 160°C.

According to eqn (2), acetic acid should be eliminated.

$$\text{—CH}_2\text{—}\underset{\displaystyle\text{OOCCH}_3}{\underset{|}{\text{CH}}}\text{—} \xrightarrow{p\text{Ts}} \text{—CH=CH—} + \text{CH}_3\text{COOH} \qquad (2)$$

The reaction can be carried out in small, closed glass tubes, which are immersed in an oil bath. After the reaction is completed, the glass tubes are broken in water. Then the liberated acetic acid is titrated potentiometrically with a 0·1 N NaOH solution. During the potentiometric titration, the bulk *p*Ts should be neutralised first. The amount of 0·1 N NaOH (a), obtained from the equivalence point differences of *p*Ts and acetic acid, allows the wt % calculation by:

$$\text{wt \% VA} = 8{\cdot}61a/m$$

where a = ml 0·1 N NaOH and m = weight sample (mg).

Although Aydin *et al.*[16] claimed the method to be applicable to any EVA type, including those in the composite materials, with a reproducibility of 0·3 wt % VA, Leukroth[17] established that it was not suitable for EVA/rubber blends. Aydin *et al.*[16] showed that there could be good agreement between results obtained by infrared spectrometry, nuclear magnetic resonance, elemental analysis and this method for VA determination. However, for EVA types with less than 15 wt % VA, this method gave considerably higher values than VA results obtained by instrumental techniques.

4.2.2. Instrumental Methods

In contrast to the chemical methods, the instrumental techniques do not depend on the reactivity of the ester group for the VA content determinations.

4.2.2.1. Elimination of acetic acid. Each method belonging to this class of analysis makes use of the thermal decomposition of EVA copolymers at elevated temperatures in an inert atmosphere, i.e. a pyrolysis (P). The pyrolysis behaviour of EVA is based on the decomposition mechanism proposed by Grant and Grassie[18] and by Grassie.[19] At temperatures above 200°C, the acetoxy groups are eliminated and form, via a six-centre transition state, unsaturation and acetic acid (eqn (3)).

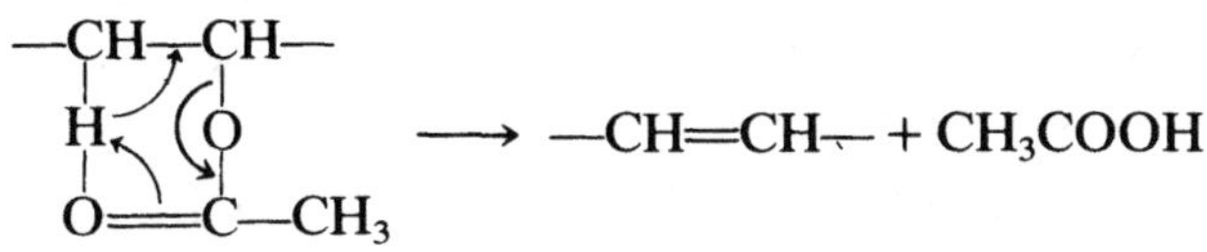
(3)

The simplest way to detect the liberated acid during a pyrolysis is by passing all gases through a 0·1 N NaOH solution.[20] By titrating the remaining NaOH, not neutralised by acetic acid, with a 0·1 N hydrochloric acid solution, using phenolphthalein as indicator, the wt % VA can be calculated by:

$$\text{wt \% VA} = (V_2 - V_1) \,.\, 860{\cdot}9/m$$

where V_1 = ml 0·1 N hydrochloric acid sample titration, V_2 = ml 0·1 N hydrochloric acid blank and m = weight sample (mg).

The optimum pyrolysis conditions are 395°C for 20 min, and EVA samples of about 1 g can be used.

This method is especially valid for examining high-pressure EVA copolymers having less than 50 wt % VA. The reproducibility is less than or equal to 0·4 wt % VA.[20] For EVA copolymers with a VA content greater than 50 wt % VA, this procedure is also applicable when the sample size is reduced and/or the concentration of the NaOH solution is increased. Unless the composition is known this method does not apply to blends or EVA copolymers containing large amounts of additives. Once the solutions are prepared and the pyrolysis initiated one experiment takes about 30 min.

4.2.2.2. Pyrolysis–gas chromatography and thermogravimetry. Pyrolysis–gas chromatography (P–GC) provides a more sophisticated method of determining the ethylene/vinyl acetate ratio. This method was developed by Barrall *et al.*,[21] in order to examine EVA/ethylene ethyl acrylate (EEA) mixtures. An absolute accuracy of 2 wt % VA can be obtained. Although no sample residence time at 475°C was mentioned, it was established that:

—no gas chromatogram can be obtained below 200°C;
—no anomalous effects are noted for sample variations (0·005–0·09 g);
—temperature variations between 350°C and 490°C do not change the amount of acetic acid formed.

Okumato *et al.*,[22] obtained for P–GC a reproducibility of less than 3·0 wt % on examining EVA copolymers containing 44·8–87·3 wt % VA. They also found that, for a 0·8–1·5 mg EVA sample held for 1 min at 550°C, the evolved acetic acid percentage could be correlated linearly with the wt % VA by:

$$\text{wt \% VA} = 1{\cdot}19 \text{ Acetic acid \%} - 4{\cdot}27.$$

A slightly different method was used by Chiu,[23] who replaced the pyrolysis oven with a device for thermogravimetry (TG). In addition to the VA content determination, TG–GC allowed the study of mass changes of the material as a function of temperature. Consequently, a better knowledge of the EVA decomposition products becomes possible.

Mlejnek *et al.*[24] used TG without GC for VA content determinations. Working with optimum conditions of 770°C for 15 s or 550°C for 10 min, the relative error for TG alone was smaller than in combination with GC. Although this method seemed to give no restrictions for the EVA types. Disselhoff[25] claimed that the method is limited to EVA copolymers having greater than 25 wt % VA. The TG method proved very valuable for examining electric cable covering, i.e. EVA copolymers containing carbon black and other ingredients.[24]

4.2.2.3. Elemental analysis.[7,26] A method commonly used by organic chemists to determine the molecular formula of hydrocarbon compounds is combustion analysis. This technique, developed by Lavoisier and Liebig, is conceptually very simple and can be used to determine the VA incorporation in EVA copolymers.

A weighed quantity of sample (a few mg) is burned in oxygen in the presence of red-hot copper oxide at about 700°C. Carbon dioxide and water are produced and are individually captured and weighed in absorption tubes. This is accomplished by passing the combustion gases through a tube containing a neutral desiccant, which absorbs the water vapour, then by leading them through a tube that absorbs carbon dioxide. The wt % carbon and hydrogen in the compound can be calculated by the stoichiometric principles of general chemistry. As the sum of these percentages is not 100, the difference is due to oxygen in the acetoxy groups. As 1 mol VA produces 2 mol atomic oxygen, the wt % VA can readily be calculated by:

$$\text{wt \% VA} = 2{\cdot}69 \text{ wt \% O}$$

Although this method has several advantages (small sample size, applicable to any EVA type, accuracy of 0·3 wt % VA), it still remains a cumbersome technique that is not specific, i.e. no composite EVA materials can be examined. Moreover, elemental analysis appears to be inaccurate at low VA content.

4.2.2.4. Proton nuclear magnetic resonance (^{1}*H-NMR*). A measuring technique that directly determines the EVA composition without destroying the sample examined is ^{1}H-NMR. By detecting the intensities of the distinguishable proton peaks of ethylene and vinyl acetate monomers, the wt % VA incorporation can be calculated. In addition, the microstructure of the EVA copolymers can be studied, e.g. sequence distribution and degree of branching. EVA copolymer spectra are readily obtained within 10 min, after scanning a less than or equal to 10% w/v copolymer solution in a suitable solvent (Fig. 7).

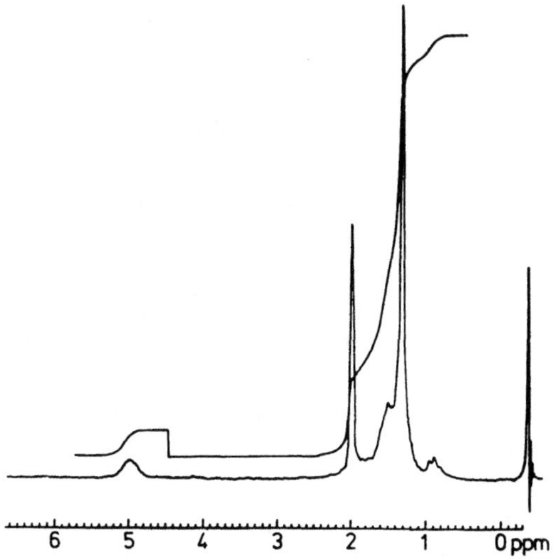

FIG. 7. A typical ^{1}H-NMR spectrum of an EVA (28·5 mol % VA) copolymer.

In the literature, several solvents are mentioned, e.g. monochlorobenzene, *o*-dichlorobenzene, chloroform, *cis*-1,2-dichloroethane. Commonly, spectra are obtained at temperatures near the boiling point of the solvent used, in order to increase the resolution.

To calculate the wt % VA content, two methods are proposed:

—The simplest procedure is to measure the ratio of the peak area characteristic for the ethylene;

$$I\,CH_3(VA) \text{ or } I\,CH(VA)/I\,CH_2$$
$$(I = \text{peak area})$$

—A more cumbersome procedure is to obtain a ^{1}H-NMR spectrum of a sample solution containing an internal standard (e.g. ferrocene). In this case, the ratio of the peak area characteristic for ferrocene and the copolymer can be used to calculate the wt % VA:

$$\frac{I_f}{I_c} = \frac{10W_f/186{\cdot}12}{X \cdot W_c} \quad \text{and} \quad X = 4(1-\alpha)/28{\cdot}05 + 5\alpha/86{\cdot}09$$

where I_f = peak area internal standard (ferrocene), $I_c = I\,CH_2(E + VA) + I\,CH_3(VA)$ = peak area copolymer (methine proton is not included)[2], W_f = weight ferrocene, W_c = weight copolymer, X = proton counts per unit weight of the copolymer and α = mol fraction VA.

Chen and Lewis[27] found this method very attractive. With increasing VA content, the changing environment of the CH_2 protons produces a less resolved CH_3(VA) proton peak, making a ratio method without using an internal standard less desirable. This complication can be anticipated by using the CH (VA) proton peak, which becomes more intense with an increasing VA content.

According to Chen and Lewis,[27] the ^{1}H-NMR method gives an accuracy of 1 wt % VA for any EVA type.

4.2.2.5. Infrared (IR) spectrometry. Infrared spectrometry cannot only be used to determine the ethylene/vinyl acetate ratio, but can also give information about the average VA sequence length[28] and the VA sequence distribution.[29] In order to obtain quantitative IR data about the VA content, a calibration curve should be available. Usually this can be obtained by determining the absorption ratio of a band characteristic for the acetoxy groups and one typical for the ethylene sequence of a standard EVA IR spectrum. When no EVA standards

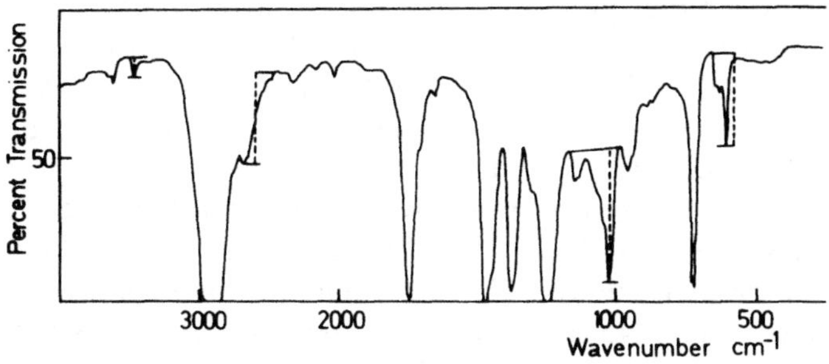

FIG. 8. Typical IR spectrum of an EVA (9·16 mol % VA) copolymer.

are available, the VA content of pure EVA resins can be determined by a method that has a higher accuracy rate than the IR technique.[13] A combination of these results and the ones from IR absorption ratios will yield a valid calibration curve.

According to Fig. 8 and Table 2, it can be seen that several IR bands allow a quantitative VA determination. Important conditions for a proper band choice, apart from being quantitative for the VA groups, are:

—good isolation (i.e., no interference from other bands) and
—satisfactory intensity (i.e. preferable base line limits between 20 and 80% transmission).

Depending on the sample preparation, different band choices might be advisable.

TABLE 2
IMPORTANT BANDS FOR THE QUANTITATIVE IR ANALYSIS OF EVA COPOLYMERS

$\bar{\nu}$ (cm^{-1})	*Band characterisation*
3 460	C=O stretching (1st overtone)
2 678	CH_2 stretching
1 743	C=O stretching (ester)
1 465	CH_2 bending
1 250	C—O stretching (acetate ester)
1 020	—COOR bending
720	—$(CH_2)_n - n \geq 4$ rocking
610	—COOR wagging

For EVA samples pressed to thin foils, Mayor and Sodomka[11] used (for the IR VA determination) the 720/610 cm^{-1} absorption ratio and found a reproducibility of 5 wt % VA for the 0–35 wt % VA range. Siryuk and Bulgakova[30] obtained a better reproducibility, 2·6 wt % VA, compared to the saponification method by using the 1743/1245 cm^{-1} absorption ratio. Koopmans *et al.*[13] proposed several possibilities (the 3460/2678 cm^{-1}, 1020/2678 cm^{-1}, and 610/2678 cm^{-1} absorption ratios), for the VA determination of high-pressure EVA copolymers. The 1743 cm^{-1} carbonyl stretching band was too intense for quantitative purposes due to the high film thickness (50 ± 20) μm. The best results, however, can be obtained using the 3460/2678 cm^{-1} absorption ratio.[19] However, for EVA copolymers with a VA content more than 33 wt % VA, the relationship of absorption ratio v. VA content does not remain linear.

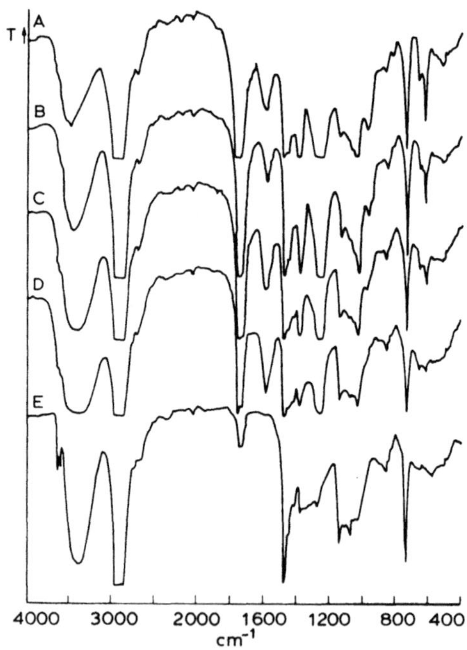

FIG. 9. IR spectra of EVAL(E)/EVALVA(A–D) polymers; obtained with a Beckman 4240 spectrometer, using thin foil samples. The parent EVA had a mol % VA of 12·0. The % VA transformation is respectively 21·4(A), 39·0(B), 63·4(C), 79·8(D) and 98·4(E).

TABLE 3
THE IR BAND CHARACTERISTICS OF EVAL CO- AND EVALVA TERPOLYMERS SPECTRA REPRESENTED IN FIGURE 9

$\bar{\nu}$ (cm^{-1})	*Vibrational mode*
3 625	Amorphous isolated OH stretching
3 595	Crystalline isolated OH stretching
3 550–3 350	OH stretching (inter- and intramolecular hydrogen bond)
3 460	C═O stretching (first overtone)
2 958	CH_3 stretching (antisym.)
2 920	CH_2 stretching (antisym.)
2 880	CH_3 stretching (sym.)
2 858	CH_2 stretching (sym.)
2 735	CH_3
2 678	CH_2
2 320	CO_2
2 010	—
1 910	—
1 735	C═O stretching (ester)
1 715	C═O stretching (hydrogen bond with OH)
1 575	C═O stretching (acetate salt; sym.)
1 470	CH_2 bending
1 450	C—CH_3 bending (antisym.)
1 410	C═O stretching (acetate salt, antisym.)
1 378	C—CH_3 bending (sym.)
1 365	CH_2 bending
1 350	CH_2 bending
1 310	CH_2 wagging and twisting
1 265	—
1 250	C═O stretching (acetate ester)
1 140	C—O stretching (tertiary OH)
1 100	C—O stretching (secondary OH)
1 065	—
1 050	C—O stretching (primary OH)
1 020	—COOR bending
960	RCH═CHR bending (out of plane)
890	R_2C═CH_2 bending (out of plane)
870	—
800	—
730	crystalline —$(CH_2)_n$—with $n \geqslant 4$; rocking
720	amorphous —$(CH_2)_n$—with $n \geqslant 4$; rocking
640	—
630	—
610	—COOR wagging

4.3. Determination of the Vinyl Alcohol Content

The quantitative determination of the vinyl alcohol (VOH) in hydrolysed EVA copolymers can be carried out indirectly by comparing the original and residual VA content. However, Koopmans *et al.*[14] developed an IR method to directly determine the percentage of VOH in partially and totally hydrolysed EVA copolymers. Considering the EVAL/EVALVA spectra in Fig. 9, it is obvious that IR spectrometry can be used to follow the ester–alcohol transformation during the EVA saponification. The VA bands gradually disappear while characteristic hydroxyl bands show an increasing absorption. The vibrational modes of the IR spectra are summarised in Table 3.

An analysis of the IR spectra reveals that quantitative mol % VOH measurements can only be performed using the broad 3550–3350 cm^{-1} hydroxyl band.[14–31] When a calibration curve is prepared from several hydrolysed EVA copolymers containing different amounts of VOH, this IR method gives fast, accurate and reproducible VOH results.

5. PHYSICAL PROPERTIES OF EVA AND THEIR DERIVATIVES

5.1. Density

Two methods can be used to establish the density of the polymer samples. For the EVA copolymers the density gradient technique (ASTM D 1505-68) was used. The water displacement method (ASTM D 792-66) was used to determine the EVAL/EVALVA polymer density because the 2-propanol/water density gradient column only has a range of $0{\cdot}88$–$0{\cdot}98 \times 10^3$ kg/m^3. Some EVAL copolymers have higher densities than $0{\cdot}98 \times 10^3$ kg/m^3. The results from both methods were comparable. The accuracy was better than 0·1%.

Figure 10 indicates an increasing density when the mol % VA or VOH becomes larger, for EVA and EVAL copolymers respectively. EVAL copolymers have a higher density than the parent EVA.

5.2. Melting Point

The melting points of the EVA and EVAL copolymers, obtained from samples with similar thermal history, are shown in Fig. 11. These measurements also allowed the percentage crystallinity to be determined (Fig. 12).[32] As the melting point is mainly determined by the

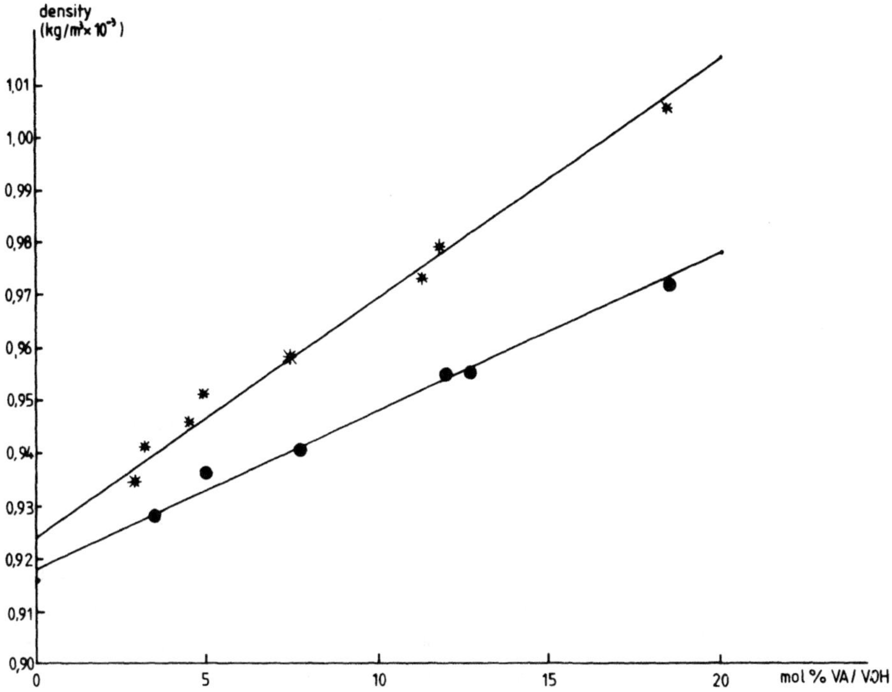

FIG. 10. Density values of EVA(●) and EVAL(∗) copolymers as a function of the mol % VA and VOH respectively.

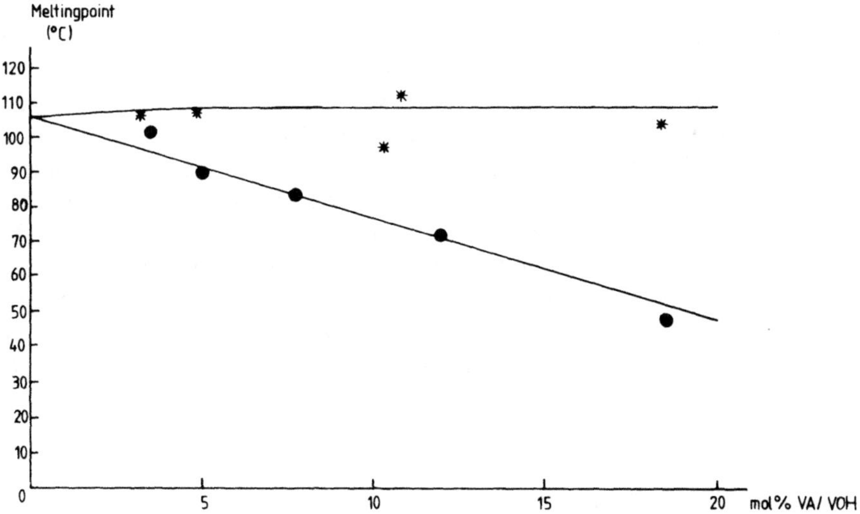

FIG. 11. Melting point values of EVA (●) and EVAL (∗) copolymers as a function of the mol % VA and VOH respectively.

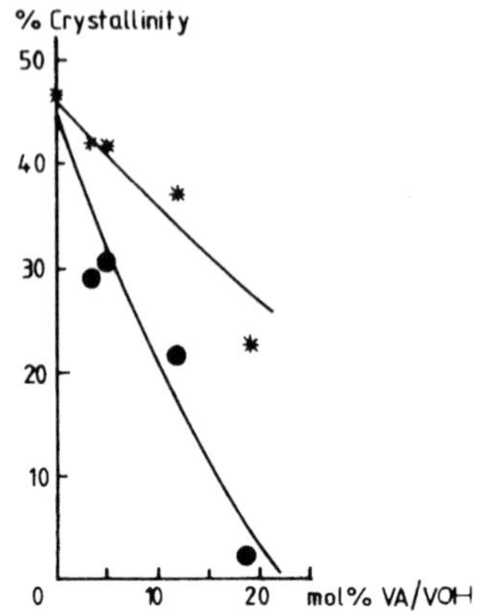

FIG. 12. Percentage crystallinity values of EVA (●) and EVAL (∗) copolymers as a function of the mol % VA and VOH respectively.

crystallinity, the decrease in melting point for EVA copolymers with increasing VA content can be expected. Indeed, the percentage crystallinity decreases in a non-linear way with increasing mol % VA or VOH. The EVAL copolymer melting points, however, are almost independent of the mol % VOH and are comparable with that of low density polyethylene (LDPE). The EVALVA melting points seem to be controlled by the residual VA content.

5.3. Vicat Softening Point

Vicat softening point determinations measure the temperatures at which polymeric substances under stress undergo a permanent change. This quantity indicates the temperature at which a polymeric material will start to flow if stressed. The Vicat softening points were obtained following the ASTM D 1525-76 method. The heating rate was 50°C/h. Figure 13 shows the Vicat softening points for the EVA and EVAL copolymers. As for the melting point observations, the EVAL Vicat softening points are higher and are less dependent on the VOH content than the EVA copolymers for which the Vicat softening points are largely VA dependent. As observed for the melting points, the EVALVA terpolymer Vicat softening values also seem determined by the residual VA content.

5.4. Mechanical Properties

The mechanical properties were obtained from compression moulded samples with an Instron Model 1122 Universal Testing Instrument.

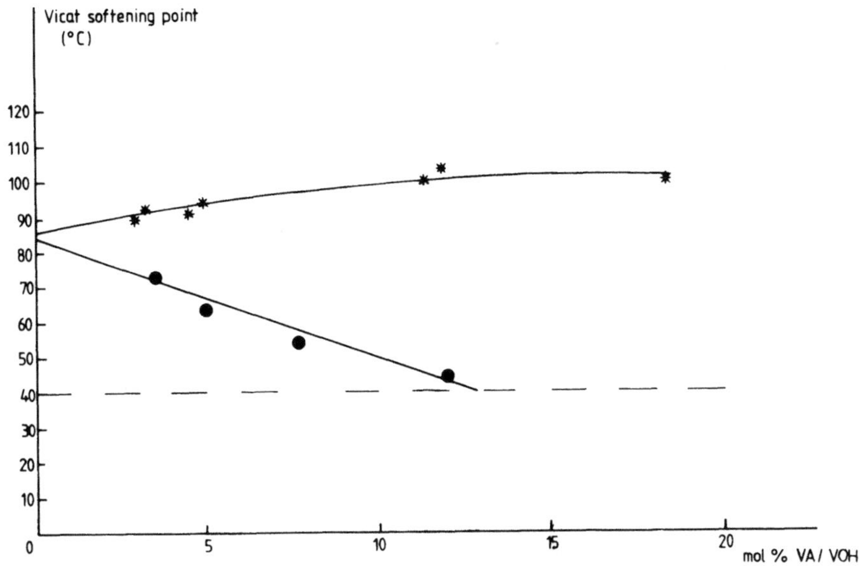

FIG. 13. Vicat softening point values of EVA (●) and EVAL (∗) copolymers as a function of the mol % VA and VOH respectively. The dotted line indicates the detection limit of the apparatus.

The speed of testing was $8{\cdot}3 \times 10^{-3}$ m/s. The stress–strain diagrams were determined in a room having a 50% relative humidity and a temperature of $23{\cdot}0 \pm 0{\cdot}5$°C. The tensile strength at break and yield as well as the percentage elongation and the modulus of elasticity could be calculated from these diagrams. The experimental procedure is described in ASTM D 638-76.

The torsional stiffness values were measured with a Clash–Berg torsion tester. The temperature interval covered, ranged from −35 to 110°C. Measurements were made following the ASTM D 1043-72 method.

5.4.1. Tensile Strength at Break

Figure 14 gives the tensile strengths at break of EVA and EVAL copolymers. EVA copolymers show a moderate increase in tensile strength values until the mol % VA is about 7. Higher VA content values give a rapidly increasing tensile strength, indicating the progressively rubberlike nature of the EVA samples. EVAL copolymers

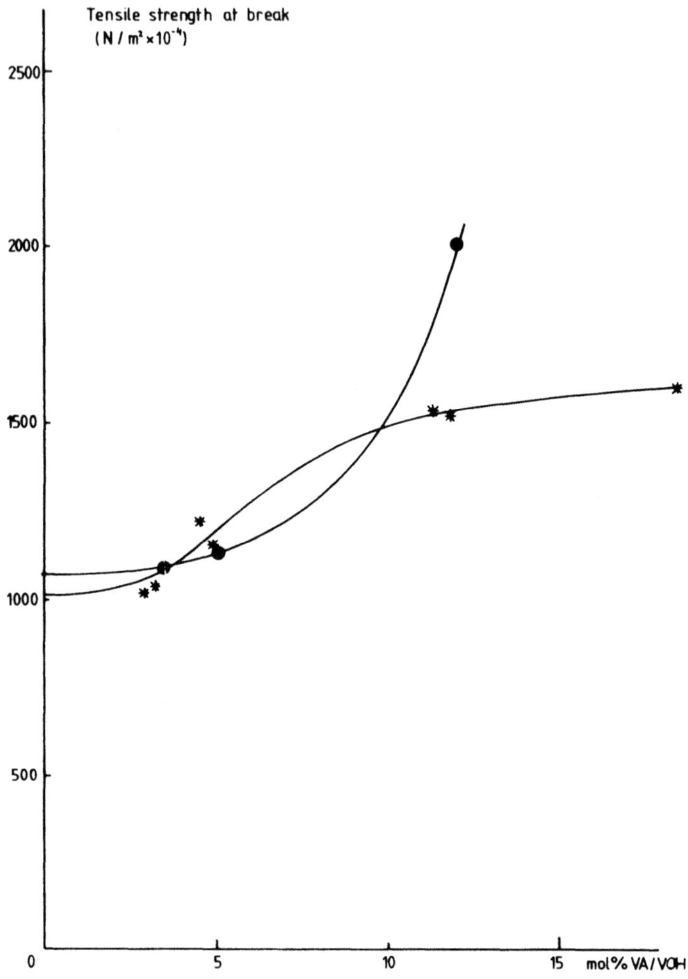

FIG. 14. Tensile strength at break of EVA (●) and EVAL (∗) copolymers as a function of the mol % VA and VOH respectively.

seemingly progress to a maximum tensile strength value, much lower than that of the parent EVA values.

The EVALVA terpolymers of different EVA types reveal higher tensile properties compared to the EVA or EVAL copolymers. The effects are more pronounced as the mol % VA of the parent EVA becomes higher.

5.4.2. Tensile Strength at Yield

With the high testing speed it was not possible to detect a yield point for EVA copolymers. The yield strength of EVAL copolymers (Fig. 15) shows an almost linear increase as the VOH content becomes larger, until a maximum, reached at about 15 mol % VOH.

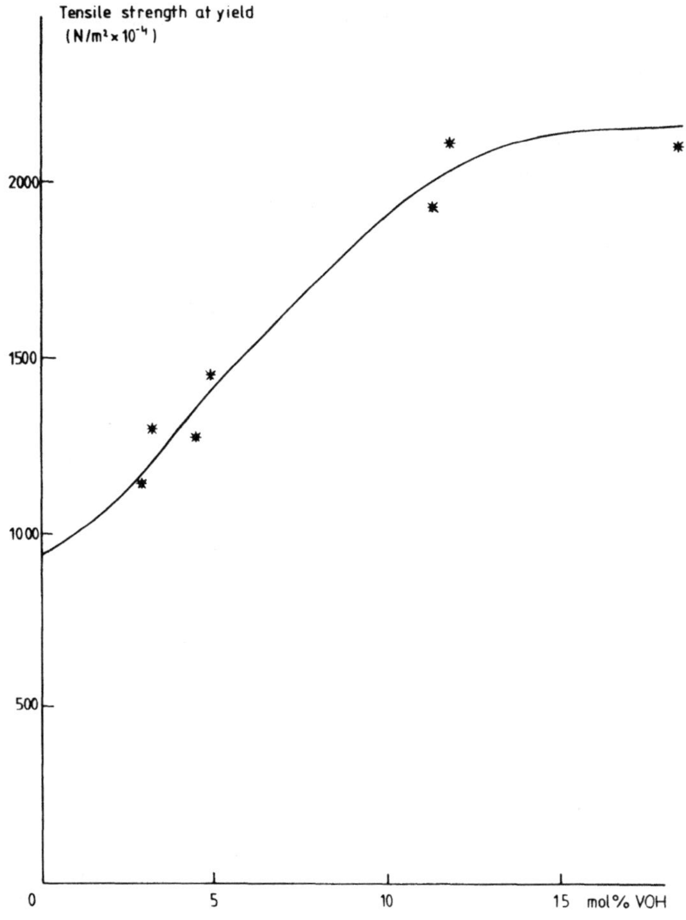

FIG. 15. Tensile strength at yield of EVAL copolymers as a function of the mol % VOH. EVA copolymers have no yield point, working with a testing speed of $8{\cdot}3 \times 10^{-3}$ m/s.

5.4.3. Percentage Elongation

EVA copolymers produce remarkably high elongation values when the mol % VA is high (Fig. 16). EVAL copolymers, however, have a very poor capacity for elongation. The percentage elongation drops almost immediately (with increasing mol % VA) to a minimum value about five times lower than for LDPE tested under the same conditions.

EVALVA terpolymers show only a moderate decrease in percentage elongation until about 80% VA transformation. Beyond this, all types show a drastic reduction in percentage elongation.

5.4.4. Modulus of Elasticity

The effect of an increasing VA content on the elastic modulus of EVA copolymers is more pronounced than a similar VOH content increase in EVAL copolymers (Fig. 17). The decreasing elastic modulus indicates a higher flexibility for EVA copolymers. Consequently, more rubberlike properties are expected at higher VA content levels.

An increase in VOH content in an EVAL polymer gradually increases the elastic modulus until a maximum value seems to be reached.

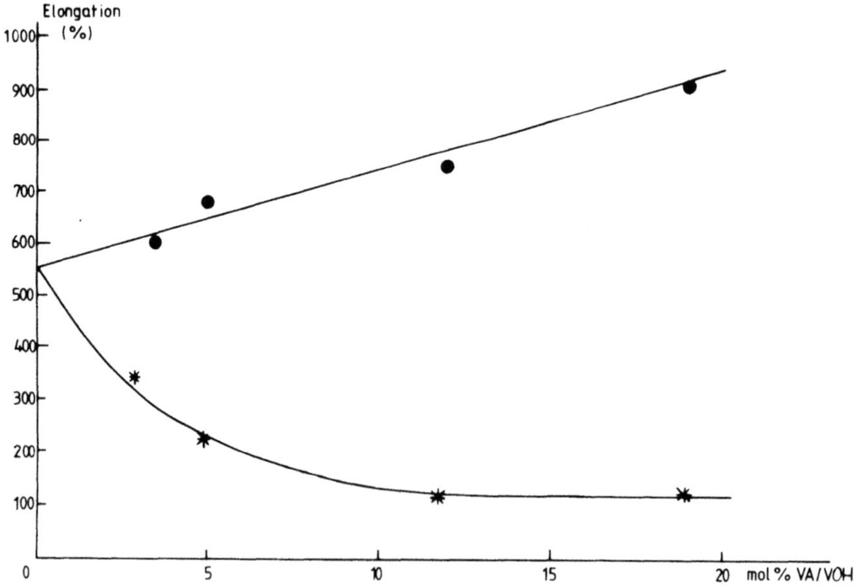

FIG. 16. Percentage elongation of EVA (●) and EVAL (∗) copolymers as a function of the mol % VA and VOH respectively.

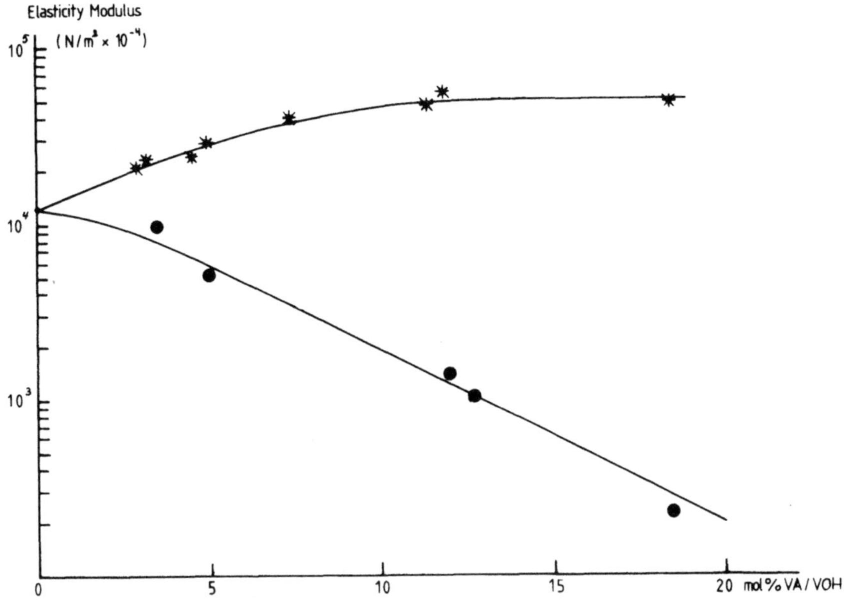

FIG. 17. Elasticity modulus of EVA (●) and EVAL (∗) copolymers as a function of the mol % VA and VOH respectively.

5.4.5. Torsional Stiffness

The torsional stiffness of EVA copolymers decreases with increase of temperature (Fig. 18). This tendency becomes more noticeable with increased mol % VA. The dependency on the VA content decreases as the temperature drops. Similar features were obtained by Emura *et al.*[33]

EVAL copolymers gave torsional stiffness values which were higher than EVA and LDPE samples. An increasing mol % VOH does not show a large temperature dependence. The EVALVA terpolymers show an intermediary behaviour between EVA and EVAL copolymers. As noted by Saito *et al.*,[34] the torsional stiffness of EVA copolymers is mainly determined by the total number of methyl groups per 1000 carbon atoms (CH_3/1000C).

5.5. Shore Hardness

Shore hardness A and D values were obtained following the ASTM D 2240-75 procedure. Figure 19 indicates that Shore D values

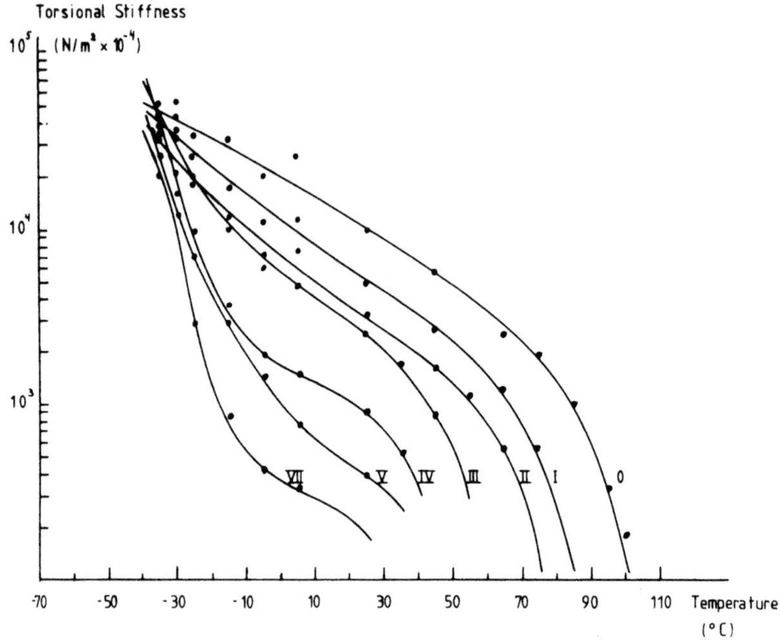

FIG. 18. Torsional stiffness of EVA copolymers as a function of temperature. The mol % VA content is respectively: 0·0(0); 3·5(I); 5·0(II), 7·7(III), 12·0(IV), 11·4(V), 12·7(VI) and 18·5(VII).

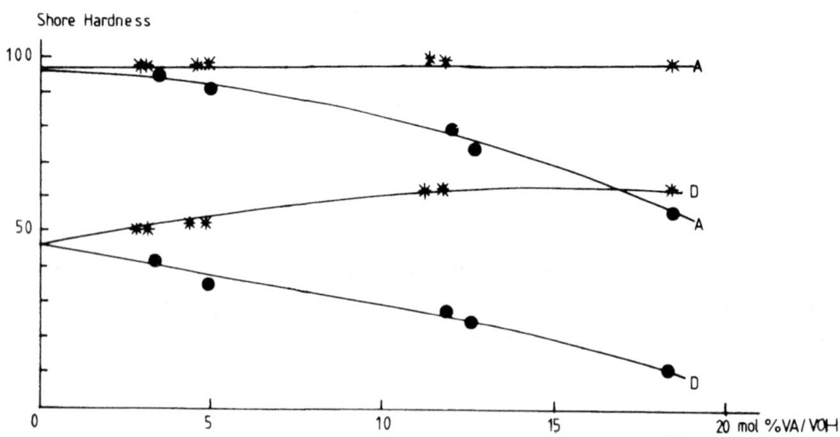

FIG. 19. Shore hardness A and D of EVA (●) and EVAL (*) copolymers as a function of the mol % VA and VOH respectively.

decrease for EVA copolymers and increase for EVAL copolymers when the mol % VA or VOH respectively becomes larger. A similar effect is noted for Shore A values. However, all EVAL copolymers seem to have reached a nearly maximum hardness. The Shore A and D values for EVALVA terpolymers are intermediary between the EVA and EVAL copolymers.

5.6. Environmental Stress Crack (ESC) Resistance

ESC experiments were carried out according to ASTM D 1693-70. The samples were put in an 10% Antarox detergent/water solution and in the pure detergent. The results can be summarised as:

(1) A melt index (MI) increase reduces the ESC resistance significantly.
(2) A high detergent concentration reduces the ESC resistance.
(3) All EVA samples had very good ESC resistance, except the one with very low VA content.
(4) All EVAL and EVALVA samples had very good ESC resistance, except the one with a low original VA content. Those with very high VOH values showed poor resistance in the pure detergent.

6. SUMMARY

Nowadays the polymer industry is concentrating its efforts on the development of new kinds of macromolecules. One of the methods of achieving this goal is to carry out chemical reactions with commercially available polymers containing reactive groups. Ethylene vinyl acetate (EVA) copolymers are of special interest in this respect. EVA not only has the potential reactivity, but can be prepared in a variety of copolymers having different vinyl acetate incorporation and molecular weight. Consequently, the partial or total transformation of the ester groups will yield a broad range of new thermoplastic co- and terpolymers with different chemical and physical properties, compared to the classical polymers such as polyethylene, polypropylene and poly(vinyl chloride).

To characterise the ethylene vinyl acetate and their acetoxyl–hydroxide transformed macromolecules, methods are described to determine the molecular weight distribution and the vinyl acetate/vinyl alcohol content.

Furthermore, different physical and mechanical properties of ethylene vinyl acetate (EVA) and their derivatives are reported and compared in order to show the influence of the polymer transformation on their behaviour.

REFERENCES

1. Koopmans, R. J., van der Linden, R. and Vansant, E. F., *Polymer Eng. and Sci.*, **22,** 1982, 645.
2. Billmeyer, F. W., *Textbook of Polymer Science,* 2nd edn, Wiley Interscience, NY, 1971.
3. Brandrup, J. and Immergut, E. H., *Polymer Handbook,* 2nd edn, Wiley Interscience, NY, 1975.
4. Fujiki, T., *J. Adh. Soc. Japan,* **5,** 1969, 313.
5. Koopmans, R. J., van der Linden, R. and Vansant, E. F., *Bull. Soc. Chim. Belg.*, **91,** 1982, 531 and **92,** 1983, 83.
6. Echarri, J., Iruin, J. J., Gurman, G. M. and Ansorena, J., *Macromol. Chem.*, **180,** 1979, 2749.
7. Salyer, I. O., Kenyon, A. S. and Ohta, M., *J. Polym. Sci. A-1,* **10,** 1972, 419.
8. Schillgalies, J., Hoffmann, H. and Fedtke, M., *Plaste u. Kautschuk,* **27,** 1980, 61.
9. Barlow, R. and Campbell, R. C., *J. Appl. Polym. Sci.*, **11,** 1967, 2001.
10. Munteanu, D., *Rev. Chem. (Bucharest),* **27,** 1976, 902.
11. Mayor, J. and Sodomka, J., *Vyzkumny Makromol. Chem. (Brno),* **26,** 1975, 601.
12. *USI Standard Test Method 287,* USI, USA, 1970.
13. Koopmans, R. J., van der Linden, R. and Vansant, E. F., *Bull. Soc. Chim. Belg.*, **91,** 1982, 531 and **92,** 1983, 83.
14. Koopmans, R. J., van der Linden, R. and Vansant, E. F., *J. Adhesion,* **11,** 1980, 191.
15. DIN 53621, Deutsches Institut für Normung ev, Berlin, 1975.
16. Aydin, O., Kaczmar, B. V. and Schultz, R. C., *Angew. Makrom. Chem.*, **24,** 1972, 1971.
17. Leukroth, G., *Gummi Asbest u. Kautschuk,* **29,** 1976, 585.
18. Grant, D. H. and Grassie, N., *Trans. Far. Soc.*, **48,** 1952, 379, and **56,** 1960, 445.
19. Grassie, N., *Trans. Far. Soc.*, **49,** 1952, 835.
20. Koopmans, R. J., PhD thesis, University of Antwerp (UIA), 1983.
21. Barrall, E. M., Porter, R. S. and Johnson, J. F., *Anal. Chem.*, **35,** 1963, 73.
22. Okumato, T., Takeuchi, T. and Tsunge, S., *J. Chem. Soc. Japan,* **73,** 1970, 702.
23. Chiu, J., *Appl. Polym. Symp.*, **2,** 1966, 25 and *Anal. Chem.*, **40,** 1968, 1516.

24. MLEJNEK, O., CVECKOVA, L., CECH, K. and GOMORY, I., *Elektroizolacna Kablova Tech.*, **30,** 1977, 15.
25. DISSELHOFF, R., *Mitt. Deutsches Kunstoff Instr. Darmstadt,* No. 12, 1972.
26. STREITWIESER, A. and HEATHCOCK, C. H., *Introduction to Organic Chemistry,* 3rd edn, McGraw-Hill, NY, 1970, p. 15.
27. CHEN, H. T. and LEWIS, M. E., *Anal. Chem.*, **36,** 1964, 1394.
28. LESCH, N., *Kolloid. Z.Z. Polym.*, **240,** 1970, 737.
29. EBDON, J. R., KANDIL, S. H. and MORGAN, K. J., *J. Polym. Sci.*, **17,** 1979, 2783.
30. SIRYUK, A. G. and BULGAKOVA, R. A., *Vysokomol Soedin Ser. B.* **19,** 1977, 152.
31. KOOPMANS, R. J., VAN DER LINDEN, R. and VANSANT, E. F., *Spectros. Int.*, **1,** 1982, 152.
32. SALYER, I. O. and KENYON, A. S., *J. Polym. Sci. A-1,* **9,** 1971, 3083.
33. EMURA, N., IMOTO, H., KOSAKA, Y., *Toyo Soda Scientific Report,* **13,** 1969, 65.
34. SAITO, M., TADOR, H. and KOSAKA, Y., *J. Polym. Sci. A-1,* **8,** 1970, 2555.

Chapter 4

ACICULAR WOLLASTONITE AS A FILLER FOR POLYAMIDES AND POLYPROPYLENE

P. J. WRIGHT

Blue Circle Industries plc, Greenhithe, Kent, UK

1. INTRODUCTION

For plastics to be used in engineering applications they must have improved mechanical strength, heat resistance, dimensional stability, low flammability and good electrical properties when compared to the commodity plastics. Many of these properties can only be raised to the desired levels by the use of fillers and reinforcements.

Wollastonite has considerable potential as a filler for plastics because of the absence of water in the mineral, its high melting point, non-toxic qualities, good colour, high strength and modulus and resistance to chemical attack. These properties, coupled with the availability of the mineral in grades with aspect ratios of up to 20:1, indicate that wollastonite can be incorporated into plastics without difficulty to produce composites with significantly improved properties.

Both polypropylene and the aliphatic polyamides are crystalline polymers having low glass transition temperatures. This type of polymer lends itself to fibre reinforcement since its ductility allows transfer of stress to the fibres when the composite is under load. Polymers which are used while in their glass state undergo failure at low strains and cannot make effective use of fibre reinforcement.

Transfer of stress from polymer to fibre requires good interfacial bonding. Coupling agents are available which give good bonding both at the wollastonite/coupling agent interface and at the coupling agent/polyamide or polypropylene interface.

2. MINERALOGICAL PROPERTIES OF WOLLASTONITE

Wollastonite is a naturally occurring calcium metasilicate ($CaSiO_3$) found in three modifications.[1] Wollastonite itself has a triclinic crystal structure forming white acicular (needlelike) crystals. Massive occurrences of wollastonite consist of this form. Parawollastonite has a monoclinic crystal structure forming prismatic or tabular crystals. Pseudowollastonite is the high temperature form, metastable at low temperatures, produced by heating wollastonite to 1120 ± 10°C, the conversion temperature being affected by the presence of foreign ions. Pseudowollastonite has a monoclinic crystal structure and forms prismatic or tabular crystals.

There are three currently available sources of wollastonite for use as a filler for plastics: Finland, New York State in the USA and Udaipur in India.

Wollastonite ore is extracted by a blasting and a primary crushing process followed, if necessary, by beneficiation to remove impurities such as silicates associated with the ore. Further milling and classification then yield a range of product sizes. The methods of production are designed to preserve the natural acicularity of the mineral. Frequently, wollastonite products are also available in surface coated forms of various types.

3. WOLLASTONITE AS A FILLER FOR PLASTICS

The important physical properties of wollastonite are given in Table 1. Of particular importance for its use in plastics are its good colour, low water sorption and ability to withstand any plastic processing temperature.

3.1. Properties Compared to Other Fillers

The modulus and tensile strength of wollastonite fibres have been measured[2] and the results are given in Table 2 along with published values for other fibrous fillers. On a modulus/strength versus cost basis wollastonite, which even in its coated forms sells at approximately one third of the price of E-glass, compares favourably with most of the fibres listed.

However, in applications where continuous reinforcement is possible, wollastonite, which has a current aspect ratio of between 10:1 and 20:1, will not be competitive.

TABLE 1
PHYSICAL PROPERTIES OF TYPICAL INDIAN WOLLASTONITE

Chemical formula	$CaSiO_3$
Molecular weight	116
Form and habit	Bladed crystal masses (fibrous)
Crystal system	Triclinic
Colour	Brilliant white
Refractive index	1·63
Specific gravity	2·9
Hardness (Mohs scale)	4·5
Solubility in water (20°C)	0·003%
Melting point	1540°C (converts to α form at 1120°C)
Coefficient of thermal expansion	$6{\cdot}5 \times 10^{-6}$ mm mm^{-1} °C^{-1}
pH of aqueous extract (20°C)	9·1

Other reinforcing fillers which will compete with wollastonite are the high aspect ratio platelet materials, talc and mica. Direct comparison with these on a single particle basis is difficult and performance comparisons in particular composites are more appropriate.

The values of modulus and tensile strength given in Table 2, along with the known aspect ratio of the Indian wollastonite and its observed orientation in fractured test pieces, can be used to predict maximum values for the strength and modulus of filled composites.

TABLE 2
COMPARATIVE STRENGTH PROPERTIES OF FIBROUS FILLERS

Fibre	*Mean diameter range (μm)*	*Tensile modulus (GPa)*	*Tensile strength range (GPa)*
Indian wollastonite	5–35	117	0·5–2·0
Crysotile asbestos	10–20	165	2·4–4·5 (3·1)
AR Glass (alkali resistant)	10–20	76	1·6–1·8
E-Glass (borosilicate)	9–15	77	2–4
Rockwood (PMF fibre)	4–6	103	0·48
Aramid (Kevlar PRD 49)	10	133	2·8
Polyacrylonitrile (Dolon 10)	18	15	0·7–0·8
Polypropylene (monofilament)	20–200	5	0·4–0·5
Poly(vinyl alcohol) (Mewlon)	6	32	2–4
Sisal	10–50	5	0·8

3.2. Predicting Composite Properties

A number of theoretical models have been proposed for predicting the elastic moduli of polymers containing fillers. The Nielsen equation, which takes into account the maximum packing fraction of the filler, has been found to provide the best prediction for the relationship between volume fraction and composite modulus for low aspect ratio fillers.[3]

$$E_{fm} = E_m[1 + (k - 1)BV_f]/(1 - B\Psi V_f)$$

where E_{fm} and E_m refer to the Young's moduli of the composite and matrix respectively, k is related to the generalised Einstein coefficient and B is defined as:

$$B = \frac{(E_f/E_m) - 1}{(E_f/E_m) + k - 1}$$

where E_f is the Young's modulus of the filler and V_f is the volume fraction of filler in the composite.

Ψ is defined as

$$\Psi = 1 + [(1 - V_m)/V_m^2]V_f$$

where V_m is the maximum volumetric packing fraction.

For nylon 6 and nylon 66 the equation predicts that the modulus of a composite containing 50% w/w (28·2% v/v) of coupled acicular wollastonite can be expected to be *c.* 370% higher than that of the unfilled nylons. The modulus of a polypropylene composite containing 40% w/w (17% v/v) of coupled acicular wollastonite should be 156% greater than that of the unfilled polymer. These increases assume perfect bonding of the filler to the polymer matrix, alignment of the fibres and measurement of the modulus in the fibre direction.

For calculating the tensile strength of a wollastonite filled composite Kelly's equation[4] for fibre reinforcement can be used. In the form given below it takes into account the observed alignment of fibres in the tensile test piece.

$$\sigma_{fm} = \tfrac{3}{8} \,.\, \tau_{fm} \,.\, l/d \,.\, V_f + \sigma'_m(1 - V_f)$$

where τ_{fm} is the shear strength of the polymer fibre interface given by the matrix shear strength (using von Mises' criterion $\tau_{fm} = \sigma_y/\sqrt{3}$),

l/d is the aspect ratio of the wollastonite fibres and σ'_m is the tensile strength of the matrix given by the tensile strength of a composite containing coupled ballotini of a similar size to the wollastonite.

For nylon 6 (PA6) the equation indicates that the tensile strength of the composite containing 50% w/w (28·2% v/v) of acicular wollastonite should be 25% above that of the unfilled nylon. For nylon 66 the tensile strength of the composite should be 38% greater than that of the unfilled nylon. For polypropylene the increase in strength should be 13%.

4. WOLLASTONITE IN PA6

The polyamides are one of the important group of thermoplastics used in engineering applications. The total world sales of polyamides are currently estimated to be 436 000 t year^{-1}, representing approximately 40% of all the plastics sold for engineering applications.[5] Of the many nylons used, nylon 66 and nylon 6 dominate the market; together they account for about 90% of the total polyamide sales.

Table 3 gives the total consumption of reinforced polyamides in W. Europe in 1983. Of the total reinforced polyamide consumed *c.* 10 000 t is mineral filled; the rest is glass fibre reinforced.

Reinforcements such as glass fibre, glass beads, talc and mica are used to increase stiffness and heat distortion temperature. A major advantage of the particulate mineral fillers lies in their ability to reduce mould shrinkage without the pronounced anisotropic effects found when the longer fibres are used. This enables components with thin sections and large flat areas to be moulded without distortion.

However, it is necessary to form a good bond between the polyamide and the mineral surface if substantial improvements in

TABLE 3
CONSUMPTION OF REINFORCED POLYAMIDES IN EUROPE, BY TYPE (1983)

	PA type 6	*Type 66*	*Other types*	*Total*
Unreinforced	76 000	40 000	14 000	130 000
Reinforced	34 000	35 000	1 000	70 000
Total	110 000	75 000	15 000	200 000

From Ref. 6.

strength are to be achieved. Silane and other coupling agents are often used for this purpose. Some experimental results which illustrate the use of such systems are reported in subsequent sections.

4.1. Materials

The nylon used was nylon 6 (MD3 grade) supplied by BIP Chemicals Ltd. The fillers used were:

—Ballotini, 17 μm mean diameter, coated with a silane coupling agent.
—Wollastonite having no significant aspect ratio and a mean particle diameter of 20 μm. This was used both uncoated and coated with 0·5% by weight of γ-aminopropyltriethoxy silane (Union Carbide A1100).
—Acicular Indian wollastonite, mean length 90 μm, having an aspect ratio of between 10:1 and 20:1. This material was also used both uncoated and coated with 0·5% by weight γ-aminopropyltriethoxy silane (Union Carbide A1100). This coating level was selected as being an optimum value after coating levels of 0·25–4% had been investigated.

4.2. Specimen Preparation

4.2.1. Filler Treatment

The silane coatings were applied by treating the filler in a high speed mixer with the silane diluted 1:1 using a mixture of methanol and water (9:1 v/v). The coated filler was then dried at 105°C.

4.2.2. Extrusion Compounding

Dry blends containing 50% by weight of each of the fillers were prepared using a ribbon blender and fed in turn to a Brabender twin screw compounding extruder. This machine contains two counter-rotating screws, each of 35 mm diameter. The length to diameter ratio is 17:1. The speed used was 15 rpm and the zone temperature was controlled as shown below:

Zone 1	268°C
Zone 2	255°C
Zone 3	255°C
Die	258°C

The extrudate was granulated and dried for 24 h at 70°C.

4.2.3. Injection moulding

Test specimens were moulded using a 35 t Szekely in-line reciprocating screw injection moulding machine. Typical moulding conditions for the 50% filled composites are shown below:

Barrel temperature, °C	
Zone 1 (rear)	230
Zone 2	240
Zone 3	240
Zone 4 (front)	250
Mould temperature	75°C
Injection line pressure (psi)	600
Back pressure (psi)	200
Injection time (s)	20
Cooling time (s)	20

Three types of specimens were moulded: tensile specimens according to BS 2782: Part 3: 320B, Charpy bars according to ASTM D256 $12{\cdot}7 \times 127 \times 6{\cdot}35$ mm and plaques $100 \times 100 \times 3$ mm used for falling weight impact tests and resistivity measurements.

The filler contents of some of the specimens were measured by ignition of the nylon at 500°C, and the recovered fibre was also used to determine the loss of aspect ratio occurring in the compounding and moulding processes. The average fibre length was found to be 40 μm and the aspect ratio had been reduced on average, by one third.

Differential scanning colorimetry (DSC) measurements showed that the presence of wollastonite, either coated or uncoated, does not cause a significant change in the degree of crystallinity of the nylon 6.[7] The glass transition temperature (T_g) also remained unchanged at 51°C.

As the mechanical properties of nylon are dependent on moisture content all the test specimens were stored over phosphorus pentoxide in a desiccator immediately after moulding.

4.3. Tensile and Flexural Testing

The testing of the injection moulded specimens was carried out according to the following test procedures:

BS 2782: 320B. Tensile strength and modulus (using a rate of grip separation of 5 mm/min).

BS 2782: 335A. Flexural strength and modulus (using a relative rate of movement of loading nose and supports of 4 mm/min).

To determine the effect of moisture on the mechanical properties of the composites, flexural and tensile specimens were conditioned in water for 15 h at 50°C. The temperature of the water was adjusted to 23°C one hour before testing. The tensile properties of the composites were also measured at 0°C and 50°C using a temperature controlled cabinet mounted around the test specimen. These results are reported in Table 6.

The tensile and flexural results given in Tables 4, 5 and 6 confirm the importance of good interfacial bonding between filler and polymer if the potential of fibre reinforcement is to be realised.

The increase in the tensile strength of the composite containing 50% of the silane coated acicular wollastonite is 18·5% compared to the unfilled nylon 6. This compares favourably with the 25% increase in tensile strength calculated using Kelly's equation (see Section 3.2), assuming perfect adhesion between filler and matrix and alignment of the fibres in the direction of test. The use of the uncoated acicular wollastonite results in a composite having poor tensile strength and a low strain at break when compared to the other wollastonites.

The tensile modulus of the composite containing the coated acicular wollastonite has increased by 122% when compared to the unfilled nylon 6; this is substantially less than predicted by the Nielsen equation (Section 3.2). Fibre alignment, fibre length and aspect ratio distribution can only be partly responsible for the discrepancy between theory and experiment, e.g. there is a need to very carefully define aspect ratio and the meaning of maximum packing.

The substantial differences between flexural and tensile modulus found for the composites containing the acicular filler are due to differences in fibre orientation distribution in the skin and core regions of the test specimens. (This statement is based on scanning electron microscopy observations.)

4.4. Heat Distortion Temperature

The heat distortion temperatures of the composites, measured according to BS 2782:121A using a load of 1·81 MPa, are also reported in Table 4.

The high values for the flexural modulus of the composites containing the acicular wollastonite are reflected in the heat distortion temperatures of 185–193°C determined for these composites. Given that nylon 6 melts at 215°C these values are near to the maximum

TABLE 4

PHYSICAL AND THERMAL PROPERTIES OF NYLON 6 FILLED WITH WOLLASTONITE AND BALLOTINI

Property	*Unfilled nylon 6*	*Silane coated ballotini*	*Non-acicular wollastonite*	*Non-acicular wollastonite, silane coated*	*Acicular wollastonite*	*Acicular wollastonite, silane coated*
Tensile strength (MPa)	81	69	44	66	65	96
Tensile modulus (GPa)	2·7	4·0	4·1	4·0	5·6	6·0
Tensile elongation (%)	60	3·4	2·4	3·3	1·3	3·0
Flexural strength (MPa)	108	112	80	108	118	166
Flexural modulus (GPa)	3·7	4·3	4·7	4·6	9·8	10·8
Heat distortion temperature (°C at 1·81 MPa)	60	104	105	114	185	193
Linear expansion coefficient (°C × 10^{-5})	10	5·0	5·0	5·0	4·0	3·0
Rockwell hardness (M scale) (ASTM D785)	83	—	—	—	—	89
Density (g/m^3) (BS 2782:620A)	1·13	1·50	1·52	1·52	1·52	1·52

TABLE 5

THE EFFECT OF WATER CONDITIONING ON NYLON 6 FILLED WITH WOLLASTONITE AND BALLOTINI

Property	*Unfilled nylon*	*Silane coated ballotini*	*Non-acicular wollastonite*	*Non-acicular wollastonite, silane coated*	*Acicular wollastonite*	*Acicular wollastonite, silane coated*
Flexural strength (MPa)						
Dry	108	112	80	108	118	166
Wet	49	61	51	67	81	116
Flexural modulus (GPa)						
Dry	3·7	4·3	4·7	4·6	9·8	10·8
Wet	2·2	2·9	3·0	2·9	5·8	6·3
Tensile strength (MPa)						
Dry	81	69	44	66	65	96
Wet	47	41	29	44	39	62
Tensile modulus (GPa)						
Dry	2·7	4·0	4·1	4·0	5·6	6·0
Wet	1·4	2·5	3·1	2·7	3·7	3·3

TABLE 6

THE EFFECT OF TEMPERATURE ON THE TENSILE PROPERTIES OF NYLON 6 FILLED WITH BALLOTINI AND WOLLASTONITE (50% by weight filler loading)

Property	*Unfilled nylon*	*Silane coated ballotini*	*Non-acicular wollastonite*	*Non-acicular wollastonite, silane coated*	*Acicular wollastonite*	*Acicular wollastonite, silane coated*
Tensile strength (MPa)						
0°C	90	76	50	76	75	111
50°C	50	38	25	38	42	66
Tensile modulus (GPa)						
0°C	2·5	3·6	3·6	3·9	4·8	5·2
50°C	1·2	1·1	1·2	1·2	2·5	3·1
Strain (%) at break						
0°C	10	2·1	1·9	2·2	1·4	2·4
50°C	—	7·2	9·0	7·0	1·9	6·1

possible with particulate reinforcement. Glass fibre reinforced nylon 6 is reported to have a heat distortion temperature of *c.* 210°C at a filler loading of 30% w/w.

4.5. Coefficient of Expansion

The linear expansion coefficients given in Table 4 were measured using the apparatus and procedure described in ASTM 696. Measurements were also made over the range −40°C to +140°C using a Mettler TMA-40 thermomechanical analysis apparatus. These measurements confirmed that the presence of the silane coating has little effect on the expansion coefficient of the composite. The reduction in value of the expansion coefficient which has occurred when the composites are compared to the unfilled nylon is directly related to the reduction in the volume of nylon present.

The reduction of the coefficient of expansion caused by the incorporation of the wollastonite results in reduced mould shrinkage. Linear mould shrinkage for the Charpy bar specimens was found to be 0·7% in the flow direction and 1·1% across the flow direction for the unfilled nylon 6. For the composites containing 50 wt% of the acicular wollastonite this was reduced to 0·4% in the direction of flow and 0·5% across the direction of flow.

4.6. Impact Behaviour

Mineral reinforcements are added to thermoplastics in order to improve such properties as stiffness, strength, hardness and creep resistance. Unfortunately, the impact strength of these plastics, already inferior to that of metals, will be further degraded by the reduced amount of tough energy absorbing material present in the composite. A detailed understanding of the impact performance of these composites, covering a wide range of notched and unnotched conditions, is required to provide information about the behaviour of the composite under practical service conditions.

4.6.1. Notched Impact

The presence of any notch has a detrimental effect on the energy required to fracture nylon. The extent of the reduction is highly dependent on notch severity. Notched impact tests were therefore performed on Izod bars according to both ASTM D256A using 6·35 mm width specimens containing the standard 0·25 mm radius

TABLE 7
NOTCHED IMPACT TESTS AT 23°C

Filler	*Izod* $(J/m)^a$	*Izod* $(J/m)^b$	*Charpy* (kJ/m^2)
None	35	29	2·1
None, containing 3% water	421	416	—
Ballotini	27	18	2·5
Non-acicular wollastonite	27	16	2·4
Non-acicular wollastonite, silane coated	28	18	2·3
Acicular wollastonite	33	22	3·4
Acicular wollastonite, silane coated	40	30	4·0
Acicular wollastonite, silane coated (nylon containing 3% water)	79	70	—

[a] No correction for energy used to remove broken part of specimen.
[b] Correction for energy needed to remove broken piece has been applied to the results.

notch, and on Charpy bars according to DIN 53453 using the more severe square notch (sometimes called a U notch).

The results given in Table 7 show that only in the presence of the more severe notch (i.e. the U notch), when the unfilled nylon is failing in a brittle manner, does the acicular wollastonite improve impact strength. Fibres can improve impact strength by impeding the propagation of the crack allowing energy to be expended in debonding and pull out processes. The acicular wollastonite fibres are too short for these processes to be significant. In the notched Izod test the impact strength of the nylon is lowered when acicular uncoated wollastonite is used because of the ease with which the crack propagates through the voids formed between matrix and filler.

4.6.2. Unnotched Impact

Unnotched impact tests were performed using an instrumented falling weight machine. The drop height was 1·3 m producing an impact velocity of 5 m/s. The total mass of the impactor was 15 kg and the impact energy 191 J. The specimens were of the Charpy type and the geometry of the impactor was identical to that of the striker used on a pendulum impact machine. The specimen support was the same as described in ASTM D256B.

A piezoelectric force transducer coupled to the impactor nose generates an electrostatic signal when the impactor strikes the specimen. The signal from the transducer is fed through an amplifier into a transient recorder where the data is stored. From the recorder the signal can be fed into an oscilloscope, flat bed plotter or a microcomputer. The signal is in the form of a force–time plot. This plot is easily converted to a force–deflection curve as the velocity of the impactor is not significantly changed by its passage through the specimen.

Typical force–deflection curves are illustrated in Fig. 1. The first peak which occurs in the curve is caused by the movement of the specimen through the supports and is not related to the fracture process. From the various values which can be obtained from the force–deflection plots the more important features of the fracture process have been selected and the results given in Table 8.

All the fillers used substantially reduce the impact strength of the nylon. Nylon 6 is a tough plastic capable of absorbing energy during fracture through deformation processes. The presence of a filler

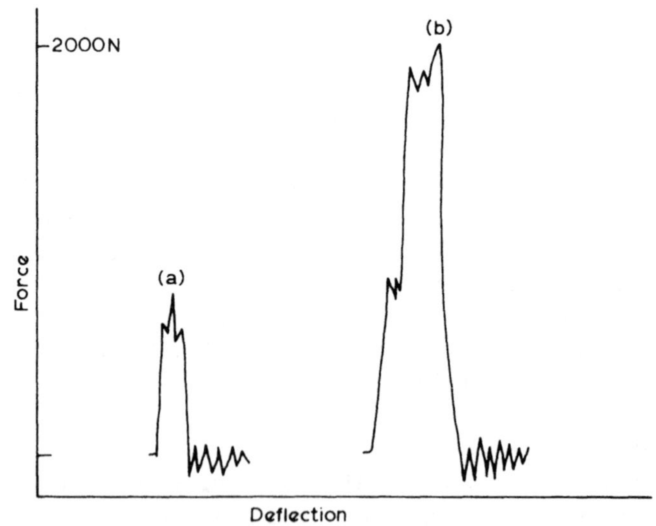

FIG. 1. Instrumental falling weight impact tests. Force–deflection curves obtained from (a) nylon 6 filled with acicular wollastonite; (b); nylon 6 filled with acicular wollastonite coated with γ-aminopropyltriethoxysilane.

TABLE 8

	Peak force (kN)	*Total deflection (mm)*	*Total energy absorbed (kJ/m²)*
Unfilled nylon 6	Does not break	—	—
Silane coated ballotini	1·4	6·2	57
Non-acicular wollastonite	0·96	4·9	32
Non-acicular wollastonite, silane coated	1·5	6·1	60
Acicular wollastonite	0·93	3·54	29
Acicular wollastonite, silane coated	1·76	7·0	88
Acicular wollastonite, silane coated (nylon contains 3% moisture)	1·36	11·2	121

reduces the amount of tough material available for these processes and therefore reduces the elongation at break and degrades the impact properties. The performance of the composites in which the filler is not bonded to the matrix is particularly poor.

4.7. Electron Micrographs of Fracture Surfaces

The fracture surfaces of tensile test and unnotched impact specimens were examined by scanning electron microscopy. The accelerating voltage was chosen to show as clearly as possible any nylon adhering to the filler surface. Figure 2 is representative of the fracture surfaces found when no adhesion exists between nylon and filler. The fracture surfaces are formed by cracks which run through the voids formed at the nylon–wollastonite interface.

When the coupling agent is present the fracture surface is smooth with few protruding fibres or holes (left by the wollastonite removed during the fracture process). Figure 3 illustrates the good bond found between nylon and the aminosilane coated wollastonites.

In this photomicrograph the wollastonite fibre has undergone transverse cleavage, an important attribute of acicular Indian wollastonite which accounts for the retention of aspect ratio when the fibre length is reduced in the compounding processes.

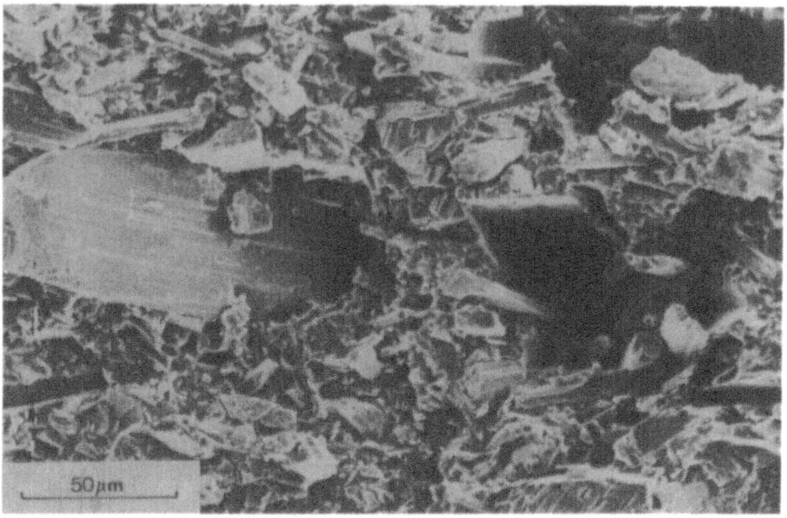

FIG. 2. Fracture surface of a tensile test specimen containing uncoated acicular wollastonite.

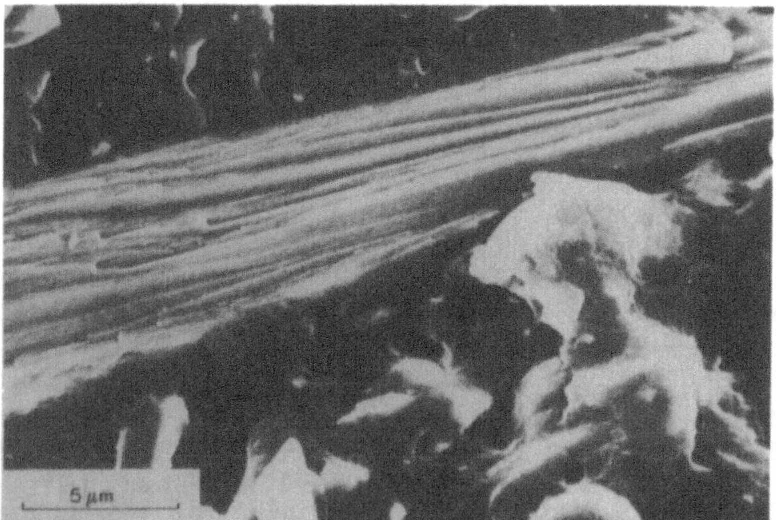

FIG. 3. Fracture of a tensile test specimen containing silane coated acicular wollastonite showing good adhesion between polymer and filler and the transverse cleavage of the fibre in the fracture plane.

Areas of ductile failure were present in the fracture surfaces of composites containing the aminosilane coated fillers. These areas of ductile failure increased when higher loadings of coupling agent were used. These higher loadings were not capable of significantly increasing the strength of the composite, nor did the use of higher loading improve properties such as thermal and impact properties.

4.8. Water Absorption

The water absorption at 50°C was determined for the unfilled nylon 6 and the composites containing uncoated and silane coated acicular wollastonites over a period of 10 days. The results were plotted as percentage of water uptake versus the square root of the time of immersion in days (Fig. 4). All the wollastonites behaved as perfectly

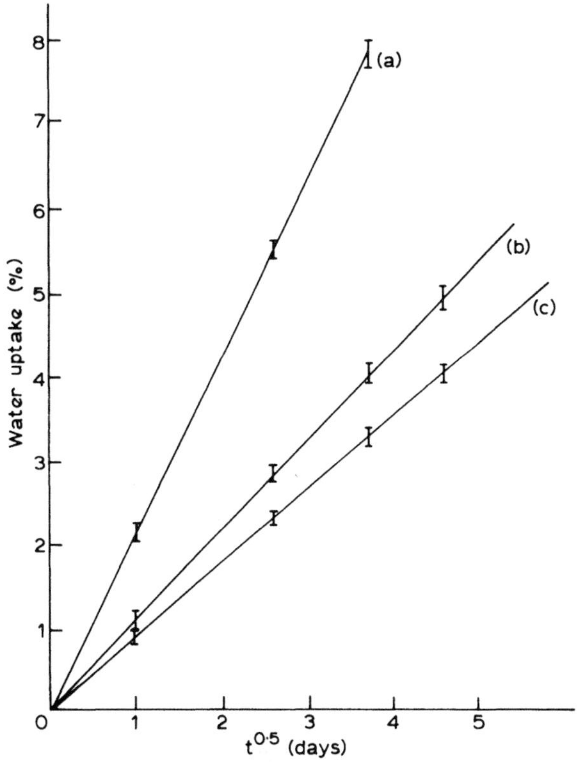

FIG. 4. The water absorption of nylon 6 and wollastonite filled nylon 6 measured at 50°C. (a) Nylon 6—slope 2·0; (b) nylon 6 containing 50 wt % of the uncoated wollastonites—slope 1·0; (c) nylon 6 containing 50 wt % of the silane coated wollastonites—slope 0·85.

inert fillers in this experiment, contributing nothing to the water uptake of the composite. When the amino silane coupling agent was present the water absorption was further reduced due to decreased absorption in the interfacial region.

4.9. Volume Resistivity

Volume resistivity was measured according to BS 2782:202A in a controlled humidity environment using plaques measuring 10 × 10 × 0·3 cm. Plaques containing uncoated and aminosilane coated acicular wollastonite were preconditioned, after painting on the graphite electrodes, in chambers containing saturated salt solutions. A relative humidity range of 12–75% was used, corresponding to a nylon water

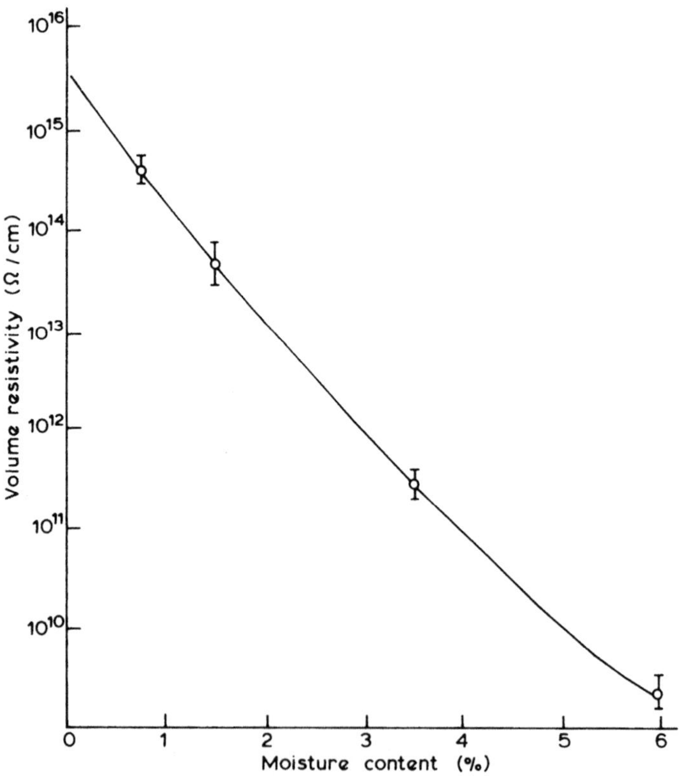

FIG. 5. Relationship between volume resistivity and moisture content of nylon in both unfilled and wollastonite filled plaques.

content at equilibrium of up to 6%. Unfilled nylon plaques were treated in the same way.

The effect of the amount of absorbed water on the volume resistivity of nylon and the wollastonite filled composites is shown in Fig. 5. The moisture content of the nylon has been calculated assuming negligible water absorption by the wollastonite. The results obtained from all the plaques, filled and unfilled, were not significantly different, showing that, as would be expected at the low volume fraction of filler used, the volume resistivity is solely dependent on the amount of nylon present.

4.10. The Flammability of Wollastonite–Nylon 6 Composites

Wollastonite has no flame retardant effect on nylon 6 in the chemical sense; it can however modify the burning characteristics by reducing the tendency to drip.

Nylon 6 burns with a blue flame from a colourless melt accompanied by much dripping of the molten nylon. The filled nylon 6 burns with a yellow flame but with no appreciable smoke. The ability of the nylon 6 to continue to burn once the flame used to ignite it is removed depends on the mobility of the melt. The presence of acicular wollastonite produces a more viscous melt which, in vertical burning tests, remains in place, allowing the ash to extinguish the flame. The oxygen index figures given in Table 9, measured according to ASTM 2863 and using specimens conditioned at 50% RH, show the increased oxygen index obtained when the acicular wollastonite is present.

In horizontal burning experiments the rate of dripping of the nylon is decreased by the presence of ballotini and wollastonite and to a greater extent by acicular wollastonite. The slow rate of dripping found in the latter case enables the specimen to continue to burn after removal of the bunsen used to ignite it. The unfilled nylon drips at such a rate that after removal of the flame the specimen is extinguished because of removal of all burning material.

TABLE 9

	Oxygen index
Nylon 6	25·0%
Non-acicular wollastonite	23·5%
Acicular wollastonite	25·8%

5. WOLLASTONITE IN PA 66

Nylon 66 was the original nylon developed in the United States and was marketed by Du Pont in 1939 prior to nylon 6 being marketed in Europe. ICI were responsible for the development of nylon 66 in the UK.

Nylon 66 offers advantages of better mechanical and thermal properties compared to nylon 6. The water absorption of nylon 66 reaches a maximum of 8% compared to 11% for nylon 6. Nylon 6 is, however, tougher and has better flexibility and surface finish.

5.1. Compound and Specimen Preparation

The compound and resultant test specimens were prepared and tested in a similar way to those containing nylon 6, so only the important difference will be described here.

5.1.2. Materials Used

The nylon used was PA 66—'Maranyl' nylon supplied by ICI PLC. The filler used was the same acicular Indian wollastonite coated with A-1100 used to prepare the nylon 6 composites.

5.1.3. Extrusion Compounding

The dry blend combining 50% by weight of the filler was fed to a Brabender twin-screw compounding extruder. A die of 6 mm diameter was used. The strand was granulated and then dried for 72 h at 80°C. The screw speed was 10 rpm and the zone temperature and torque were as shown in Table 10.

The blend was difficult to process; it did not feed well and the machine found the blend hard going.

TABLE 10

	Temperature (°C)				*Torque (N m)*
	1	*2*	*3*	*4*	
PA 66	270	280	280	270	130/150
PA 66 + acicular wollastonite, silane coated	270	280	280	270	180/200

TABLE 11
INJECTION MOULDING CONDITIONS

Sample type	*Injection time (s)*	*Cooling time (s)*	*Line pressure (psi)*
ISO tensile			
PA 66	10	18	400
PA 66 + acicular wollastonite, silane coated	10	18	500
Charpy bar/Flexural bar			
PA 66	20	20	350
PA 66 + acicular wollastonite, silane coated	20	20	400
Plaques			
PA 66	14	20	400
PA 66 + acicular wollastonite, silane coated	14	20	650

5.1.4. Injection Moulding

Test specimens were moulded using a 35 t Szekely in-line reciprocating-screw moulding machine. The barrel temperatures were:

Zone 1	rear	265°C
Zone 2		275°C
Zone 3		285°C
Zone 4	front	285°C
Mould		75°C

The other conditions were as given in Table 11.

5.2. Tensile and Flexural Tests

The tensile and flexural results given in Table 12 confirm the ability of the silane coated acicular wollastonite to provide significant reinforcement when used as a filler for polyamides. The 28% increase in tensile strength caused by the presence of 50% of the coated wollastonite is close to the 38% increase calculated using the Kelly equation if allowance is made for fibre orientation in the test piece. The increase in tensile modulus of 122% is, as with PA 6, much less than predicted by the Nielsen equation. The increase in tensile strength is similar to that achieved by coated mica at a comparable loading but is inferior to

TABLE 12

THE PROPERTIES OF NYLON 66 FILLED WITH 50% W/W SILANE COATED ACICULAR WOLLASTONITE (DETERMINED AT 0% RH)

Property	*Unfilled nylon 6·6*	*Nylon 66 + silane coated acicular wollastonite*
Tensile strength (MPa)	83	106
Tensile modulus (GPa)	2·7	6·0
Tensile elongation at break (%)	50	3·0
Flexural strength (MPa)	119	186
Flexural modulus (GPa)	3·1	10·5
Heat distortion temperature (°C at 1·81 MPa)	80	225
Linear expansion coefficient (°C × 10^{-5})	7	2–3
Rockwell hardness (M scale)	90	98
Density (g/m^3)	1·14	1·62
Notched Izod impact (J/m) (6·3 mm thick specimens)	12	22
Water absorption (24 h, 23°C) (ASTM D570)	1·2	0·5
Oxygen index (% O_2)	23	28

TABLE 13

THE EFFECT OF TEMPERATURE ON THE TENSILE PROPERTIES OF NYLON 66 FILLED WITH SILANE COATED ACICULAR WOLLASTONITE

Property	*Unfilled nylon 6·6*	*Nylon 6·6— + acicular wollastonite*
Tensile strength (MPa)		
0°C	96	127
50°C	67	80
Tensile modulus (GPa)		
0°C	3·0	7·0
50°C	2·0	5·0
Strain (%)		
0°C	17	3·0
50°C	—	4·0

TABLE 14

THE EFFECT OF WATER CONDITIONING ON NYLON 66 FILLED WITH SILANE COATED ACICULAR WOLLASTONITE

Property	*Unfilled nylon*	*Nylon 66— + acicular wollastonite*
Flexural strength (MPa)		
Dry	119	186
Wet	79	151
Flexural modulus (GPa)		
Dry	3·1	10·5
Wet	2·6	7·7
Tensile strength (MPa)		
Dry	83	106
Wet	62	79
Tensile modulus (GPa)		
Dry	2·7	6·0
Wet	1·4	3·6

that given by glass fibre at a loading of 20% w/w. The increase in flexural modulus is similar to that achieved by mica and equivalent to the use of glass fibre at a loading of 33% w/w.

The effects of temperature on the tensile properties of nylon 66 filled with the acicular wollastonite are shown in Table 13. The improved tensile strength of the composite demonstrated throughout the temperature range (0–50°C) with the expected decrease in the difference between filled and unfilled material as the temperature increases. The rate of change in tensile properties of the composite reflects the rate of change of the tensile properties of the unfilled nylon 66.

To determine the effect of moisture on the flexural and tensile strength of the composite specimens were conditioned in water for 15 h at 50°C; the results are given in Table 14. As with nylon 6–wollastonite composites the value of the silane coating agent is demonstrated by the preservation of the strength of the composite, compared to the unfilled nylon, under these adverse conditions. If uncoated wollastonite is used the loss of strength by the composite under these conditions is much greater, resulting in a composite having a lower tensile strength than the unfilled nylon.

As the results given in Table 12 for the water absorption of the PA 6/acicular wollastonite composite show, the wollastonite behaves as a perfectly inert filler contributing nothing to the water uptake of the composite.

5.3. Impact Behaviour

The notched Izod impact results given in Table 12 show that the presence of the acicular wollastonite improves impact performance more in nylon 66 than in nylon 6. This reflects the greater notch sensitivity of the plastic. The magnitude of the notched impact results given here are difficult to compare with other quoted values because of the effect of the thickness of the specimen (6·3 mm) and the manner in which the notch has been introduced (notches were made by using a high-speed milling machine equipped with an appropriate cutter).

The unnotched impact performance of the composite was determined using the instrumented falling weight machine described previously. Specimens in the form of Charpy bars and plaques measuring 5 cm × 5 cm × 3·0 mm were used.

In contrast to the bars of nylon 6 which were only bent by the falling weight machine, bars of nylon 66 could be broken.

The shape of the load–deflection curve obtained when the Charpy bars are broken is identical to that obtained from the nylon 6–acicular wollastonite composite. The increased load at break and the decreased elongation at break are to be expected from the comparison of the tensile properties of nylon 6 and 66. The results given in Table 15 show that the addition of acicular wollastonite does not improve the impact strength of nylon 66. The presence of the filler stiffens the polymer, decreasing the deflection at break and the total deflection. The loss of energy absorbing polymer by the addition of a high modulus filler can only be compensated for by energy expended in pulling out long fibres. The length of the wollastonite fibres is inadequate for this process and does not make any significant contribution to the impact strength of the composite.

5.4. Heat Distortion Temperature

The heat distortion temperature of 225°C is higher than recorded for most nylon 6·6 composites containing a similar loading of filler. It is slightly less than is quoted for some nylon 6·6–mica composites, *c.* 235°C, and for nylon 66–fibre glass composites, for which a value of >240°C is given (m.p. nylon 6·6: 264°C).

5.5. Fibre Orientation

The mould used to prepare the rectangular plaques used an edge gate designed to give minimum orientation of polymer or fibre. To obtain

TABLE 15
UNNOTCHED IMPACT TESTS AT 20°C

Material and specimen type	*Peak force (kN)*	*Peak deflection (mm)*	*Fail deflection (mm)*	*Peak energy (N m)*	*Fail energy (N m)*	*Stiffness (N/mm¹)*	*Total energy absorbed (kJ/m²)*
Charpy bar							
Unfilled nylon 66	1·5	—	11·1	—	9·0	—	113
Nylon 66–Snowfort SN4	2·2	—	6·1	—	7·78	—	98
Plaque							
Unfilled nylon 66	1·54	2·47	5·47	1·15	2·39	562	—
Nylon 66–acicular wollastonite, silane coated	1·20	1·38	4·06	0·44	2·13	870	—

Impactor velocity 5 m/s. Total weight of impactor 15 kg.

TABLE 16

ORIENTATION IN PLAQUES WITH EDGE GATE

Mode of testing	*Stiffness* $E90°/E0°$	*Strength* $\sigma 90°/\sigma 0°$
Tension	1·26	1·15
Flexure	1·20	1·13

an indication of the degree of anisotropy that can be expected when wollastonite–polyamide composites are produced using a well designed mould, measurements were made of the flexural and tensile strength of samples cut at 90° and 0° to the major flow direction. The results are given in Table 16.

The results obtained for the unfilled mouldings of nylon 66 showed no anistropy in flexure or tension. The degree of anistropy in the moulding containing the acicular wollastonite is small when compared to the differences which can be found in similar mouldings containing glass fibre.

6. INDIAN WOLLASTONITE AS A FILLER FOR POLYPROPYLENE

Polypropylene (PP) has been commercially available since the early 1960s. It now makes up the largest tonnage of mineral filled thermoplastics used in injection moulding applications. Table 17 gives the total consumption of reinforced PP in Europe in 1983. Polypropylene offers the best growth prospect of any thermoplastic as it is used in increasing quantities by the automotive industry.

TABLE 17

CONSUMPTION OF REINFORCED POLYPROPYLENE IN EUROPE, BY TYPE (1983)

Talc-reinforced	105 000 t	58%
Calcium-carbonate reinforced	60 000	33%
Glass-fibre reinforced	12 000	7%
Other reinforcements	3 000	2%
Total	180 000 t	100%

From Ref. 6.

6.1. Filled PP

A wide range of polypropylene homopolymers and copolymers are available. By control of molecular weights, molecular weight distribution, tacticity and the inclusion of comonomers a wide range of physical properties can be obtained. The intrinsic chemical resistance of polypropylene is outstanding and stabilisation systems are available to enable it to be used in a wide range of environments.

Despite these favourable properties, polypropylene cannot be considered as a high performance engineering plastic unless its low stiffness, mechanical strength, heat distortion temperature and coefficient of expansion can be improved.

Particulate reinforcement such as talc, calcium carbonate, mica, etc. are commonly added to polypropylene, both to improve the physical properties of the moulded part and to improve moulding characteristics such as shrinkage and cycle time. Talc filled polypropylene components are favoured by the automotive industry when increased stiffness is the main requirement. The highest level of stiffness and the ability to maintain useful mechanical properties at elevated temperatures are given by mica containing composites.

A major disadvantage when talc or mica are used is the poor impact performance of the composite, particularly in notched tests.

Compared to mica and talc little work has been published about the use of wollastonite as a filler for polypropylene. Some results have appeared[8] showing that the use of azidosilane coupling agents substantially increases the tensile and flexural strength of wollastonite containing composites. The use of the organosilicon chemicals (OSCs) developed by Union Carbide, which do not have the safety, handling and environmental problems associated with the azidosilane, has only been reported in mica containing composites.[9] The use of these coupling agents, in conjunction with polypropylene, was therefore studied.

The polypropylenes used were Hercules 6524 (MFI 4·0) and ICI 'Propathene' HWM 25 (MFI 4·0). The filler was the same acicular Indian wollastonite used to prepare the nylon 6 and nylon 66 composites. The wollastonite was not pretreated with the organosilicon chemical; in this case the polypropylene and the wollastonite were placed in a ribbon blender and Union Carbide Ucarsil PC1A, followed by PC1B, were added while mixing. 2% of PC1A and 0·5% of PC1B were used, based on the filler weight. This is referred to as

TABLE 18

PROPERTIES OF 40% W/W WOLLASTONITE FILLED POLYPROPYLENE COMPOSITES

Property	*Hercules 6524 unfilled*	*Hercules 6524 + uncoated acicular wollastonite*	*Hercules 6524 + wollastonite integral treatment*	*ICI 'Propathene' HWM 25 unfilled*	*ICI 'Propathene' + wollastonite integral treatment*	*ICI 'Propathene' + talc*
Tensile strength (MPa)	31	28	35	29·5	34	30
Tensile modulus (GPa) (1% Secant)	1·1	2·2	2·0	1·2	2·9	2·7
Tensile elongation (%)	>600	22	20	>600	12	20
Flexural strength (MPa)	40	44	58	37	56	47
Flexural modulus (GPa) (1% secant)	1·8	4·3	3·9	1·8	4·0	3·6
Izod impact (J/m)	20	15	18	21	17	13
Izod impact 1·0 mm radius notch (J/m)	52	24	40	61	31	23
Heat distortion temperature (°C at 1·81 MPa)	62	80	85	60	83	79
Density (g/m³)	0·91	1·25	1·25	0·91	1·25	1·24
Melt flow index (g/10 min)	—	—	—	7·0	11·0	7·0

integral treatment. (PC1A and PC1B are organosilicon compounds manufactured by Union Carbide.)

6.2. Extrusion Compounding

Blends containing 40% w/w of the fillers were fed in turn to a Werner and Pfleiderer ZDSK twin-screw compounding machine containing one kneading zone. The temperatures of all the zones and the die were adjusted to 220°C. The screw speed was 76 rpm. The extrudate, in the form of strands, was cooled by passing through a water bath and then granulated. The integrally treated blends of wollastonite and polypropylene were easily processed when compared to the blend containing the uncoated wollastonite. The unfilled polypropylene was also passed through the compounding machine.

The melt flow indices of the composites based on 'Propathene' were measured after compounding; the results are given in Table 18.

6.3. Injection Moulding

Test specimens were moulded using a 35 t Szekely in-line reciprocating-screw injection moulding machine. Typical moulding conditions for the 50% filled composites are given in Table 19.

The injection line pressure and injection time were changed according to the sample type; the flexural/HDT bars required long injection and cooling times to prevent void formation.

6.4. Mechanical and Thermal Tests

The testing of the injection moulded specimens was carried out using the same procedure and conditions described for PA 6 and PA 66.

TABLE 19
INJECTION MOULDING CONDITIONS

Zone 1	225°C
Zone 2	230°C
Zone 3	240°C
Zone 4	225°C
Melt temperature	220–224°C
Mould temperature	25°C
Injection line pressure (psi)	300–600
Back pressure (psi)	150
Injection time (s)	4–50
Cooling time (s)	16–40

6.4.1. Tensile and Flexural Tests

The values of tensile strength given in Table 18 for the composites prepared using the organosilicon chemicals are close to the values predicted by Kelly's equation (Section 3.2) if it is assumed that there is good bonding between the polymer and filler. These values for tensile strength are inferior to the reported values of 40–50 MPa which can be obtained using glass or surface treated high aspect ratio mica at similar loadings. The flexural strength of the composites also show the effect of the silane treatment.

When the flexural and tensile moduli of the composites containing the uncoated and silane treated wollastonite are compared the presence of the silane appears to have caused a slight reduction in stiffness, possibly by reducing the crystallinity or changing the crystalline morphology of the polypropylene.

The elongation to break of the polypropylene has been greatly reduced by the presence of the wollastonite; this is reflected in a reduction in impact strength.

6.4.2. Impact Behaviour

The notched impact results given in Table 18 show the expected reduction in impact strengths, when the composites are compared to the unfilled polypropylene. The presence of the wollastonite fibres has reduced the amount of energy absorbing polymer available and the fibres are not long enough to impede the crack running out from the notch.

The unnotched impact tests were performed using the equipment described in Section 4.6.2. The specimens were in the form of plaques measuring 5×5 cm $\times$ 3 mm supported on a cylindrical base containing a hole of 32 mm diameter. The impactor was hemispherical, having a diameter of 16 mm.

The results obtained for the composites based on the 'Propathene' homopolymer are given in Table 20. None of the filled plaques showed any signs of ductile failure, breaking into four roughly equal pieces. The behaviour of the polypropylene–uncoated wollastonite composite was inferior to that of the unfilled polypropylene, having a reduced load and elongation at break. The formation of voids at the interface between the filler and the polymer matrix weakens the polymer. When the wollastonite is bonded to the polymer by using the organosilicon chemical the load and deflection at maximum load are increased and although the total deflection and the energy absorbed after the peak

TABLE 20
UNNOTCHED IMPACT

Property	*Polypropylene unfilled*	*Polypropylene + uncoated wollastonite*	*Polypropylene + wollastonite, silane coated*	*Polypropylene + talc*
Peak force (kN)	0·49	0·34	0·62	0·66
Peak deflection (mm)	1·55	1·3	1·80	1·94
Fail deflection (mm)	10·4	5·3	5·45	5·38
Fail energy (N m)	0·19	0·15	0·48	0·55
Total energy absorbed (N m)	2·05	1·1	2·0	2·27
Stiffness (N/mm)	314	342	346	340

force is reached are decreased, the net result is that the total energy absorbed remains virtually unchanged.

6.4.3. Heat Distortion Temperature
The increase in heat distortion temperature obtained when the silane treatment is used is disappointing when compared to reported values for glass and mica filled composites of up to 130°C. It is evident that a fibre of much higher aspect ratio is required to obtain a useful increase in the heat distortion temperature of polypropylene composites.

6.5. Electronmicrographs of Fracture Surfaces
The fracture surfaces of all the composites, as obtained from tensile and impact tests, were examined by scanning electron microscopy. Only those composites prepared from the silane treated mixture of polypropylene and wollastonite showed any evidence of bonding

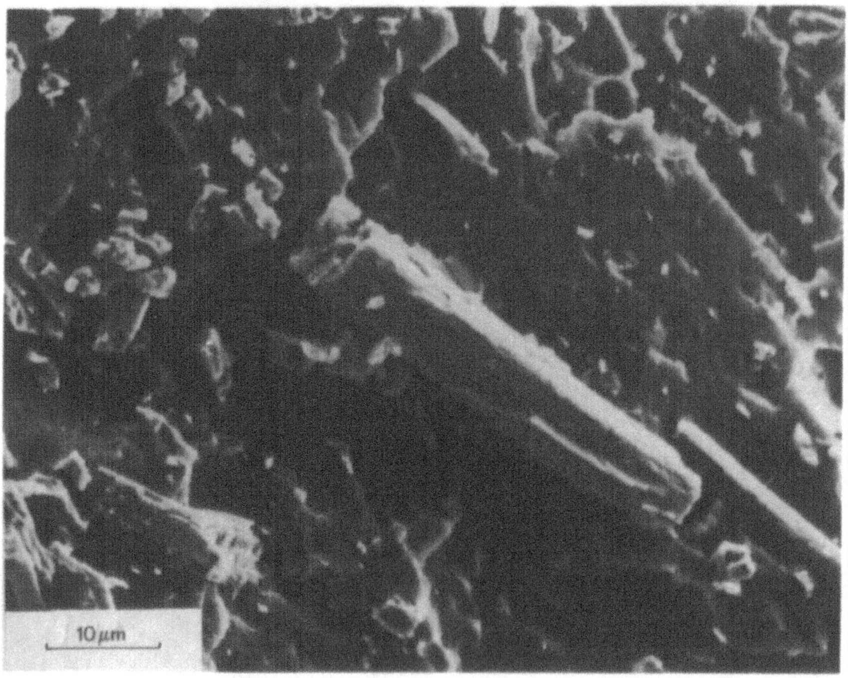

FIG. 6. Fracture surface of a notched Izod specimen containing uncoated wollastonite.

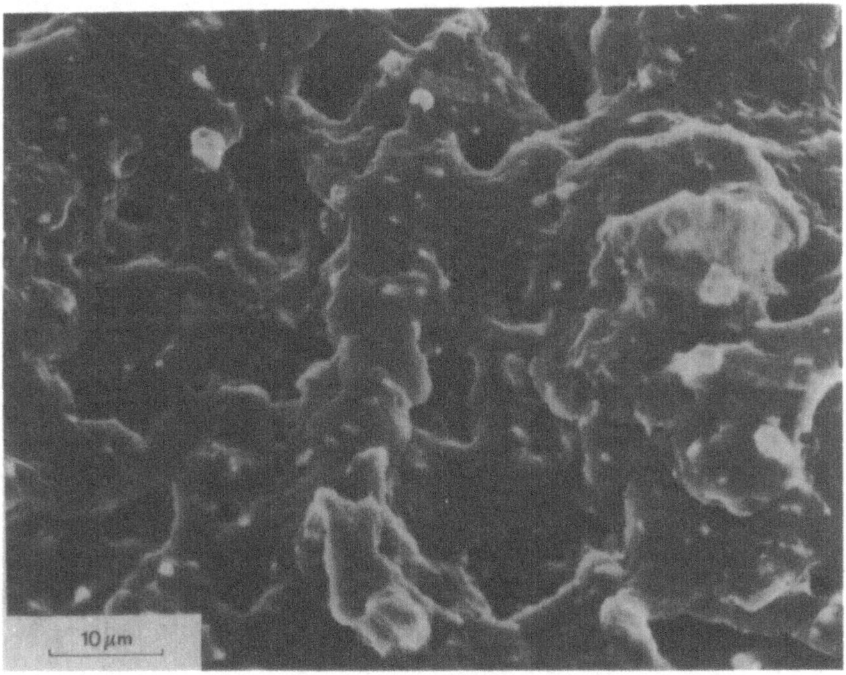

FIG. 7. Fracture surface of a notched Izod specimen containing acicular wollastonite. The wollastonite and polypropylene blend having been treated with silanes PC1A/PC1B.

between matrix and filler. In these composites the fracture plane runs through the plastic, revealing few of the wollastonite fibres. In Fig. 6 the fracture surface of an Izod specimen shows no sign of adhesion, many of the wollastonite fibres having been pulled out or lying on the surface. In Fig. 7 the fracture surface of the Izod specimen prepared from the silane treated mixture of polypropylene and wollastonite shows how well the wollastonite is wetted by the polymer matrix, with practically all of the failure taking place in this matrix.

Figures 8 and 9 show fracture surfaces produced by the unnotched falling weight test; again the differences in wetting and adhesion can be clearly seen. There is evidence in these photographs of the transverse cleavage of the wollastonite fibres found in the fractured nylon 6 and 66–wollastonite composites.

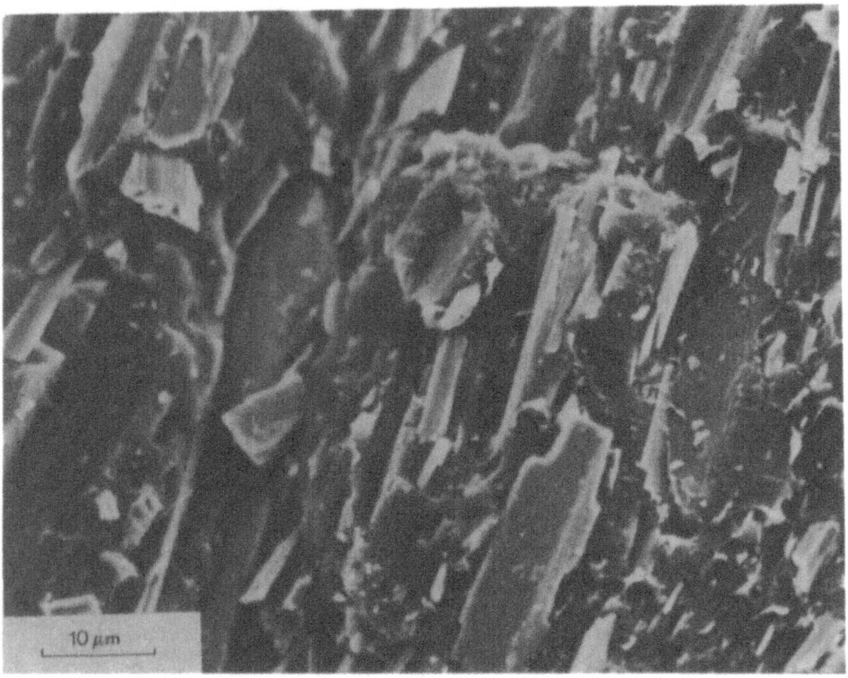

FIG. 8. Fracture surface of an unnotched impact test specimen containing uncoated acicular wollastonite.

7. CONCLUSION

Silane coated acicular wollastonite can be used as a filler for both nylon 6 and 66 to provide composites having significantly improved tensile strength, both dry and wet, compared to the unfilled nylons. These increases in strength can only be obtained when a coupling agent has been used to provide good adhesion between the filler and the polymer matrix.

Heat distortion temperatures are increased to values near to the maximum obtainable with particulate fillers, and other important properties such as the coefficient of thermal expansion and mould shrinkage are also improved to a similar degree.

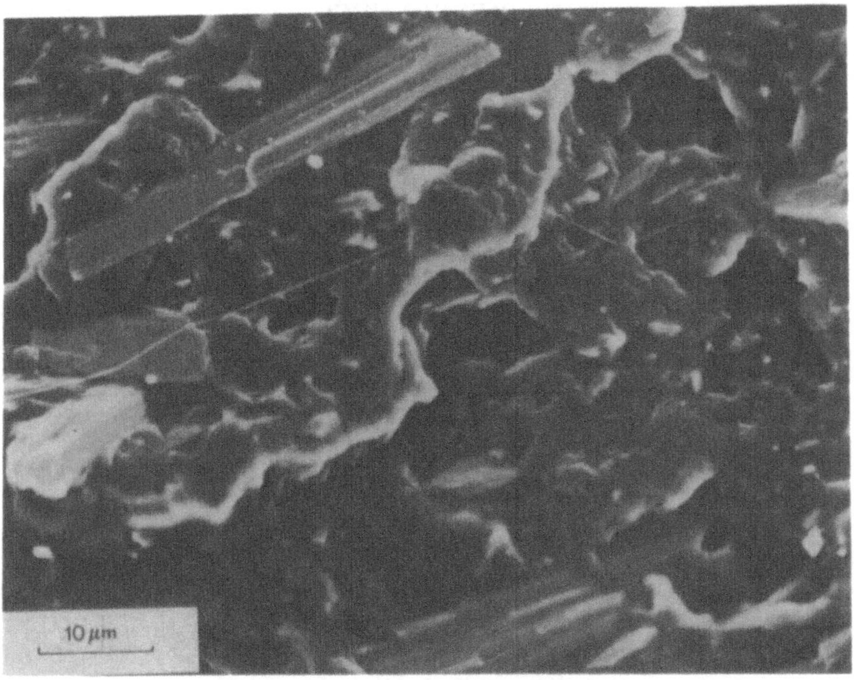

FIG. 9. Fracture surface of an unnotched impact test specimen containing acicular wollastonite. The wollastonite and polypropylene blend having been treated with silanes PC1A/PC1B.

The only property of nylon 6 and 66 adversely affected by the presence of acicular wollastonite is the impact strength. The impact strength of nylon will, however, be diminished by the presence of any particulate mineral; only longer fibres such as chopped strand glass can offer some improvement in notched tests, but at higher cost and with a great increase in anisotropic behaviour.

When used as a filler for polypropylene acicular wollastonite does not seem to offer any appreciable mechanical or thermal advantage when compared to other filllers. The use of the organosilicon chemical PC1A/PC1B has resulted in a composite having good bonding between filler and polymer matrix, but the increases in tensile strength and heat distortion temperature are not sufficient to justify the cost of the treatment.

REFERENCES

1. Andrews, R. W., *Wollastonite,* Institute of Geological Sciences, HMSO, London, 1970.
2. Harwell Laboratory, Oxfordshire. Private communication.
3. Nielsen, L. E., *J. Appl. Phys.* **41** (11), 1970, 4626.
4. Kelly, A., *Proc. Royal. Soc.* **A282,** 1964, 63.
5. Celanese Engineering Resins, quoted in *Plast. Rubb. Int.* **9** (4), 1984, 22.
6. Plastic Consult srl, Milan, 1984.
7. Perera, D. Y., Coatings Research Institute, Belgium, Private Communication.
8. McFarren, G. A., Sanderson, T. F., and Schappell, F. C., *Polym. Eng. Sci.* **17** (1), 1977, 46.
9. Godlewski, R. E., *SPI 38th Annual Technical Conference 13-E,* 1983.

Chapter 5

FATIGUE AND LONG-TERM STRENGTH OF THERMOPLASTICS

K. V. GOTHAM

*Rubber and Plastics Research Association, Shawbury, Shrewsbury, UK**

1. INTRODUCTION

Thermoplastics are complex materials in their response to applied loads. They do not obey Hooke's law, being non-linear in their relationship between stress and strain. They are also viscoelastic in nature. This property is seen either as a time dependent response in stress (stress relaxation), when held under conditions of constant strain, or as a time dependent response in strain (creep), when subject to a constant continuously applied load. As a general rule their mechanical properties are more dependent upon temperature than metals which they seek to displace in engineering components. As a class of materials they have, in their unmodified form, comparatively low modulus and low strength. In addition to this complicated response to continuously applied loads, which results in a progressive decrease in strength with elapsed time under load, thermoplastics still suffer from 'fatigue' under cyclic loads. Hence the qualification implied within the title. At one time[1] these responses were known as static and dynamic fatigue, but this terminology led to some confusion with engineers. Consequently, as a matter of convention nowadays, the term 'fatigue' is restricted to the effects of cyclic or intermittent loads only, as is the case for metals, and 'long-term strength or creep rupture' to the effect of continuously applied loads. Both phenomena

* Present address: 1 Kensington Gardens, Minsterley, Shrewsbury, UK.

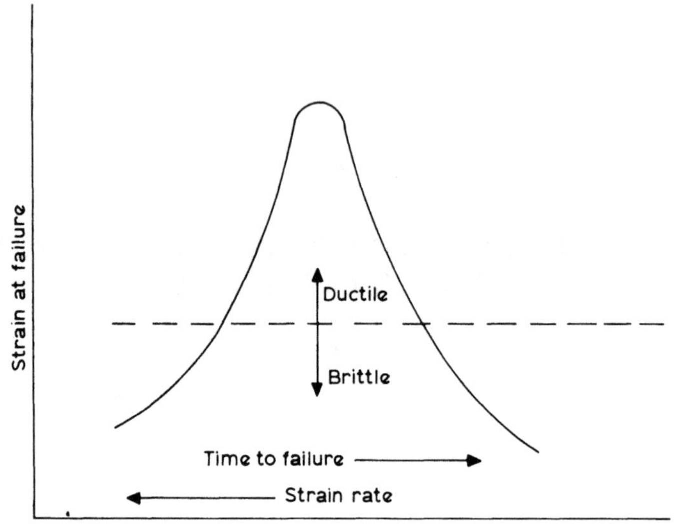

FIG. 1. The ductility peak (schematic).

produce a progressive weakening of a component and so the general term 'durability' collectively describes fatigue and long-term strength.

Although thermoplastic materials tend to be inherently ductile they respond so only within a limited range of operating conditions. Typically, as shown in Fig. 1, there is a tendency for ductility, measured as strain or energy to break, to maximise at moderate input rates of mechanical energy density. Thus, by association, there is a tendency for brittle failure to occur under conditions that lead to very short (impact) or very long (fatigue) times to failure. These two regimes of embrittlement, which correspond to service usage, *a priori* are not necessarily caused by the same failure mechanisms. Further, it is important to appreciate that polymers, which in their natural state or by modification are resistant to failure, say under impact loads, are not necessarily also resistant to fatigue. The implication here is that the mechanisms involved in crack initiation are not the same for very short and very long-term failure.[2]

There are various forms of failures that can be observed in thermoplastic materials. The most obvious is that of macroscopic fracture which can be catastrophic in nature and, as has already been mentioned, can be either ductile or brittle. However, it may not be appropriate or desirable to allow for macroscopic fracture as a design criterion. As a consequence, such incipient failures as local stress crazing or cracking and stress whitening, which are also observed, can serve as natural design limits or boundaries. These phenomena can arise as a result of material or processing variables or quite simply as inherent molecular structural defects or contamination.

The importance of defect initiation should not be underestimated. On the one hand it may be inextricably linked with and result from the process by which the component has been produced. Yet on the other, these defects can develop during component lifetime as a direct consequence of preceding service history. Thus failure under impact loads most probably stems from defects already present within the material, whereas fatigue failure may result from defects conceived and grown during component lifetime.

Because of the large difference in rate at which a crack grows at initiation and towards the end of its lifetime, the whole failure process can be conveniently separated into two main components, crack initiation and propagation. Unfortunately, much attention over the years has been spent by workers in studies concerned with the crack propagation component, virtually to the exclusion of initiation. A possible reason for this is that crack propagation is that part of the failure spectrum which is amenable to analysis by linear elastic fracture mechanics. It can, therefore, be presented mathematically and enables apparently easy materials comparison. However, since the major part of service lifetime, probably in excess of 95%, is consumed in conception and initiation of a labile defect, the importance of this component cannot be ignored. The exclusive examination of crack propagation, and a comparison of material fatigue resistance simply on this count, can be misleading. A traditional fatigue characteristation provides a straight forward and comparatively quick insight into brittle failure of the type encountered in service.

Attention has already been drawn to the subject of 'fracture mechanics'—the analysis of crack propagation from existing crack-like defects in a component. Under laboratory test conditions, it is necessary to induce a crack artificially by the use of machined notches with very sharp tip radii ($<10\,\mu m$). This is a special notch geometry. More usually, the effect of a notch is characterised in terms of 'notch sensitivity' and stress concentration factors.[3] Such characterisations provide useful data for the engineer and a number of illustrations are provided later in the chapter. Their presence will be seen to induce premature brittle failure under both continuous and cyclically applied loads.

In a broad appreciation of the durability of plastics there are other factors known to increase the likelihood of brittle failure. For instance, the nature of the environment in which a component may be called to operate. Although generally more resistant to chemical environments than metals, thermoplastics are still subject to not only direct chemical attack but also environmental stress cracking. This phenomenon is in a number of respects akin to stress corrosion in metals. Essentially, environmental stress cracking is a physical/chemical interaction. In the absence of stress/strain, there is no effect and the component functions satisfactorily. However, if stress/strain is either externally applied or generated internally at a sufficiently high level as a result of processing, crack initiation followed by propagation can occur.

Other factors also known to increase the propensity of the material to brittle failure are: increase in the complexity of the applied stress field and poor processing resulting in the generation of material defects. Decrease in temperature also will cause a general increase in the likelihood of brittle failure for all polymers. As mentioned earlier, at 20°C most thermoplastics are ductile but as temperature is reduced a point termed the 'brittle point' is reached at which there is a change in nature of the failure. This point varies from polymer to polymer. Thus, for example, in a standard tensile test bulk cast poly(methyl methacrylate) will become brittle at around 0°C whereas low density polyethylene with a density of 0·915 Mg/m^3 will not become brittle until its temperature is reduced to about −50°C.

For totally different reasons we shall see that incorporation of stiff glass fibre reinforcing elements in an otherwise ductile polymer, e.g. nylon 66, causes a complete change in the stress/strain relationship. The yield point disappears and the strain to break is reduced from a very high value for this 'natural' fibre forming thermoplastic to the order of 4%. Finally, degradation of the polymer, for any reason, leading to a reduction in molecular weight, particularly at the surface of the component, initiates the onset of brittle failure.

Since fatigue is the main consideration of this chapter, the effect of such variables as frequency, waveform, mode of deformation, etc. on the phenomenon are discussed. In addition, although the instances in which thermoplastics components are called upon to resist classic fatigue may be few, the importance of characterisation for service performance of this phenomenon will be appreciated.

2. FAILURE UNDER CONTINUOUS LOADS

Under continuously applied loads thermoplastics deform but the amount of deformation is not constant; it continues to increase with the elapsed time that the component is under load. This phenomenon is referred to as creep and although the essential concern of the chapter is long-term failure, creep is a very important and integral part of it. As time under load increases certain things may happen to the material. If the applied stress is of sufficient magnitude, the level of strain accumulated may increase to such an extent that incipient local failure such as crazing, cracking, stress whitening or microvoiding can occur. In the limit, of course, terminal failure could take place. The rate at which these various types of failure develop is dependent upon the magnitude of the applied stress, the temperature and the environment in which the component is operating. These various aspects of failure are considered in this section in terms of their practical importance.

2.1. Creep

The deformational behaviour of thermoplastics is classed as viscoelastic. Originally, this term was meant to imply a combination of viscous and elastic properties, but should not be interpreted in this way for plastics. The reason for this is chiefly because viscous flow rarely contributes towards the deformation below the yield point. It is important to realise and appreciate that the strain developed during creep in thermoplastics is nearly always fully recoverable.[4] There are, of course, exceptions to this general rule. For example, if during creep severe damage has been caused then complete recovery is unlikely. In addition, complete recovery is only possible if the unloaded time is much longer than the time under load during which the creep developed. A rough guide to the relationship between recovery and creep is that, for a creep period of t seconds, full recovery at the same temperature, on removal of the load, would occur after $10t$ seconds.

There is a further aspect of deformational behaviour which needs to be considered, i.e. stress relaxation. This is a complementary process to creep and takes place when a thermoplastic is held under constant deformation as opposed to constant load. When the material is first deformed the instantaneous stress peaks and then gradually and continuously decreases with increase in elapsed time. Both the rate of creep under a continuously applied load and the rate of stress

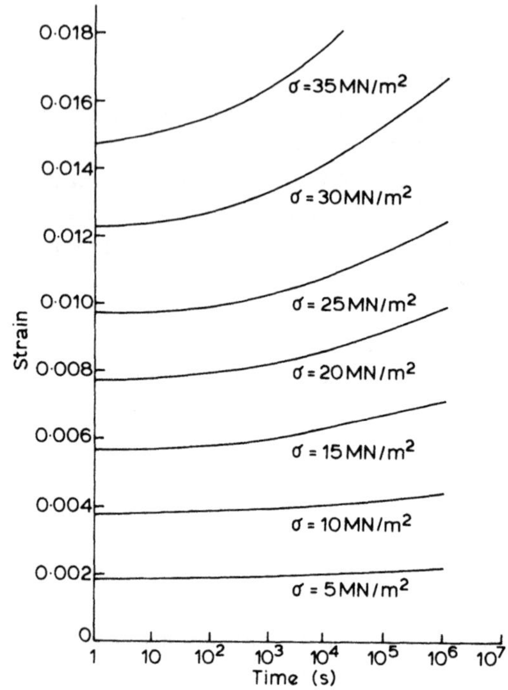

FIG. 2. Creep in polycarbonate at 20°C—tension.

relaxation under constant deformation are important parameters to consider during design.

Figure 2 shows a set of creep curves for polycarbonate, generated under a series of tensile stresses at 20°C. Polycarbonate is a transparent, amorphous thermoplastic, water white in colour and used in a number of applications which demand an inherently stiff, impact resistant material.

Although Fig. 2 is a standard presentation for creep data, there are other ways in which the data can be represented which are important for plastics. A useful reference in this context is BS 4618.[5] One particular presentation which proves useful in the context of pseudo-elastic design is the 'creep modulus'. Figure 3 shows such a presentation for polycarbonate, as a function of the level of tensile strain.

In service it is quite possible that the load will be intermittently rather than continuously applied. This leads to a situation in which both creep and recovery can take place. The level of accumulated

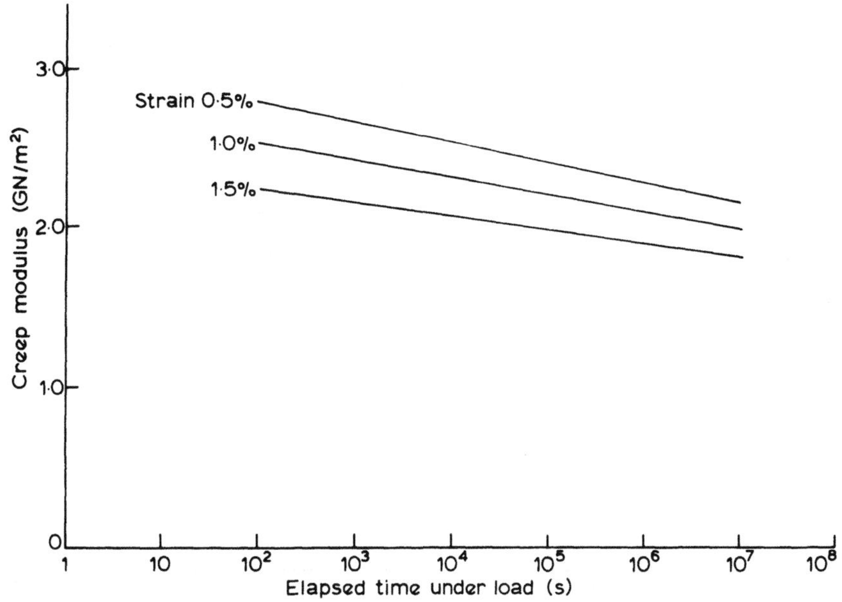

FIG. 3. Creep modulus for polycarbonate at 20°C, as a function of strain—tension.

strain will naturally depend upon the frequency at which the 'load', 'unload' cycle occurs but will lie below that for the continuously loaded case. An example of this sort of response is shown in Fig. 4 for a transparent grade of uPVC. The implications of premature failure under intermittent load are discussed in Section 3.

2.2. Limits of Durability

The user of plastics will appreciate that there are a number of limiting points to which the materials can be used. Often the limit is defined by the application itself. The most obvious and unambiguous one is that of macroscopic fracture. This can be either ductile or brittle in nature.

Brittle fracture can be defined as that which occurs with no observable plastic deformation, in the body of the plastic near or on the fracture surface itself. With ductile fracture, on the other hand, there is associated irreversible deformation. Dependent upon the application, macroscopic fracture may not be an acceptable limit for the design. For example, by the time ductile failure approaches, strains in excess of 5% will have developed and such deformational changes are unlikely to be able to be accommodated within the design.

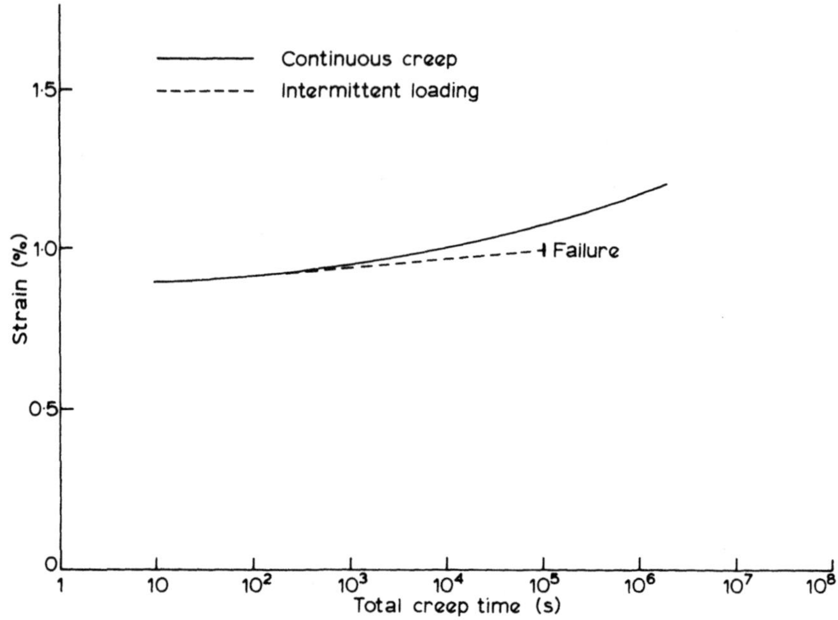

FIG. 4. The response to intermittent load (uPVC transparent grade 20°C). Mode—tension.

At such high working levels of strain also, the likelihood of fracture intervening for one reason or another cannot be ignored.

2.2.1. *Crazing*

The phenomenon of crazing has long been associated with the organic glasses, poly(methyl methacrylate), polystyrene, polycarbonate and styrene acrylonitrile. It is seen in the transparent grades as fine hair-like flaws mainly in the surface of the material. They can also develop, however, internally as a result of processing conditions.[6] Until the work of Spurr and Niegisch[7] and in particular the later work of Kambour and Robertson[8,9] crazes, initially observed in ceramic, had been treated as though they were true cracks. The fact that the craze contains polymeric material, interlinking the surfaces of the craze and oriented parallel to the direction of the stress field, is important since it affects the load bearing capability of the material. The strength of the craze itself is important because of its potential influence on the nature of subsequent failure.

2.2.2. Cracks

A crack can now, therefore, be defined as a flaw in which two new surfaces are created as a result of the stress field. These surfaces, although not bridged by polymer as a whole, may on creation have had localised linking of the two surfaces near the crack tip. Because of the unstable nature of a labile crack, these linked surfaces break down and permit the crack to propagate.

2.2.3. Stress Whitening

This is an omnibus term which is used to cover a range of associated phenomena in plastics. It arises as the result of local changes in refractive index of the thermoplastic. It can, for instance, be caused by the creation of microvoids which *en masse* appear white as the result of light scattering from regions well below the limit of visual resolution. Alternatively, in heterogeneous thermoplastics such as glass fibre reinforced plastics, it may occur at the onset of breakdown of the bond between the glass fibres and surrounding matrix. Similar phenomena can be seen in rubber reinforced composites such as the ABS class of plastics. In this instance Bucknall[10] has attributed the whitening to the formation of microcrazes around the rubber particles. In the same context, these microdefects could be cracks rather than crazes.

2.3. Creep Rupture—Tension

Creep rupture or static fatigue is a phenomenon observed in glasses as well as plastics. Only at high temperatures, however, is anything similar observed in metals. It is seen as a progressive reduction in load bearing capability as time under load increases. Most thermoplastics are inherently ductile[11] at room temperature and continue to fail as such at long times. There are exceptions to this generalisation: for example, injection moulded grades of polystyrene and poly(methyl methacrylate), and transparent grades of unplasticised poly(vinyl chloride) and glass reinforced thermoplastics. These materials break before the yield point has been reached. The creep rupture data are presented as graphs of applied stress vs. logarithm of the time under load. Figure 5 shows a typical characteristic for polycarbonate. The ductile failure is seen as the creation of a neck in the specimen. In the case of polycarbonate, ultimate failure is preceded by craze initiation which is also time dependent. Ductility will persist until at some stage crack-like defects develop, crack propagation ensues and failure is premature in that it precedes what would have been predicted by

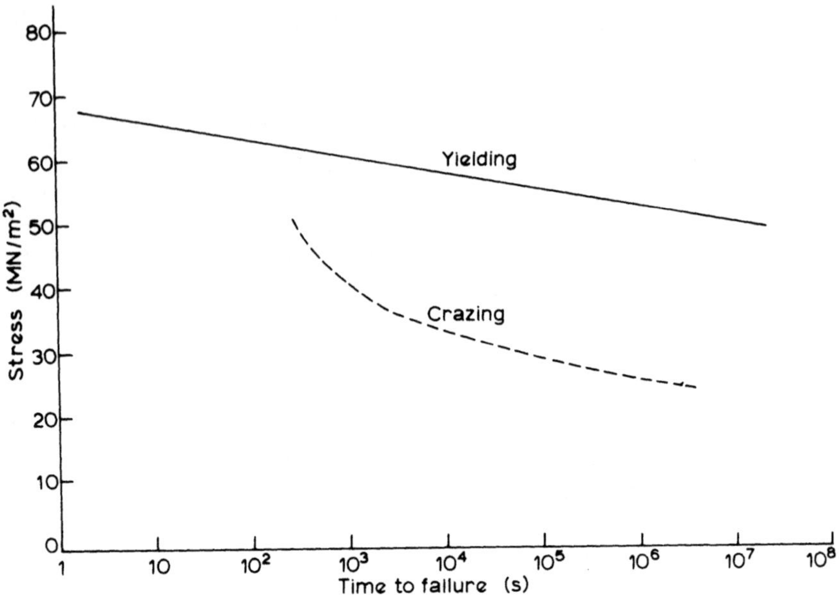

FIG. 5. Creep rupture in polycarbonate under continuous load at 20°C. Mode—tension.

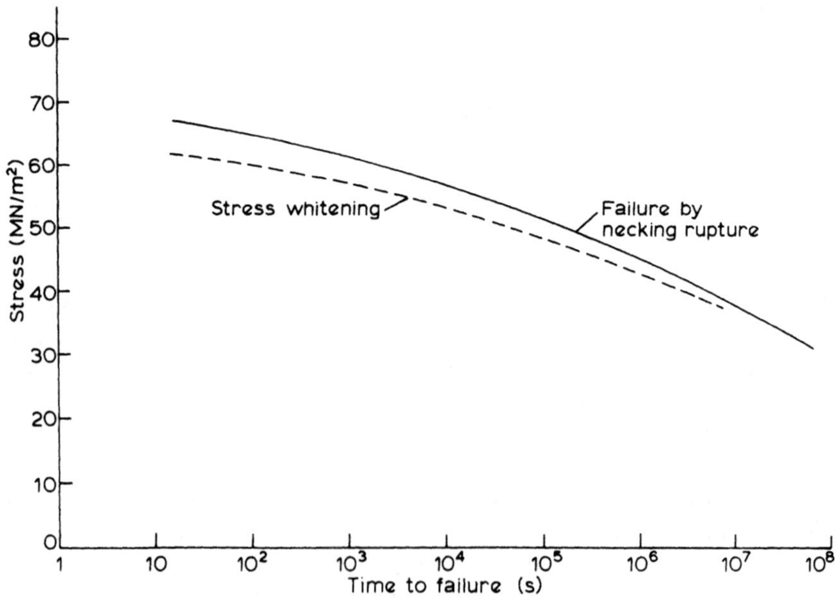

FIG. 6. Creep rupture in an acetal copolymer under continuous load at 20°C. Mode—Tension.

extrapolation of the ductile regression curve. As a matter of convenience in design, Fig. 5 essentially represents the time dependent yield stress curve for polycarbonate at 20°C. Thus, after one year under continuous load, the yield stress (σ_y) at 20°C for polycarbonate will have fallen from 65 MPa to approximately 50 MPa.

A further example of creep rupture is shown in Fig. 6 for an acetal copolymer. In this case the precursor to failure is stress whitening and the time dependent characteristic for this occurrence is also included.

The usefulness of this type of data is enhanced if they are set against a background of isometric strain curves. The integration of deformation and failure data provides a format in which trends in the failure data can be seen or interpreted. This function is better considered in the next sub-section dealing with the ductile/brittle transition.

2.4. The Ductile/Brittle Transition

The problem of brittle failure in the recognised thermoplastics used in engineering applications is a continual source of annoyance for the designer. Figure 7 is a schematic representation of durability in thermoplastics. Under continuously applied loads, failure is controlled

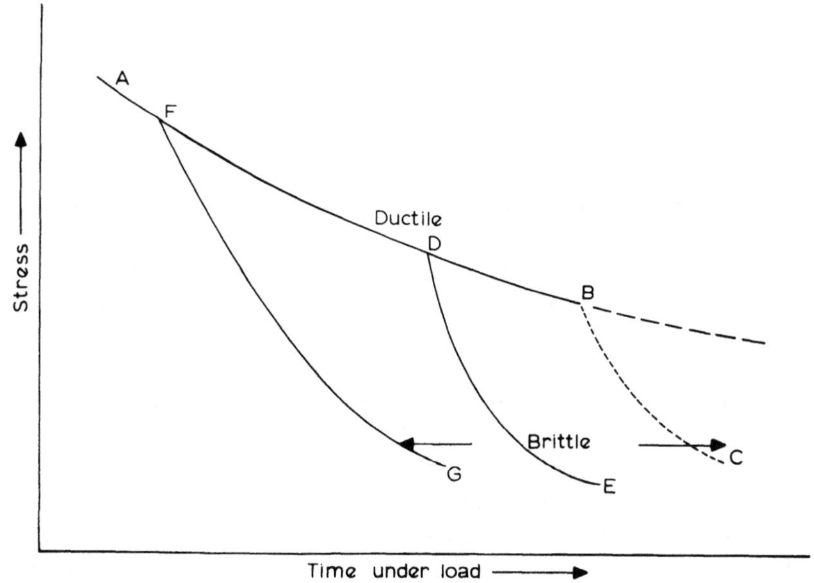

Fig. 7. The lifetime of a thermoplastic subject to continuously applied loads (schematic). Mode—tension.

by two main mechanisms, yielding and cracking. If the yield processes are dominant, then ductile failure occurs as signified by *AB* in Fig. 7. Materials, however, are imperfect so that although ductility may persist, at some stage the influence of cracking begins to take over and brittle fracture intervenes as depicted by *BC*. This transition from ductile to brittle may precipitate failures over a much wider period than the idealised Fig. 7 suggests.

Knowing the position of this transition in the multivariable surface (stress, strain, time, temperature, environment) is important and essential to good design practice. Avoidance of brittle failure, therefore, requires an intimate understanding and appreciation of factors that are known or are suspected of inducing brittle failure. For instance, if there is a very sharp change in section in a component or a deep scratch in its surface, then the transition can be initiated. Produced under laboratory test conditions it could take the form of curve *DE*. Alternatively, the component under load may become contaminated with an active environment which accelerates the effect of the notch even further. This is depicted by the curve labelled *FG*.

As a general rule then, any factor or influence likely to accelerate the rate at which a labile crack grows in the stressed component precipitates the transition earlier. Throughout the remainder of the Chapter it will become apparent that there are a number of different factors which cause brittle failure to happen during service and even sometimes in storage.

3. FATIGUE: FAILURE UNDER CYCLIC LOADS

Components in service will not be exposed always to continuously applied loads. The load may be intermittent or even cyclic and, although phenomenologically the response of plastics is similar to that of metals, there are some notable differences. As regards the data obtained in fatigue tests, however, these are presented in a similar fashion to that for metals, viz. as S/N curves. These are plots of the number of cycles (N_f) required to cause failure in a specimen at a particular stress level (S) and are presented on linear stress/log N_f axes. Figure 8 shows such a characteristic for a pipe grade of uPVC.

The response of plastics to cyclically applied loads is in general more complex than that of metals. This complex behaviour stems from certain properties of these materials: their high mechanical hysteresis,

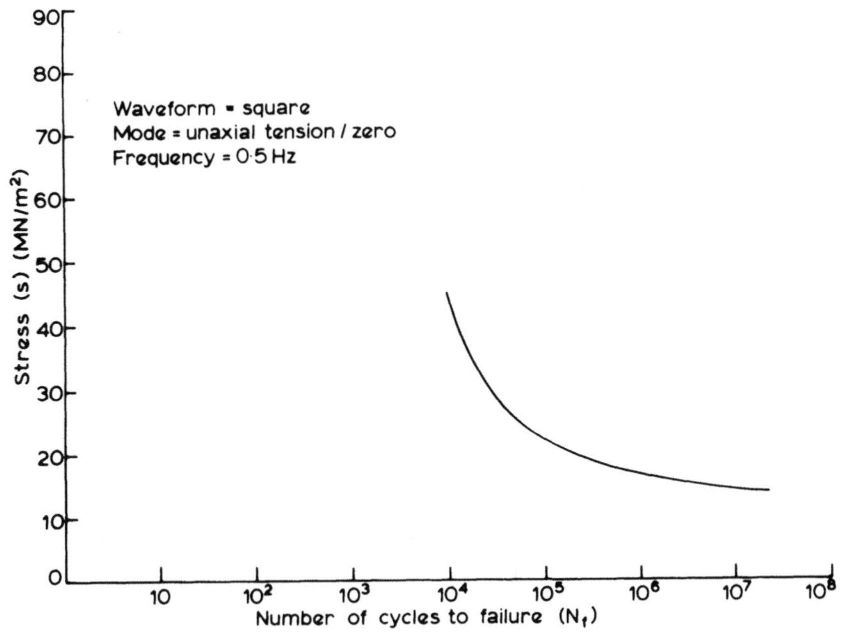

FIG. 8. Fatigue in uPVC (pipe grade) at 20°C.

non-linear viscoelastic response and low thermal conductivity coupled with their marked temperature dependence. These properties interact to cause the temperature of the specimen or component to rise dramatically when the frequency is too high. Thus, whereas frequencies of load cycling of up to 100 Hz are often used in testing metals, much lower frequencies need to be used in characterising plastics. Early work in this field by Riddell, Koo and O'Toole[12] demonstrated the problems involved. Failure by 'thermal softening', therefore, imposes a limit to the stress or strain amplitude that can be applied to a component if it is to be used at high frequencies. Forced cooling at the surface of the component does of course help to extend the range of use. Figure 9 shows curves of the temperature rise observed in PMMA (poly(methyl methacrylate)) under fatigue loading in flexure with and without air cooling at the surface.

Although there is a similarity between the shape of the fatigue curves for a number of plastics and those obtained for metals, the existence of a 'fatigue limit' for instance, observed in ferrous metals,

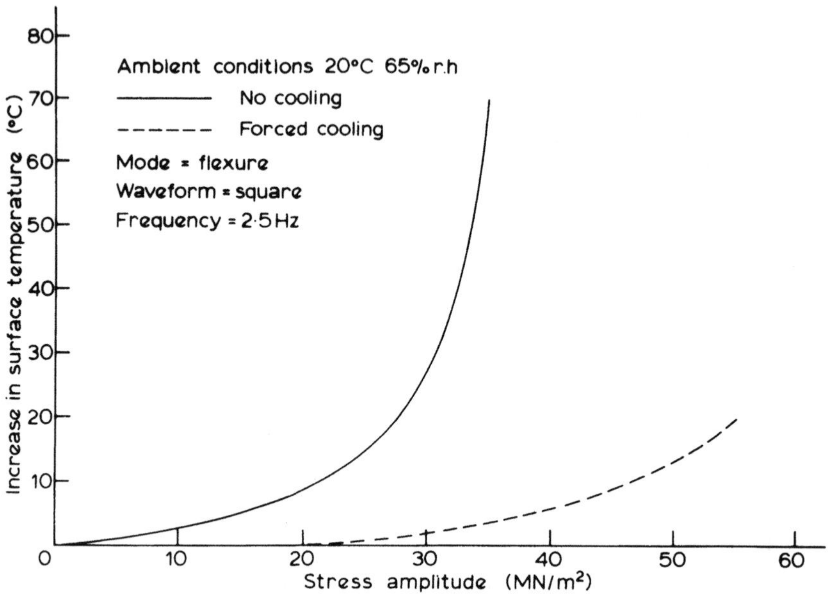

FIG. 9. Thermal effects in PMMA under cyclic loading.

has yet to be established for plastics. (The fatigue limit is a value of the stress amplitude below which a material may successfully endure an infinite number of cycles.)

So for plastics, the meaning of a fatigue test can be obscured if the frequency is too high, the mean stress level $\sigma_{mean} \neq 0$ and of course if the test conditions are not isothermal. Of course, if particular applications demand that the component operate at high frequencies etc., then it is necessary to establish satisfactory working levels of strain etc., by direct component testing or at least ensuring that laboratory test conditions simulate those required for actual service.

Failure as a result of externally applied forces is essentially a continuous process from beginning to actual end. As a matter of convenience, phenomenologically it can be split into two main phases, crack initiation and crack propagation. The crack propagation phase has been of particular interest to a number of workers because it responds to mathematical treatment and analysis. However, the fundamental importance of crack initiation cannot be overemphasised. The reason for this is that the initiation phase occupies the major part of component lifetime with the initiated crack growing very slowly until a critical size is reached. At this juncture, growth accelerates and the process continues until failure occurs. The two component parts of the failure process, crack initiation and crack propagation, are considered in more detail in the following sub-sections dealing with plain and notched specimens respectively.

3.1. Plain Specimens

3.1.1. The S/N Curve

The need to reduce test frequencies has encouraged the study of the effect of different waveforms upon fatigue life. In particular, the use of a cyclic step input or square wave has enabled comparisons to be made between lifetimes of thermoplastics exposed to continuous or cyclic loads.[1] Although data can be and frequently are presented in their simplest and most easily recognised form, i.e. as conventional *S/N* curves, in the case of square wave loading the data transpose conveniently to real time characteristics. This is achieved by summing the time under tensile load over the N_f cycles to failure. The resulting stress/time characteristic can then be compared with the corresponding creep rupture characteristic for the material. Figure 10 shows failure in uPVC at 20°C under continuous and cyclic loads ($\sigma_{mean} = 0$). This comparison shows that short-term failure under either continuous or cyclically applied loads is dominated by yield mechanisms and is ductile. Then as the stress applied is reduced with increasing lifetimes, a stage is reached where the two curves diverge. Creep rupture failure remains ductile but, under cyclic loading, crack initiation followed by crack propagation intervenes, causing premature failure. This then is the onset of the ductile/brittle transition. The position on the time scale and the severity of the transition depends upon a number of factors but the important thing is that it is materials based. However, for this particular uPVC grade, it can be seen that at the 10^6 cycles/s section, the stress required to cause failure has fallen to about $\frac{1}{3}$ of the creep rupture stress. Additionally, inspection of the fracture surface shows it to be brittle.

In the case of metals, the presence of a non-zero mean stress does not give rise to creep during the test except at high temperatures. For plastics, however, if $\sigma_{mean} \neq 0$, the presence of such a stress can well dominate the strain history resulting in accumulated increasing creep strain as N increases. The nature of failure is also clearly affected, but this point is more conveniently discussed in the next section.

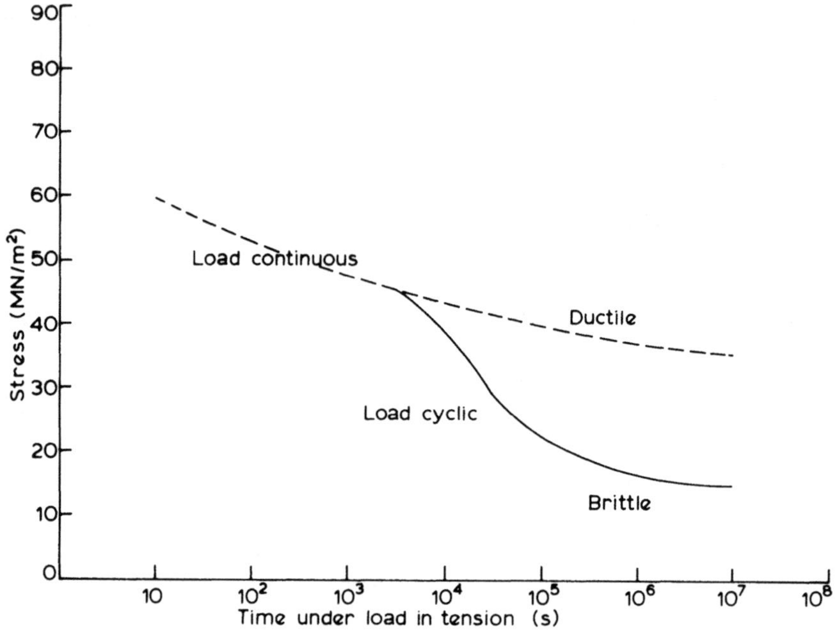

FIG. 10. Creep rupture and fatigue in uPVC at 20°C.

3.1.2. Crack Initiation

Under continuously applied loads at 20°C, thermoplastic materials such as uPVC, polyamide (nylon), polypropylene and polyethylene are ductile in their response.[11] Consequently, such regression curves as those of Figs 5 or 6 are, to a good approximation, time dependent functions of the yield stress. Thermoplastics, like many other materials, are imperfect and, as intimated previously, brittle failure will intervene at some stage in their lifetime under load. This brittle failure results from crack initiation at defects in the material thereby causing premature failure. Under continuously applied loads there is a tendency for cracks so initiated to blunt at the tip. This is caused by the high stresses, in excess of the yield stress, developed at these stress concentrations. However, when the applied load is interrupted, or is intermittent or cyclic, the plastic deformation formed at the tip of the defect relaxes during unload periods.[13] By this mechanism, the growth of plastic deformation is controlled and the onset of premature brittle failure is accelerated. Even at such a low frequency as that of diurnal application of the load, reductions occur in load bearing lifetime.[14]

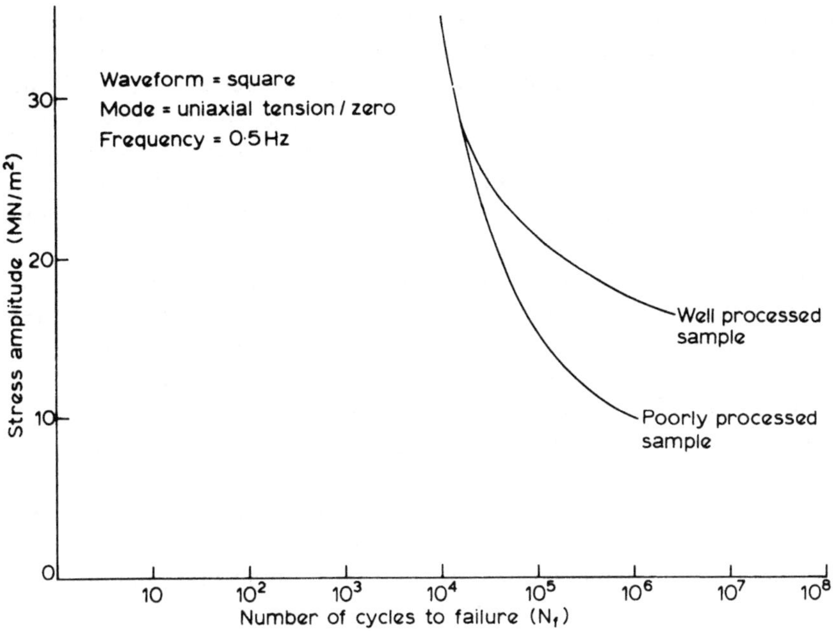

FIG. 11. Fatigue: The effect of processing level on uPVC.

The ease of crack initiation depends upon whether flaws or imperfections are present in the material. Consequently, fatigue resistance will be affected by processing history—extrusion or injection moulding. It has been shown for example in uPVC that good processing of this particulate material is essential if good fatigue resistance is to be realised. Figure 11 shows the response of a uPVC pipe grade which has been processed at two different levels. The obvious benefits of good processing (e.g. good dispersion and proper degree of gelation—see Section 5.1) can be readily appreciated. In other instances, however, the main concern has been the presence of residual stresses/strains and how these influence fatigue resistance by affecting craze/crack initiation.[15] Clearly, either breakdown of particulate materials or avoidance of residual stresses/strains by correct processing effectively improves the resistance to fatigue failure of the resulting components.

3.1.3. Crack Propagation

The energy required for crack initiation consists of the energy absorbed in deforming the material elastically and in some cases plastically, to the point at which cracks begin to propagate. Notches reduce both the elastic and plastic contributions by accelerating crack initiation. In this sense, flaws and imperfections act as notches. Hence with very severe notches, deep and with sharp tip radii, and crack-like defects, the crack initiation component of failure is very much reduced. In fact in classically brittle materials like polystyrene and poly(methyl methacrylate) it may disappear altogether, with crack propagation occurring upon load application.

The notch sensitivity of a thermoplastic material under fatigue loading can be quantitified as for any other material.[16] There is a real need to know how notches or stress concentrations reduce fatigue life, for two reasons. First, however meticulous component design may be, it is inevitable that some degree of stress concentration be there, simply because the changes in section, grooves, holes, etc. are essential to the design and function of the component. Thus it is important to minimise such effects by using generously radiused corners wherever possible, avoiding sharp changes in section etc. Further, where metal inserts are used it is important that these be incorporated using the best known practices.[17]

Secondly, a study of the effect that notches have upon the fatigue resistance of a material provides the designer with a measure of its toughness.

3.2. Notched Specimens

Figure 12 shows a series of S/N_f curves for both plain and notched specimens of bulk cast poly(methyl methacrylate). The notched specimens have different notch geometries effected by different notch tip radii. Fatigue was in uniaxial tension/compression with a mean load of zero. The frequency was 0·5 Hz and the waveform square. The results show a steady trend with the net stress decreasing as notch tip radius is decreased. All fractures were brittle.

The data of Fig. 12 have been adjusted by the appropriate stress concentration factor[3] to give corresponding peak stresses (Fig. 13). The individual results are now seen to unify well with the exception of those for the plain specimen and those for the very sharply radiused notch tip ($r \cong 10\,\mu$m). In the case of the plain specimens, their response can be rationalised if the material itself is assumed to contain defects responding at a particular level of stress concentration.

In the case of the very sharp tip radiused notches, if the net stresses are adjusted by the apparent stress concentration factor ($\cong$16), quite

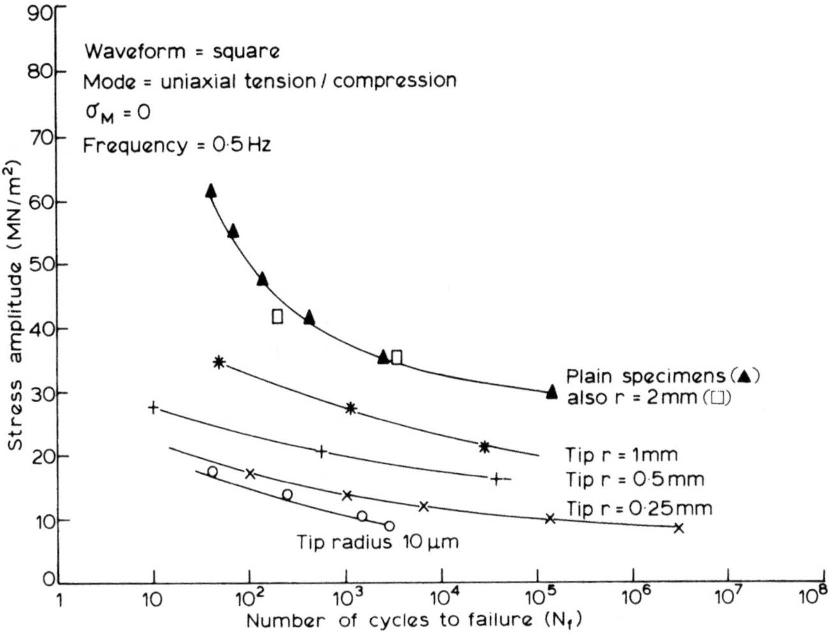

Fig. 12. The effect of notch tip radius on fatigue of bulk cast PMMA at 20°C.

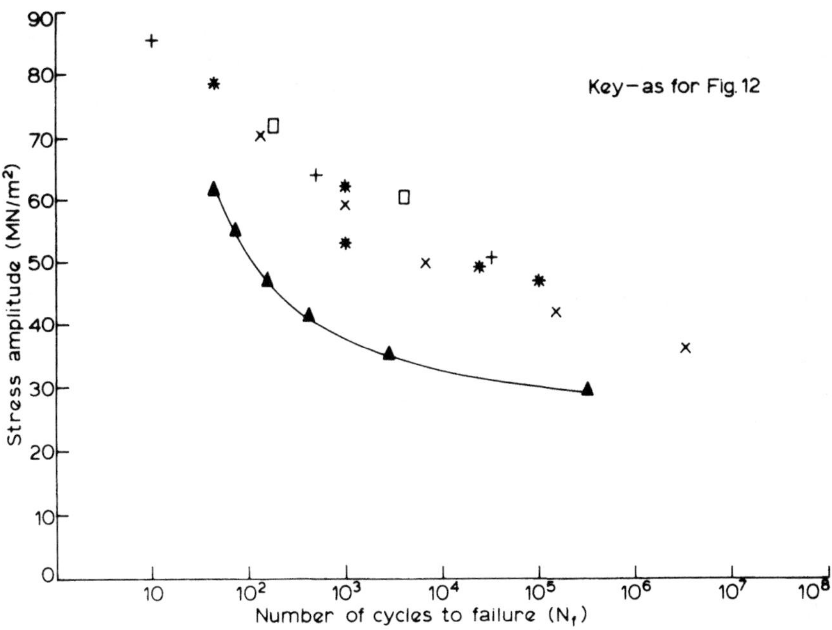

Fig. 13. The data of Fig. 12 adjusted to take account of stress concentration factor.

clearly simple stress concentration theory breaks down. The concentrator is acting as though it were a labile crack and, consequently, its response can only be described by linear elastic sharp crack fracture mechanics analysis.[18]

3.2.1. Fracture Mechanics

Over the past 20 years the concept of fracture mechanics has been increasingly implemented into the investigation and analysis of problems concerning failure of thermoplastic components. In the early days, its use was very much restricted to the analysis of fracture in classically brittle materials such as PMMA and PS (polystyrene).[19] It was then extended to modelling crack growth in environmental stress cracking situations[20] and later to the analysis of impact failure.[21] Essentially, linear elastic fracture mechanics is concerned with the propagation of brittle cracks in a solid medium. It is not the function of this particular section to elaborate upon or present in detail the subject of fracture mechanics since this information can be readily obtained from a number of sources, e.g. Refs 18 and 22. Suffice it to say that the fundamentals of fracture mechanics stem from the Griffith theory,[23] which states that a brittle material can contain small defects which allow a stress to become concentrated sufficiently for brittle fracture to occur. The concept was formulated into the well known equation

$$\sigma_F = \left(\frac{2E\gamma}{\pi c}\right)^{1/2}$$

where σ_F = fracture strength (brittle fracture), E = modulus, γ = surface energy of fracture surface and $2c$ = intrinsic material defect length.

Plastic materials are rarely if ever truly brittle, but rather quasi-brittle, in which plastic deformation occurs along the surface of the crack. To take account of this Orowan[24] reviewed the theory and rewrote the Griffith equation as follows:

$$\sigma_F = \left(\frac{E(2\gamma + \gamma_p)}{\pi c}\right)^{1/2}$$

where γ_p is now the energy expended during plastic flow at a crack tip.

Since 2γ is very small compared with γ_p it can be ignored and the equation simplifies to:

$$\sigma_F = \left(\frac{E\gamma_p}{\pi c}\right)^{1/2}$$

The next step in evolution of the theory was made by Irwin[25] who related the surface energy γ and plastic flow energy γ_p to a parameter K, called 'the stress field intensity parameter', where

$$K^2 = E(2\gamma + \gamma_p)$$

Irwin also postulated that catastrophic unstable fracture might occur at a critical value of K, i.e. at K_C such that

$$K_C = \sigma_F(\pi c)^{1/2}$$

This is the relationship for a sheet of infinite size. For smaller finite specimens, the equation is modified to take account of this by a geometrical factor Y such that:

$$K = \sigma_R(\pi a)^{1/2}Y$$

In this equation σ_R, the stress remote from the crack, replaces σ_F and $2a$ is the crack length. The simple principle is that, having established a value for K using, say, standard specimen geometry, it is possible to predict the stress required to cause crack propagation in a component or specimen of entirely different geometry.

Under fatigue loading Paris[26] established that the rate of crack growth per load cycle was related to the stress range $(\sigma_{max} - \sigma_{min})$ and that the fatigue crack propagation process itself was controlled by the stress field intensity parameter K.

$$\text{Crack growth rate } \frac{d2a}{dN} = A\Delta K^n$$

where ΔK = stress intensity factor range $(K_{max} - K_{min})$ and A and n = are material constants.

A typical relationship is shown for a number of polymers in Fig. 14. Considerable attention has been paid to the subject of crack propagation in thermoplastics under cyclic loading and this has been particularly well illustrated and discussed by Hertzberg and Manson.[27]

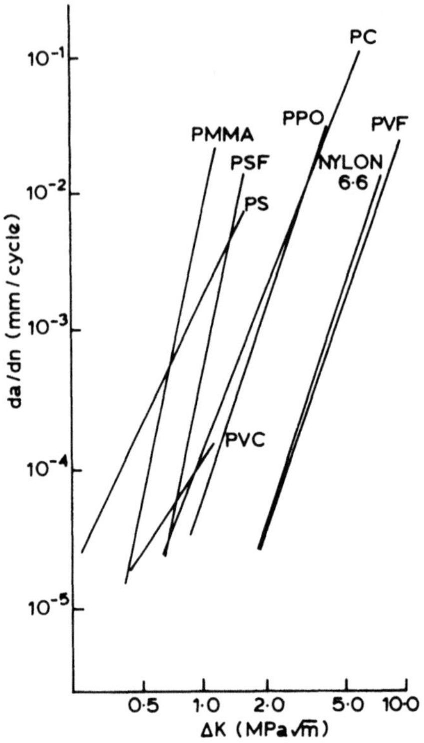

FIG. 14. Fatigue crack propagation behaviour in selected crystalline and amorphous polymers. (Reproduced by kind permission of R. W. Hertzberg and J. A. Manson.)

4. FACTORS AFFECTING FATIGUE RESISTANCE

Resistance to fatigue failure of a thermoplastic material is affected by the common operating variables such as frequency, load waveform, stress complexity and temperature. As a general rule, increase in: frequency, severity of the load waveform, complexity of the applied stress field and temperature all reduce fatigue resistance.

4.1. Frequency

In the first instance, increase in frequency of the load application can lead to failure by thermal softening. Figure 11 illustrated the temperature rise observed in poly(methyl methacrylate) under fatigue loading in the flexure mode.

A number of workers over the years have modelled the thermal softening phenomenon. These models are prescribed in terms of stress or strain, component geometry, waveform, cooling conditions, mechanical losses and, of course, frequency.[28,29]

The importance of the interaction between the variables listed above cannot be overlooked as regards formalised characterisation of fatigue by *S*/*N* curves. If the temperature varies then the resultant *S*/*N* curve is not an isothermal relationship and becomes difficult to use and interpret. In order to try and avoid this problem, characterisation frequencies for thermoplastics have been reduced and generally are of the order of 1 Hz. The influence of frequency upon fatigue life, in the region where failure is now dominated by crack initiation and crack propagation, is more predictable. Consider the data obtained for a pressure pipe grade of uPVC shown in Fig. 15. The conditions under which these data were obtained were as follows: the load was cycled between tension and zero at frequencies of 0·5 Hz and 5 mHz and the

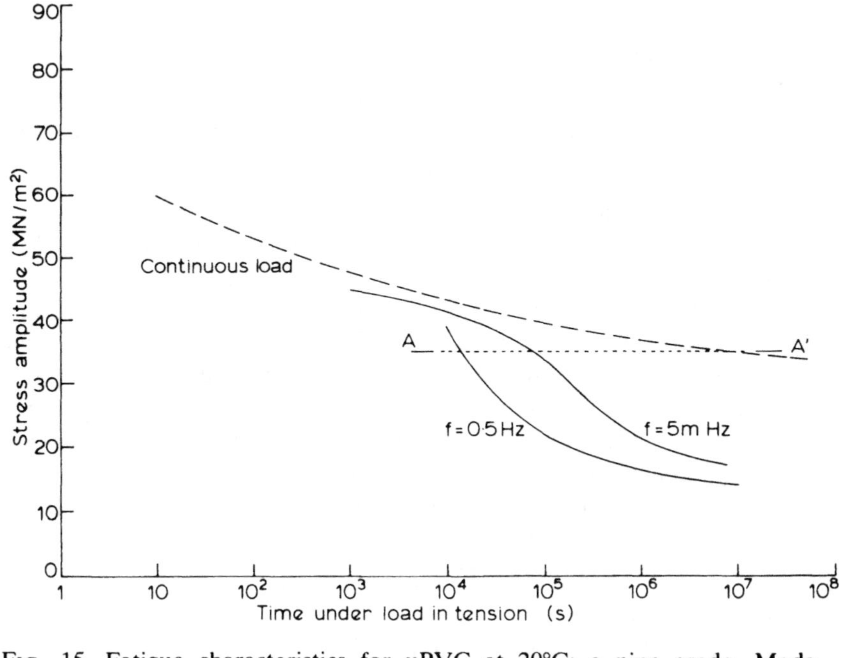

FIG. 15. Fatigue characteristics for uPVC at 20°C: a pipe grade. Mode—tension.

waveform was square. The data have been presented in terms of real time rather than cycles to failure, enabling both fatigue and creep rupture failure to be presented together. Consider a constant stress section such as AA' which identifies the extremes of lifetime under cyclic and continuous loads at the intersection points of the fatigue (A) and creep rupture (A') curves. At such a section, a series of tests was done for a stress level of 35 MN/m^2. Failure varied from brittle under cyclic loading to ductile under a continuously applied load. Frequency correspondingly varied from 0·5 Hz to that of the continuously loaded state. The frequency for the latter can be determined using the Hamon transform.[30] This states that

$$\text{Frequency} = \frac{1}{10t}$$

where t = time to failure under continuous load. The relationship between frequency and lifetimes, specified as real time and plotted on log/log axes, approximates to being linear, with time to failure increasing as frequency decreases. Thus if the two 'fixed points' A and A' can be identified then this simple relationship provides a convenient means of interpolation. Figure 16 shows the relationship for uPVC at a stress level of 35 MN/m^2.

It is generally true to say that, as the frequency of load application is reduced, load bearing lifetime increases, but correspondingly the number of cycles required to cause failure (N_f) decreases.

4.2. Load Waveform

Much fatigue testing has been performed with sinusoidal loading. For metals, in the region of the fatigue limit there is little information as to the effect of waveform shape. However, since there is no significant effect of frequency, over the range 1–200 Hz it is not expected that variations in waveform shape will influence lifetime to any marked extent either.[16] For thermoplastics, however, waveform shape does affect lifetime. For example, Fig. 17 compares the S/N curves for square and sinusoidal wave loading. Essentially, the larger the energy input per cycle at a particular stress/strain amplitude, the shorter lifetime becomes. Incidentally, this trend is also borne out by the analysis for thermal softening by Opp *et al.*[29]

4.3. Stress Complexity

The general effect of increasing the complexity of the stress field imposed on a component from simple uniaxial to biaxial to triaxial

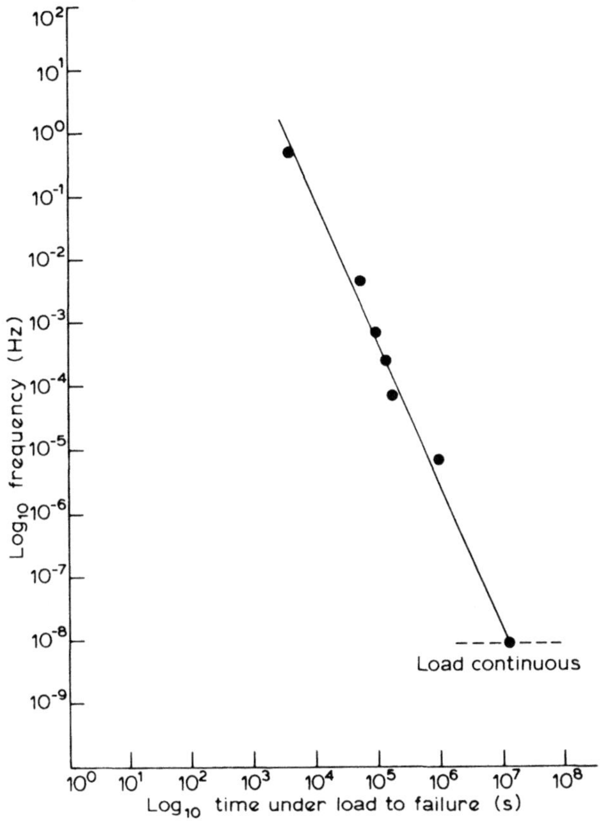

FIG. 16. The effect of frequency on the load bearing lifetime of uPVC at 20°C.

tension is to increase the likelihood of brittle failure. Thus the simple fact that the introduction of a notch into a uniaxial field increases the possibility of brittle failure is wholly consistent with this trend, since the stress field becomes distorted around the notch tip by introducing a biaxial component. In the case of very thick material this stress field becomes even more complex, by the generation of a through thickness triaxial component. More generally, any change in the nature of the stress field which causes an increase in the ratio of dilational/shear stresses is accompanied by an increase in potential embrittlement. This effect is illustrated very neatly in Fig. 18 by the response of polycarbonate to changes in the stress complexity under fatigue loading.

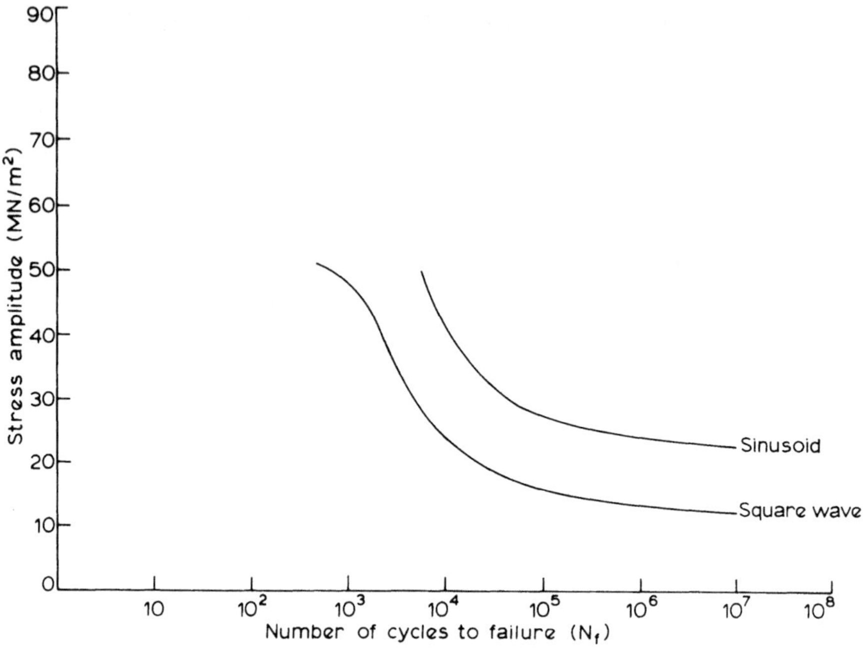

FIG. 17. The influence of waveform shape on the lifetime of uPVC under fatigue loading.

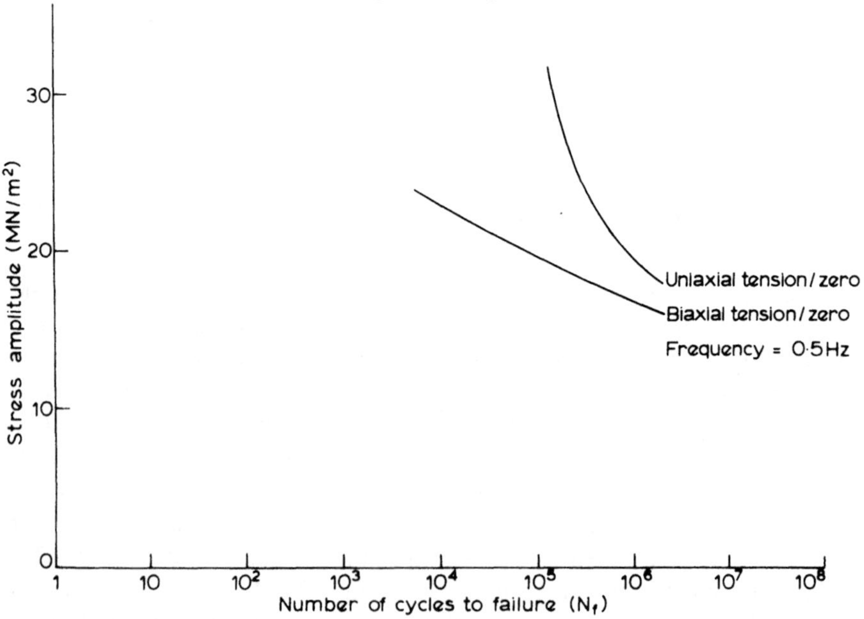

FIG. 18. The resistance of polycarbonate to fatigue loading under uniaxial and biaxial stress fields.

In this particular example, we see that a change from a uniaxial to a biaxial field causes a reduction in fatigue resistance. A further increase in the complexity of the simple uniaxial field to include a compressive component also causes a reduction in fatigue resistance. The reasons for these effects and interactions are discussed in more detail in Ref. 31.

It is unlikely that in service a component will only be exposed to a simple uniaxial field. For one thing the geometry of the component is rarely that of an ideal prismatic beam or tensile bar, which is used to establish the fatigue characteristics of materials under ideal conditions. Additional information about notch sensitivity is very useful but often there is no real substitute for actual component testing. Care needs to be taken in this instance that the procedures used to accelerate failure do have practical significance.

Pressure pipe is one such application which lends itself to component testing. It also provides an instance in which complex stress fields are generated and met quite naturally under service operation. When under load, a hoop stress is generated in the wall of the pipe and this is accompanied by an axial stress at right angles to and along the length of the pipe. Further, it is not uncommon for bending stresses to develop as a consequence of actual pipe installation. Even higher local stresses can be caused by poor installation with pipes grounded upon sharp stones.

Pressure pipe also provides an ideal example in which the load waveform can become complicated. In service, it is unlikely that pipe will be subjected to a load with a profile similar to that depicted in part (a), (b) or (c) of Fig. 19. It is more likely that the pipe will operate with a variation in stress superposed upon a continuously applied background or static pressure (Fig. 19(a)). Under ideal conditions, such a load waveform or profile would look something like that of Fig. 19(d).

There is a convenient and most powerful way in which data of this sort can be presented: the SMITH diagram. In this particular presentation of fatigue response, the information is plotted on maximum and minimum stress axes, with tension positive, compression negative, vs. mean stress. Hence as Fig. 20, the SMITH diagram for uPVC, shows, it ranges from pure dynamic loading with $\sigma_{mean} = 0$, i.e. the fully reversed loading case, to a pure static load for creep rupture. Each curve represents a section of fatigue life; Fig. 20 a 10^7 s lifetime. In between the extreme states of pure dynamic to creep

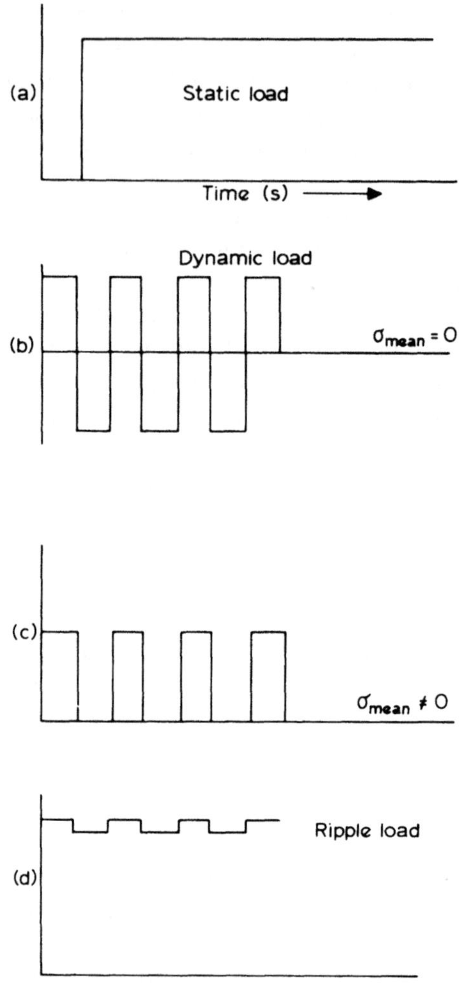

FIG. 19. Load cycle shape.

rupture loading, the amplitude of the oscillatory load is seen to decrease as σ_{mean} increases until it looks much like the ripple load of Fig. 19(d). There are variations of this type of presentation which have been used for metals, e.g. Goodman and Gerber diagrams[32] but the SMITH diagram[33] has a particular appeal and usefulness.

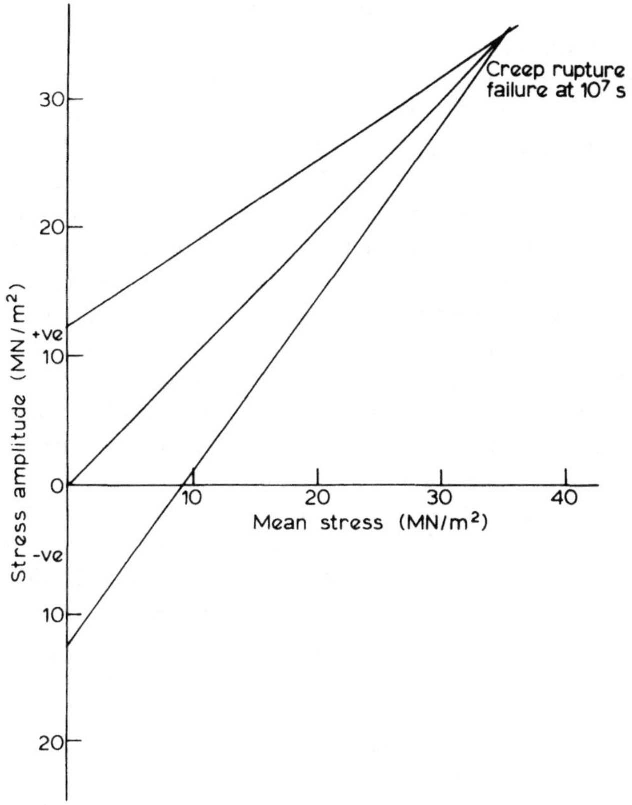

FIG. 20. SMITH diagram for uPVC: 10^7 s section.

5. SOME OTHER FACTORS AFFECTING FATIGUE LIFETIME

Factors other than the traditional ones discussed under Section 4 can affect fatigue lifetime. For instance, such material variables as molecular weight, crystallinity, the effect of polymer modification by rubber or stiff element reinforcement, additives generally and, not least, processing.

5.1. Processing

It is understandably important for any thermoplastic component that the raw material be well processed. It should be injection moulded or extruded under optimum conditions. The ease or difficulty with which a thermoplastic material can be so processed is chiefly a property of the plastic. Thus processing can range from easy, with such materials as polystyrene and polyethylene, to difficult, with rigid PVC, polycarbonate and some of the recent high temperature thermoplastics like poly(ether ether ketone). Implicit in these comments regarding processing is the resultant state of the moulded material. An easy material to mould is less likely to give rise to high levels of undesirable residual stresses/strains than a difficult one. An example of the effect that level of processing can have upon fatigue resistance was shown earlier in Fig. 11 for well and poorly processed uPVC. This thermoplastic is particulate and as such needs to be processed adequately in order to break down the complex structure of the PVC particle.[34] With inadequately processed PVC more of the original structure persists and it is possible in some samples to identify granules of polymer which still retain their identity.[2] Further, such granules can act as flaws and, as is illustrated in Fig. 11, 'crack initiation' becomes significantly easier in the poorly processed sample, leading to a reduction in fatigue resistance. Some early examples of rotationally cast polyethylene components, before the process had been fully developed, exhibited poor fatigue properties. This behaviour was attributed to poor processing.

Naturally there is an interaction between processing and mould design. The latter is of equal importance for the resulting component since the position and severity of weld lines, for instance, with relationship to the service forces, can be crucial in reducing fatigue lifetime.

5.2. Molecular Weight

The influence that the molecular weight of a material has upon its properties are often not given sufficient consideration in 'design for service'. It is generally true to say that as molecular weight is increased so does mechanical performance improve. Ideally, the molecular weight of the grade selected should be determined by a balance between ease of processability and good toughness. Too often it is ease of processing which seems to be the major consideration. Changes in molecular weight have little direct effect upon moduli and yield stresses but decreasing molecular weight too much eventually leads to embrittlement. Increasing molecular weight, on the other hand, assuming no adverse interactions with processing, leads to an

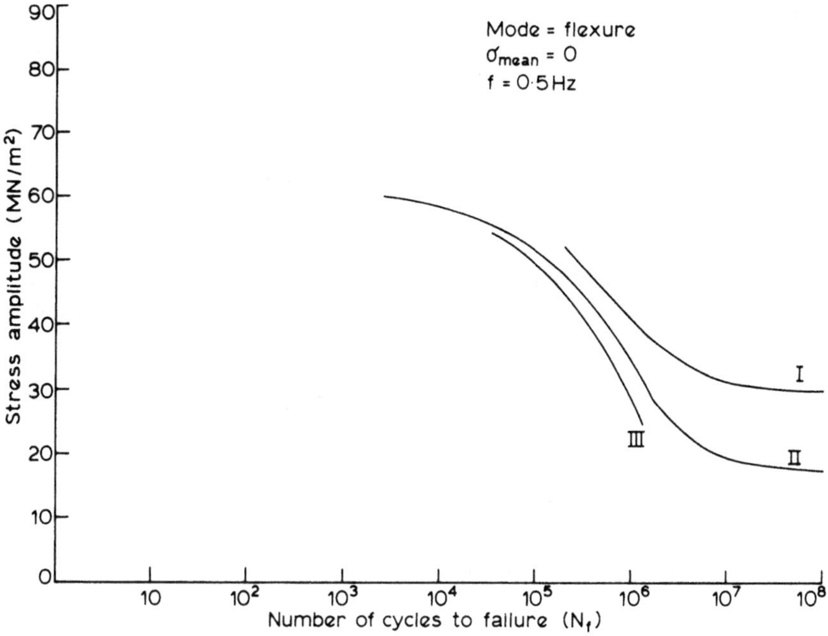

FIG. 21. Fatigue characteristics of a acetal copolymer.

improvement in toughness and fatigue resistance. This is exemplified by the response of three different grades of an acetal copolymer, shown in Fig. 21. The grade coded (I) is more often used for extrusion but, if it can be injection moulded satisfactorily, leads to the best fatigue resistance of the three grades illustrated. It has the highest molecular weight. The normal injection moulding grade (II) is less resistant and clearly the performance of (III), an easy flow grade with the lowest molecular weight of the three, is the worst.

Reduction in molecular weight, besides reducing fatigue resistance, increases the propensity to brittle failure in impact and reduces resistance to environmental attack. Conversely, too high a molecular weight may cause processing difficulties coupled with an increase in the degree of anisotropy within the component because of higher viscosity. The disadvantage of this trend in anisotropy is that it can outweigh the potential advantages imparted by higher molecular weight grades. As a consequence, the net result can be that the higher molecular weight material produces a component having inferior service performance.

5.3. Crystallinity

The relationship between crystallinity and mechanical performance is complicated and has not been entirely resolved. The reason for this is that simple measures of percentage crystallinity are inadequate in themselves as a characterisation. The size, shape and perfection of both crystals and spherulites are clearly important but above all intraspherulitic strength, as evidenced by the tie molecules between units, is crucial to mechanical toughness.

Considered as a group, however, semi-crystalline polymers tend to have better fatigue resistance than do amorphous ones. Consider the data of Table 1 which details the strain amplitude various thermoplastics will support for a lifetime of 10^6 cycles at 20°C. The mode of deformation is uniaxial with a constant applied force varying from tension to zero, at a frequency of 0·5 Hz.

These data illustrate three important points. First, semi-crystalline polymers, as a group, are more resistant to fatigue than amorphous ones. Second, when these materials are reinforced with glass fibres, this superiority of semi-crystalline materials is still retained. Third, the

TABLE 1
FATIGUE: FAILURE STRAIN

Polymer	*Strain %* (10^6 *cycles*)
Amorphous	
Polyacrylate	0·35
Polycarbonate (PC)	0·55
Poly(ether sulphone) (PES)	0·45
Poly(methyl methacrylate) (PMMA)	0·45
Unplasticised PVC (uPVC)	0·3
Acrylonitrile–butadiene–styrene (ABS)	0·4
Semi-crystalline	
Acetal (POM)	0·75
Nylon 66 (PA 66)	1·0
Polypropylene (PP)	1·0
Glass fibre reinforced thermoplastics	
PES	0·2
Poly(ether ether ketone) (PEEK)	0·6
PA	0·6
PP	0·55

addition of glass fibre reinforcement, however, results in an overall reduction in the strain amplitude that the materials can support. This point is considered in more detail in the next section.

5.4. Polymer Modification

Thermoplastics can be modified in a number of ways: by blending with other polymers, by plasticisation, by copolymerisation and by the addition of particles or fibres. In the space available it is not possible to consider the influence all these have upon fatigue resistance. We shall examine two of the major modifications, copolymerisation and the addition of particles or fibres. The very versatile polymer, polypropylene, offers a good example in which these modifications have been effected.

5.4.1. Copolymerisation

The simultaneous polymerisation of propylene and ethylene provides an interesting range of materials. The general effect of the co-monomer, ethylene, is to reduce yield stresses by softening but this is accompanied by a counteracting toughening effect in impact. The major advantage of rubber–resin blends is that they can have superior resistance to the propagation of fast cracks as in the case of ABS.[35] This advantage is not necessarily extended to slow cracks. Dynamic fatigue tests on ABS, for instance,[36] have shown that it does not have particularly good resistance to slow cracking under cyclic loading conditions. The general effect then of copolymerisation is to reduce yield stresses and moduli. For polypropylene under fatigue loading (Fig. 22), it can be seen that the S/N curve for the homopolymer is depressed to lower stress levels when copolymerised with ethylene. The counteracting effect of the reduction in yield stresses and moduli upon fatigue response is an increase in ductility for the copolymer as shown in Fig. 23.

5.4.2. Glass Fibre Reinforcement

Figure 24 shows the fatigue characteristic of a 25% w/w glass fibre reinforced homopolymer. Obviously the stress bearing characteristics of this material are better than those of the unreinforced matrix but, as Fig. 25 shows, this improvement is gained only at the expense of ductility. The main attribute that the incorporation of glass fibre confers upon the resultant composite is that it stiffens the matrix. It does not produce corresponding improvements in fatigue resistance.

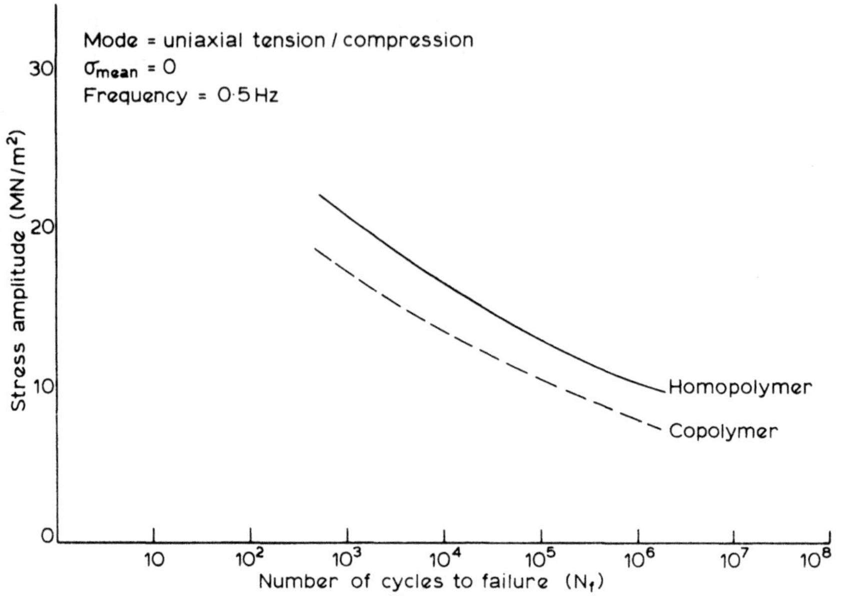

FIG. 22. The effect of copolymerisation on the fatigue properties of polypropylene.

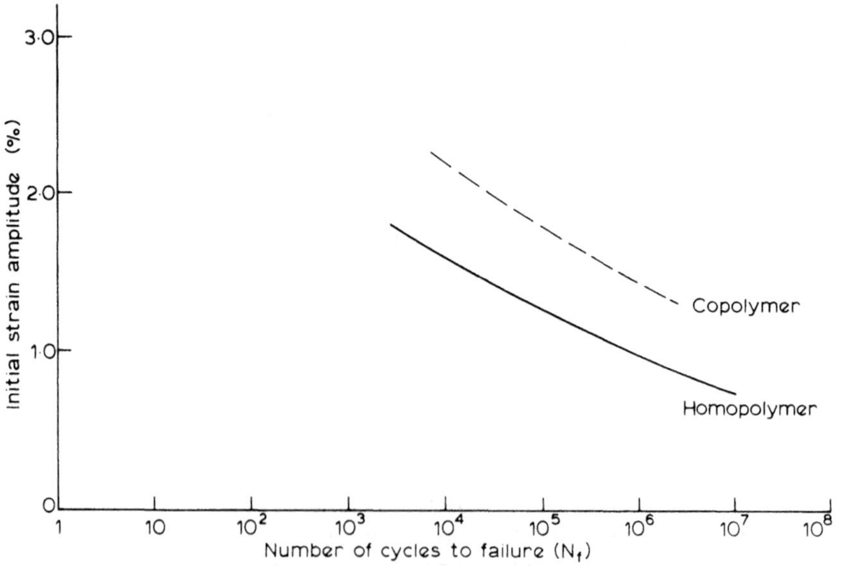

FIG. 23. The data of Fig. 22 represented in terms of strain.

There are two further important points about the data of Figs 24 and 25. These concern first, the anisotropy in fatigue resistance which is imparted by the glass fibres. The data were obtained on waisted specimens prepared from a 115 mm diameter, edged gated, injection moulded disk, 6 mm thick. The 0° direction is in line with the sprue. From Fig. 24 it is estimated that the degree of anisotropy is quite marked, amounting to a factor ≅1·4 (ratio 90°/0° directions). As Fig. 25 shows, the data of Fig. 24 are completely rationalised when considered in terms of strain. This result found for glass fibre reinforced (GR) polypropylene is not singular. The concept of failure being strain rather than stress dominated has wider applicability, certainly for glass fibre reinforced thermoplastics. It had previously been observed by the author for nylon 66 GR[36] and also more recently in a programme of work on high temperature polymers.[37] The matter is given further consideration in Section 6.

The second point concerns anisotropy. It is significant that the 90° direction is stiffer than the 0° which is parallel to the flow direction. The reason for this lies in the balance between the properties of the

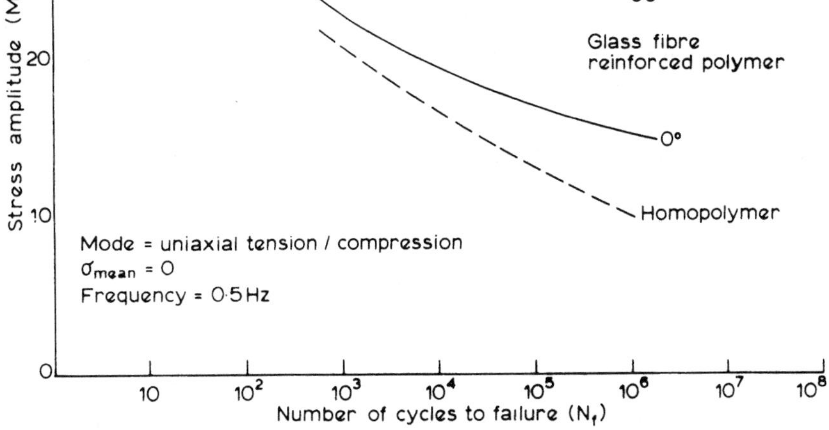

FIG. 24. Fatigue data for a glass fibre reinforced polypropylene homopolymer.

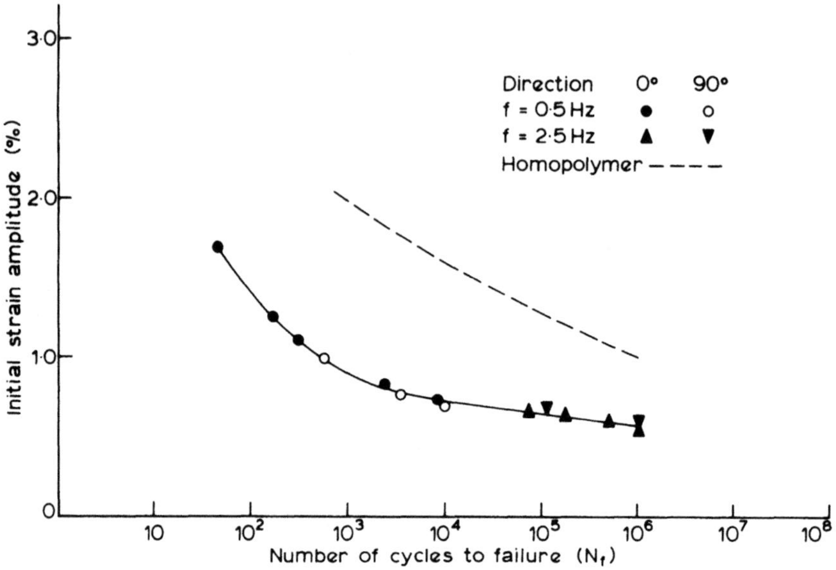

FIG. 25. Strain dependence fatigue curve corresponding to Fig. 24.

core and skin.[38] On injection, the skin cools first and glass fibre alignment is in the 0° direction. The core, cooling more slowly, however, tends to reorient itself in the orthogonal direction. The balance between the two directions depends upon thickness. In thin mouldings, e.g. <3 mm, the skin will dominate and the stiffest direction will be in the 0° direction. As thickness is increased, the core will begin to take over and the stiffest direction becomes the 90°. A word of warning needs to be given at this juncture. The data under consideration were generated in the uniaxial mode. It is conceivable that if the composite were to have been tested in flexure, then the skin and near substrate would be dominant in fatigue response, irrespective of thickness.

6. ENVIRONMENTAL EFFECTS

The mechanical performance of a thermoplastic component can be affected, usually in an adverse manner, by the environment in which it is called upon to function. Of particular importance with respect to lifetime is the effect that outdoor weathering has upon the performance of thermoplastic components. Generalisations can be dangerous, and the reader is strongly recommended to consult a recent publication by Davis and Sims.[39] Even the general atmospheric conditions can influence the response of certain thermoplastics in quite a dramatic way. The nylon family of thermoplastics in particular is perhaps the most sensitive to the water content of the air in which it is being used. It is not intended to discuss this very important problem for nylon components in this section and since it is considered in detail by the raw materials manufacturers, the reader is recommended to refer to their literature. For example, Refs 40 and 41 consider the effect of water uptake on dimensions and toughness and Ref. 42 considers the subject of nylon plastics more generally.

In this section, only two main environments are discussed, temperature and chemical.

6.1. Temperature

The influence of temperature upon the mechanical response of thermoplastics is greater than that of metals over the same temperature range. Major changes take place from low temperatures to the upper limit of working. A good guide to the general response of any material is provided by a plot of tensile strength vs. temperature. For example Fig. 26 shows the dependence of tensile strength upon temperature of poly(ether sulphone) (PES), both natural and glass reinforced grades. Under rates of straining in monotonic loading, ≅0·17%/s, PES remains ductile over a wide range of temperature and failure occurs beyond the yield point. At temperatures below −100°C failure changes from ductile to brittle whereas at the other end of the spectrum, in the vicinity of the rubber/glass transition ($T_g = 225$°C), PES has a much reduced strength. The effect of incorporating glass fibres is to provide a general reinforcement to the strength of PES and also increase its stiffness. It does not, however, materially influence the trend with temperature, but it does, of course, severely reduce ductility. Breaking strains under monotonic loading are of the order of 2–3% which are significantly less than the yield strain of the base resin, 6–7%. As a matter of interest, the upper safe working temperature for PES is ≅180°C which is well below the T_g. So much then for the general effects, but how is fatigue affected by the temperature?

Fatigue data as a function of temperature are generally scarce, probably because of the shear effort and time involved in generating a

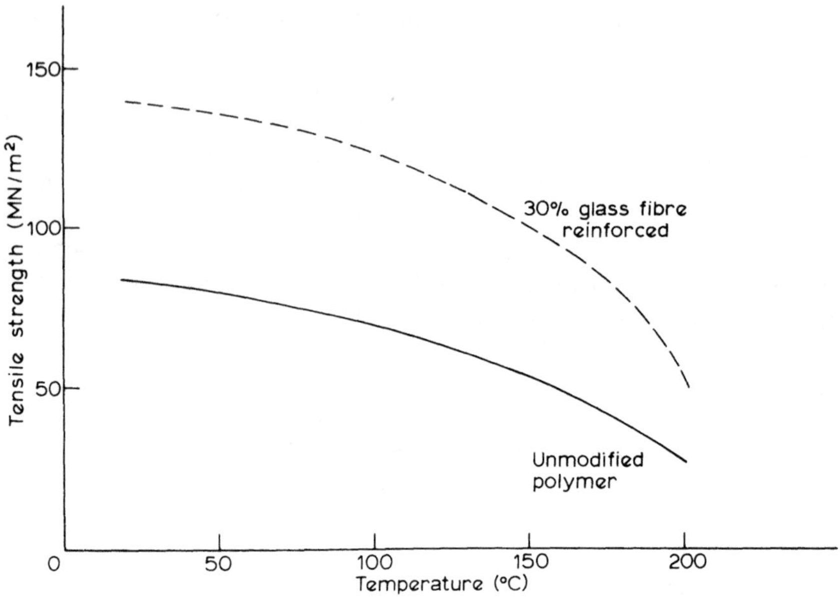

Fig. 26. The temperature dependence of the tensile strength of poly(ether sulphone).

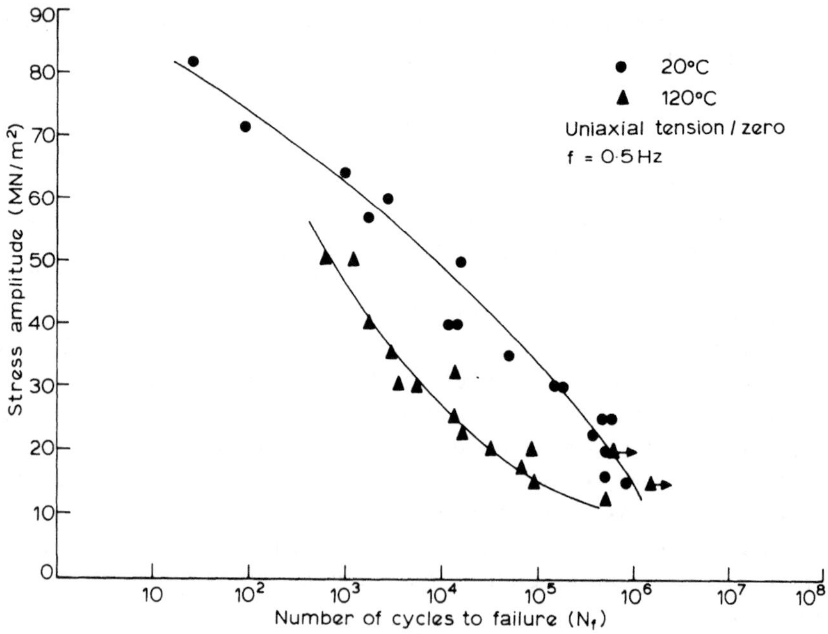

Fig. 27. Fatigue in poly(ether sulphone) (GR) at 20°C and 120°C.

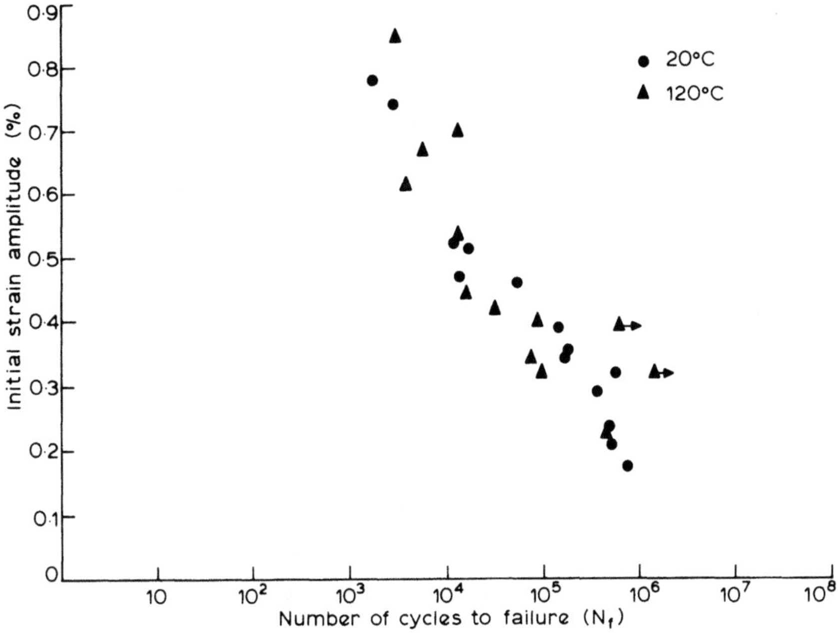

FIG. 28. The data of Fig. 27 transposed to strain/N_f axes.

series of S/N curves. In order to reduce this effect, one possible approach is to characterise fatigue response at selected temperatures and then interpolate for temperatures in between. Figure 27 shows the fatigue characteristic for PES (GR) at 20°C and 120°C. It can be seen that increasing temperature reduces fatigue resistance, interpreted as stress bearing capability. Transposing the data to strain vs. log N_f axes, by using the appropriate stress/strain relationship, i.e. for a frequency of 0·5 Hz, the data are now seen to coalesce (Fig. 28). This result is further evidence that fatigue failure is more likely to be a strain rather than a stress dominated phenomenon. Certainly, the response of poly(ether ether ketone) (PEEK) and poly(phenylene sulphide) (PPS) over the same temperature range, shown in Figs 29 and 30, would generally support this even though the data are somewhat scattered.

6.2. Chemical Effects

6.2.1. General

Plastics are now widely used in applications which demand good resistance to chemical attack. Consequently, an understanding of the

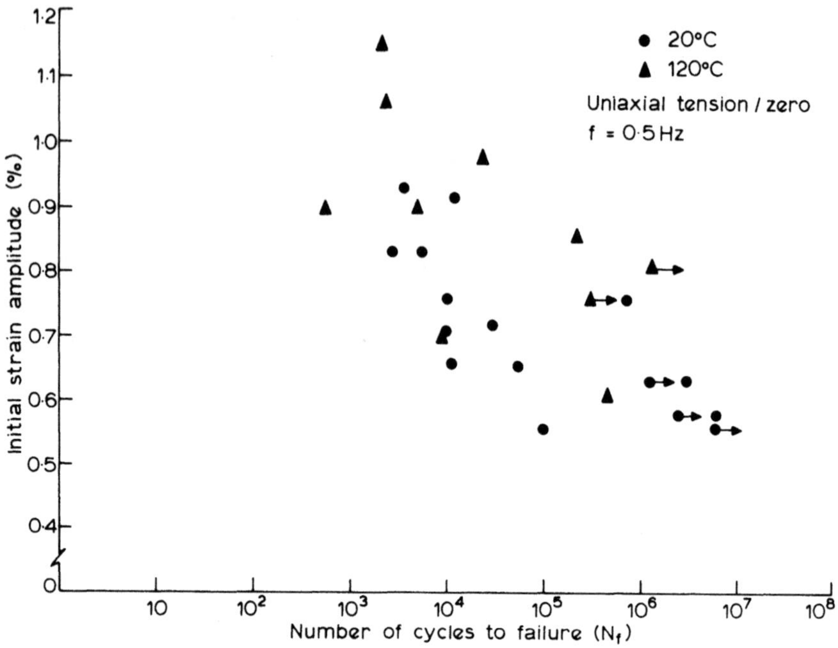

FIG. 29. Fatigue in poly(ether ether ketone) at 20°C and 120°C.

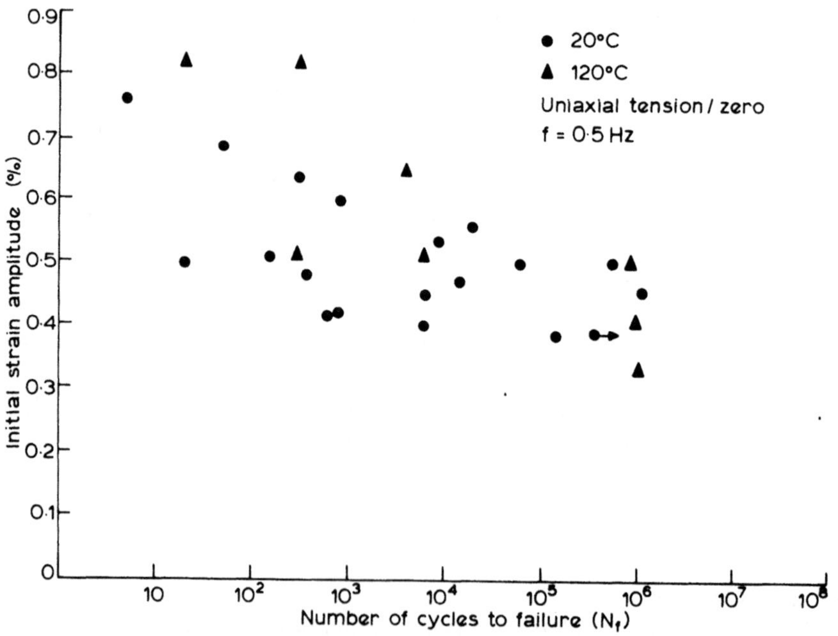

FIG. 30. Fatigue in poly(phenylene sulphide) at 20°C and 120°C.

effects of chemical media upon the physical and mechanical properties of thermoplastics has achieved some priority. There are in fact two major aspects to the problem, which, although related, demand different types of information. The first concerns the effect that chemical media have when the component is exposed to externally applied stresses. This category is generally referred to as 'environmental stress cracking' (e.s.c.). The second involves a much less complex interaction, being concerned with the direct effect chemical media have upon the material itself in the absence of applied stresses. This can be considered as straightforward chemical resistance. Both categories are commonly met in service. Unfortunately, although they are related, it does not necessarily follow that, if there is no chemical attack for a particular material/environment combination, there will be no environmental stress cracking hazard. The effects of direct chemical attack are generally fairly obvious, causing oxidation, solvation, or molecular degradation. Detection of the more subtle physical-chemical interaction of e.s.c. and the accommodation of this within a design specification are generally more difficult.

6.2.2. Chemical Attack

For information about this particular category of chemical interaction, the reader is recommended to consult such definitive sources as Butt and Wright,[43] who discuss the use of plastics in chemical plant construction.

6.2.3. Environmental Stress Cracking

Observed in thermoplastics, this phenomenon bears a strong resemblance to that of 'stress corrosion' in metals. As a general rule, however, the main difference is that in metals it is considered for its effect upon fatigue whereas for thermoplastics the main studies have been restricted to its effect upon creep rupture strength. Very little data have been generated for its interaction with fatigue.

It first came to prominence in the 1950s and 60s for the case of low density polyethylene (LDPE) which was finding a growing market in the packaging of liquid environments in blow moulded polyethylene bottles and containers. The reader is recommended to consult the earlier work of J. B. Howard[44] who was largely instrumental in defining the problem for polyethylenes. As a simple example, Fig. 31 shows the response of LDPE MFI 20 (injection moulding grade, low

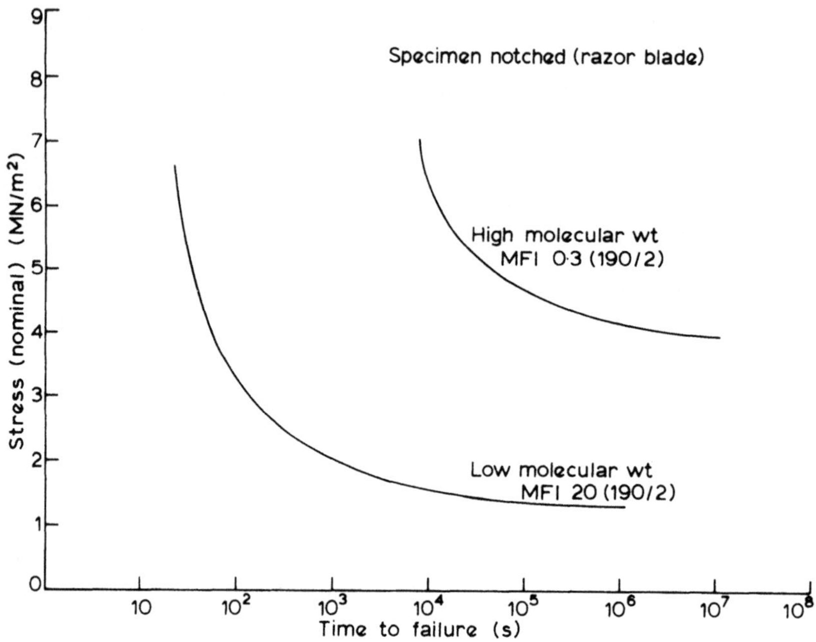

FIG. 31. Environmental stress cracking in LDPE in silicone oil 200 cSt: 20°C. Mode—tension.

mol. wt) and MFI 0·3 (extrusion grade, high mol. wt). The environments are air and a silicone oil of viscosity 200 cSt, and the temperature 20°C. The higher resistance of the MFI 0·3 polymer is consistent with the improvements in mechanical performance offered by increasing molecular weight (see Section 5.2). The need for a wider investigation into this field of polymer/liquid interaction has grown with time and with the continued exploitation of plastics in load bearing applications. There is a variety of ways in which the potential hazard and the likelihood of environmental stress cracking being initiated can be studied. The influence on creep rupture is clearly an obvious one, but in a number of instances a more sophisticated approach can be helpful. Such an approach is provided by monitoring the influence an environment can have upon the creep response of a material. The philosophy is a simple one. A comparison is made between the creep response in the environment under investigation with that observed in air, under the same externally applied stress. The deeper phenomenological considerations of this technique have been discussed[45] but Fig. 32 illustrates the principles.

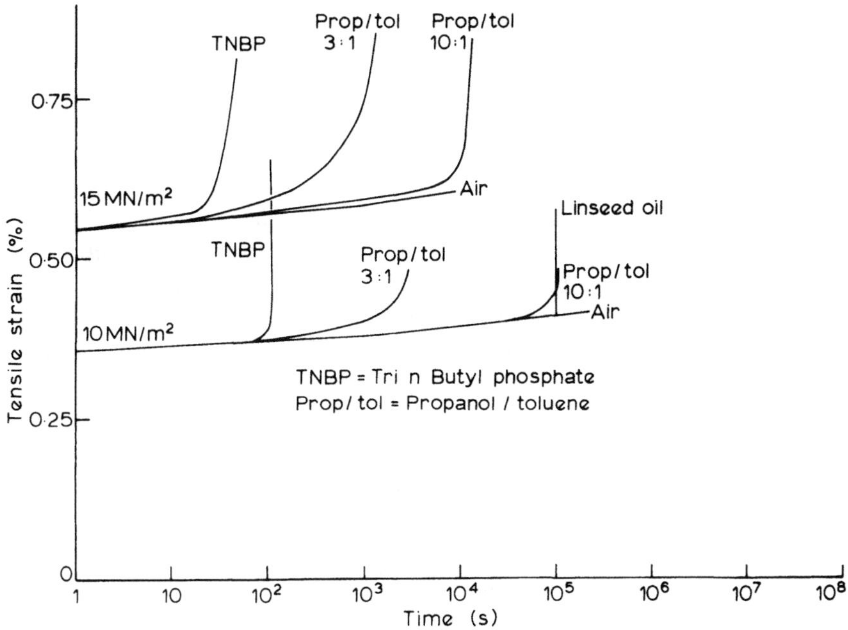

FIG. 32. Creep response at 20°C of a modified poly(phenylene oxide) (PPO) in a number of different environments. Mode—tension.

The simple conclusion drawn from this experiment is that tri-*n*-butyl phosphate (TNBP) is more reactive than propanol/toluene since it causes damage initiation to occur earlier, leading to the onset of 'run away' creep. All three environments are significantly more interactive with poly(phenylene oxide) (PPO) than is air alone. Clearly the role of the environment, particularly in combination with an applied stress, can be a complicated one. It is seen in Fig. 32 to accelerate the rate of growth of a crack or defect once initiated. As to the question whether it promotes easier initiation than air by itself, this remains essentially unsolved. The consensus would favour that initiation takes place at a time independent of the nature of the environment present and that it is only the defect/crack growth rate that is accelerated. Understandably, the situation becomes more complex when the load is interrupted or cycled at a specific frequency. Some experiments on polycarbonate, exposed to biaxial fatigue loading in flexure ($f = 0{\cdot}5$ Hz), suggested that load cycling plus environmental interaction promotes faster crack growth. The effects were additive and perhaps this is the most important aspect to bear in mind in any design specifications which require data for fatigue resistance in interactive media.

7. LONG TERM STRENGTH

Two principal aspects of mechanical behaviour need to be considered with regard to the long term strength of thermoplastics. These are deformation and durability. In the design of a component to be load bearing, either or both these aspects of mechanical behaviour become important and need to be accommodated by the design. Thus a realistic practical and detailed component specification is vital. Given the relevant specification requirements and availability of corresponding mechanical data, an appropriate stress or strain limit can be agreed. Dependent then upon factors such as temperature, environment, type of load application, continuous or intermittent, if the material is still expected to remain ductile, a design limit based upon deformation would be satisfactory. If, however, these or any other factors encountered in service are known or thought likely to cause brittle failure within the planned lifetime of the component, then these design limits, based upon deformation, will need to be reduced to that level which will avoid brittle failure.

Although thermoplastics as a class of materials, with some exceptions, are inherently ductile, it will be unlikely that design can be based solely upon deformation and creep rupture strength. The reasons for this have been seen to be various. The existence of adventitious faults or defects within the molecular structure, or simply within the body of the plastic, can cause a transition in the nature of failure at some time. A knowledge of when this transition, from ductile to brittle, does take place is important in assessing the inherent toughness of a thermoplastic. It has been demonstrated, however, that the propensity to brittle failure is dependent upon material variables such as molecular weight and processing history. Superposed upon these are the effects of other factors such as thermal history and heat ageing.[37]

Again, since as a class of materials thermoplastics have low modulus and strength compared with the materials they often seek to displace, correct material selection and good design becomes essential if they are to provide an effective competitive alternative. To offset certain deficiencies in the properties of plastics, modification by incorporation of stiff reinforcing elements has found increasing favour in engineering components. Although these modifications lead to significant increases in modulus and strength, this is not accompanied by corresponding improvements in fatigue resistance and general durability. This follows as a natural consequence of fatigue and long term strength being deformation dominated phenomena.

Emphasis has also been placed upon the role played by crack initiation in long-term durability. As a general principle, any modification or treatment likely to delay crack initiation will extend fatigue lifetimes. Hence the benefits of good processing and the copolymerisation of propylene with ethylene which causes a general increase in ductility both under short-term (impact) and long-term (fatigue) loading. Semi-crystalline thermoplastics, with their crystal reinforcement, are generally more fatigue resistant than amorphous polymers.

In conclusion, although inferior in certain areas of mechanical performance, thermoplastics offer an interesting range of materials with an attractive profile of properties. To make effective use of them as engineering materials requires sound characterisation, good materials selection and intelligent and innovative component design.

REFERENCES

1. Gotham, K. V., 'A formalised experimental approach to the fatigue of thermoplastics', *Plastics and Polymers,* Aug. 1969, 309–19.
2. Gotham, K. V. and Hitch, M. J., 'Factors affecting fatigue resistances in rigid uPVC pipe compositions', The *Brit. Poly. Journal,* **10,** March 1978, 47–52.
3. Peterson, R. E., *Stress Concentration Factors,* John Wiley and Sons Inc., New York, 1974.
4. Turner, S., '*Mechanical Testing of Plastics*', George Godwin, London, in association with The Plastics and Rubber Institute, London, 1983.
5. BS 4618, *Recommendations for the Presentations of Plastics Design Data,* British Standards Institution, London, 1970.
6. Haward, R. N. and Mann, J., *Proc. Roy. Soc. A,* **282,** 1964, 120–36.
7. Spurr, O. K. and Niegisch, W. D., *J. App. Poly. Sci., Part A-2,* **6,** 1962, 1083.
8. Kambour, R. P., 'Structure and properties of crazes in polycarbonate and other glassy polymers', *Polymer,* **5** (3), March 1964, 143.
9. Kambour, R. P. and Robertson, R. E., Chap. 11, in Vol. 1, *Polymer Science,* A. D. Jenkins, Ed., North-Holland, Amsterdam, 1972.
10. Bucknall, C. B. and Smith, R. R., 'Stress whitening in high impact polystyrenes', *Polymer,* **6** (8), 1965, 437–46.

11. GOTHAM, K. V., 'Long-term strength of thermoplastics: the ductile–brittle transition in static fatigue', *Plastics and Polymers,* April 1972, 59–64.
12. RIDDELL, M. N., KOO, P. G. and O'TOOLE, J. L., 'Fatigue mechanisms of thermoplastics', *Poly. Eng. Sci.,* **6,** 1966, 363.
13. VINCENT, P. I., 'Localised plastic deformation and fracture', *TR No. 97,* Division of Polymer Science, Case Western Reserve University, Cleveland, Ohio.
14. GOTHAM, K. V. and HITCH, M. J., 'Design considerations for fatigue in uPVC pressure pipelines', *Pipes and Pipelines International,* Feb. 1975.
15. MANDELL, J. F., SMITH, K. L. and HUANG, D. D., 'Effects of residual stress and orientation on the fatigue of injection moulded polysulfone', *Poly. Eng. Sci.,* **21** (17), mid Dec. 1981, 1173–80.
16. FROST, N. E., MARSH, K. J. and POOK, L. P., *Metal Fatigue,* Clarendon Press, Oxford, 1974.
17. BAYER, *Technical Literature Makrolon* (*Polycarbonate*), 1975.
18. KNOTT, J. F., *Fundamentals of Fracture Mechanics,* Butterworth, London, 1973.
19. VAN DEN BOOGAART, A. and TURNER, C. E., 'Fracture mechanics': review of principles, *Trans. J. Plastics Institute,* **31,** Aug. 1963, 109–17.
20. MARSHALL, G. P., LINKINS, N. H., CULVER, L. E. and WILLIAMS, J. G., 'Environmental stress cracking in polyolefins', *SPE J.* **28** (9), Sept. 1972, 26.
21. MARSHALL, G. P., WILLIAMS, J. G. and TURNER, L. E., 'Fracture, toughness and absorbed energy measurements in Impact tests on brittle materials', *J. Materials Sci,* **8,** 1973, 949–56.
22. WILLIAMS, J. G., *Fracture Mechanics of Polymers,* Ellis Horwood Ltd, Publishers, Chichester, Sussex, 1984.
23. GRIFFITH, A. A., *Phil. Trans. R. Soc.,* **A221,** 1920, 163.
24. OROWAN, E., *Trans. Inst. Engrs. Shipbuilders Scotland,* **89,** 1945, 165.
25. IRWIN, G. R., *9th Inter. Congr. Appl. Mech. VIII,* Paper 101 (11), University of Brussels, 1957, p. 245.
26. PARIS, P. C., *Proc. Sagamore Army Materials Res. Conf., 10th,* 1964, p. 107.
27. HERTZBERG, R. W. and MANSON, J. A., *Fatigue of Engineering Plastics,* Academic Press, New York, 1980.
28. CONSTABLE, I., WILLIAMS, J. G. and BURNS, D. J., 'Fatigue and cyclic thermal softening of thermoplastics', *J. Mech. Eng. Sci.,* **12** (1), 1970, 20–9.
29. OPP, D. A., SKINNER, D. W. and WIKTOREK, R. J., 'A model for polymer fatigue', *Proc. 24th ANTEC Conf., SPE,* 1968, 149.
30. HAMON, B. V., *Proc. IEE,* **94** (IV), 1952, 151.
31. GOTHAM, K. V. and WRIGHT, D. C., 'Fatigue in polycarbonate', *Plast. and Rubber Processing and Applications,* **4** (1), 1984, 43–51.
32. BS 3518, *Methods of Fatigue Testing,* British Standards Institution, London, 1970.
33. TAPROGGE, R., 'D 53441', *Kunststoff-Rundschau,* **9,** Sept. 1968, 439–45.

34. Burgess, R. H., *Manufacture and Processing of PVC,* Applied Science Publishers Ltd, London, 1982.
35. Fletcher, K., Haward, R. N. and Mann, J., 'Rubber reinforced polystyrene copolymers', *Chem. and Ind.,* **45,** 1965, 1854.
36. Bucknall, C. B., Gotham, K. V. and Vincent, P. I., 'Fracture—II, 'The empirical approach', Chap. 10, in Vol. 1, *Polymer Science* A. D. Jenkins Ed., North-Holland, Amsterdam, 1972.
37. Gotham, K. V. and Hough, M. C., *The Durability of High Temperature Polymers,* RAPRA, Shawbury, 1984.
38. Stephenson, R. C., *Plast. & Rubber Materials Applications,* May 1979, 45–51.
39. Davis, A. and Sims, D., *Weathering of Polymers,* Applied Science Publishers Ltd, London, 1983.
40. Imperial Chemical Industries PLC, *Maranyl–Nylon 66 Compounds,* Tech. Service Note N104, London.
41. Dupont, *Zytel Nylon Resins Design Handbook.*
42. Kohan, M. I., *Nylon Plastics,* John Wiley and Sons, New York, 1973.
43. Butt, L. T. and Wright, D. C., *Use of Polymers in Chemical Plant Construction,* Applied Science Publishers Ltd, London, 1980.
44. Baer, E., *Engineering Design for Plastics,* SPE Polymer Science and Engineering Series, Reinhold Publishing Corp., New York, 1964.
45. Gotham, K. V. and Wright, D. C., 'Solvent crazing criteria', *Poly. Eng. Sci.,* **23** (3), Feb. 1983, 135–9.

Chapter 6

COMPUTER-AIDED ANALYSIS OF SOME THERMOPLASTICS PROCESSING OPERATIONS

J. F. T. PITTMAN

Department of Chemical Engineering, University College, Swansea, UK

NOTATION

C	specific heat
d	mould cavity thickness
g	heat release rate, by phase change
G	factor for time dependence of flow into a mould
h	heat transfer coefficient
H	enthalpy
k	thermal conductivity
K	Avrami rate constant
m	Avrami index
n	power-law index
P	pressure
Q	volumetric flow rate into mould
r	radial coordinate
R	residence time
S	defined in eqn (30)
t	time
T	temperature
v	velocity
$\bar{v}$	mean velocity over the thickness of the mould cavity
X	interface position in freezing

$\dot{\gamma}$ shear rate
θ relative crystallinity
μ viscosity
μ_0 consistency coefficient, in the power law
ρ density
ϕ defined in eqn (9b)
ϕ' defined in eqn (9a)
ψ stream function

1. INTRODUCTION

This contribution is concerned with the use of computers to provide information about what happens to the material during various thermoplastics processing operations, and the use of this information to help improve the operation of existing equipment and to assist with the design of new machinery. The information is provided by solving numerically the relevant differential equations governing flow, heat transfer, etc., or in some cases by using the computer to implement analytic solutions. Numerical solutions are generally necessary because realistic modelling of the processes results in very difficult mathematical problems; specifically, the governing differential equations are non-linear, and often highly so, as a result of the non-Newtonian flow properties of melts, the temperature dependence of their physical properties, the phenomenon of viscous heat generation and the resultant coupling between energy and momentum conservation equations. In addition, the often complicated geometry of the domains within which flow and heat transfer occur makes analytic solutions very difficult to obtain.

1.1. Simulation Benefits

The benefits of these computer simulations are that they can provide extremely detailed and complete information on conditions within the processed material—much more detailed than can be obtained even on highly instrumented research machines. In addition, numerical experiments are generally cheaper than physical ones, and this is increasingly so with the present rapid advances in computer technology, as well as in numerical methods. In many instances, numerical experiments can replace, or at least supplement, physical experiments, allowing the engineer to look into a process in a way which would otherwise be impossible. Through this, the reasons for unsatisfactory processing can often be identified, and improved operating conditions arrived at. New designs can be tested by numerical simulation, thereby decreasing development times and expensive trial and error.

Numerical solution of the equations governing melt flow, heat transfer and other relevant phenomena can be termed computer-aided analysis. This is a prerequisite for, and merges into, the design phase where the computer interacts with the engineer, allowing him rapidly to explore design variations, together with a range of other functions which can include property data retrieval, plastic material selection, economic assessments, mechanical design and part selection and, finally, the creation of digital information to control machining of engineering components.

1.2. Industrial and Academic Interest

In the following pages the discussion centres mainly on computer-aided analysis, but also touches upon the design activity, through mention of some commercially available programme packages. Industrial interest in these is increasing, and the volume of research literature is expanding dramatically.

Of the 200 or so references cited in the present article, about half originated during the past five years. Discussions of computer-aided design methods in the trade journals are also becoming more frequent. This trend is part of a broader one throughout industries concerned with forming processes, notably metal forming, and the state of the art in this wider sphere is indicated by recent publications.[1,2] The impression from these is that despite a significant growth of interest, the plastics industry is lagging behind the metal forming industries in the application of computer methods for analysis and design. It therefore seems very appropriate to attempt to summarise what has been done in this area, and what present capabilities are. Within certain limits, this is the objective of the present contribution.

1.3. Topics Covered

In the space available it will not be possible to cover all the thermoplastics processing operations; however, an attempt has been made to treat a number of the most important ones fairly comprehensively. These largely coincide with the areas where, to date, there has been the most work applying numerical methods, and they fall quite neatly into the category of processes, or process aspects, involving *internal flows and heat transfer.*

Specifically, the main topics covered are plasticating extrusion, flow and heat transfer in ducts (flow channels) and dies, and injection moulding (including the cooling phase). The cooling phase of blow moulding is also included because of its close similarity. Work on analysis of melt thermocouple errors, and on calendering is also mentioned briefly. Numerical treatments of an important class of problems to do with the experience of the material during processing are also described: the class includes analyses for residence time fields, degradation, laminar mixing and orientation.

Important topics omitted are those where the presence of free surfaces (melt–air interfaces) has a dominant effect on the flow, namely the parison inflation stage of blow moulding, film blowing, melt spinning, coating and die swell. These are areas where there has been significant numerical work, but by relatively fewer research groups, whose work is well reviewed in a recent book[3] to which the reader is referred for details. A further comprehensive review of melt spinning is available.[4]

A further restriction in the present article is the omission of a discussion on the treatment of visco-elastic effects. This is largely consistent with the omission of free surface phenomena, where the effects are most important. Internal flows are very often modelled using the lubrication approximation,[5a] which broadly applies in ducts of slowly varying cross-section. Within this approximation elastic effects do not influence the kinematics of the flow, though they will influence stresses, and useful solutions can often be obtained assuming purely viscous melt flow. Abrupt contractions, where elasticity would significantly influence flow fields, are generally avoided if possible in processing equipment, because of the danger of material hold-up and degradation. A further reason for the general omission of elastic effects in modelling internal flows is simply the difficulty of including them. Despite the fact that numerical simulation of visco-elastic flow has been an active research area for several years, present solution methods all break down, through convergence failure, at quite modest degrees of elasticity. In terms of the Weissenberg number, the product of a characteristic material relaxation time with a characteristic shear rate, an increase of one or two orders of magnitude is required before many practical melt flows can be simulated. Excellent and up-to-date accounts of this area are available.[6,7]

In the following pages, then, the objective is to summarise work in the specified areas and also to illustrate briefly some present capabilities by showing a few previously unpublished results from the writer's group. There will not be space to provide a fully critical account: rather, the intention is to indicate how the literature resource may be used as an aid to those who wish to make a more detailed study, and to develop and apply the methods described in order to further realise the economic benefits which computer-aided analysis and design can undoubtedly achieve for the plastics industry.

2. FLOW AND HEAT TRANSFER IN DUCTS AND DIES

2.1. Developed Isothermal Flow in Ducts

At the simplest level, flow in parallel-sided ducts may be approximated as isothermal and fully developed. Using the lubrication approximation,[5a] this representation can also be applied to gradually tapering or expanding channels. In practice, melt flows are seldom sufficiently close to isothermal for the temperature dependence of viscosity to be neglected. Some examples of the consequences of this approximation are shown by Pittman and Nakazawa.[8] However, provided that the Nahme and Brinkman numbers for the flow are sufficiently small, the simplification can be applied, and the equation of motion based on the lubrication approximation becomes

$$\frac{\partial P}{\partial z} = \frac{1}{r}\frac{\partial}{\partial r}\left(r\mu\frac{\partial v_z}{\partial r}\right) \tag{1}$$

for axisymmetric flows, or

$$\frac{\partial P}{\partial z} = \frac{\partial}{\partial x}\left(\mu\frac{\partial v_z}{\partial x}\right) + \frac{\partial}{\partial y}\left(\mu\frac{\partial v_z}{\partial y}\right) \tag{2}$$

in Cartesian form, with viscosity expressed as a function of shear rate.

For the one-dimensional, axisymmetric case, such as flow in an annulus, the first integration of eqn (1) is analytic. The second integration involves the constitutive equation for viscosity as a function of shear rate, and requires numerical quadrature for non-Newtonian flows (see also Section 5.2). For two-dimensional cases, such as flow through rectangular ducts, the (linearised) equation of motion is elliptic and is solved by standard finite difference methods, with

TABLE 1
NUMERICAL ANALYSES OF DEVELOPED ISOTHERMAL FLOWS

Reference	*Duct cross-section*	*Flow model*	*Solution method*
9 (1961)	Rectangular	Power-law	Variational principle/ Ritz method
10 (1965)	Rectangular	Power-law	Finite differences with over-relaxation
11 (1965)	Rectangular	Power-law	Finite differences with over-relaxation
12 (1966)	Annulus	Ellis model	Quadrature by Simpson's rule
13 (1968)	Annulus	Eyring model	Quadrature by Romberg method
14 (1968)	Equilateral triangle	Power-law	Variational principle/ Ritz method
15 (1969)	Rectangle, isosceles triangle	Sutterby model	Variational principle/ Ritz method
16 (1970)	Square, trapezoidal, star-shaped	Power-law	Weighted residuals method

iteration to take account of shear-dependence of viscosity. For less simple cross-sections finite difference methods are not so convenient, and solutions for trapezoidal, triangular and star-shaped cross-sections have been obtained using Ritz and weighted residuals methods. A number of works in this category are summarised in Table 1. Flow-rate/pressure-drop relationships for various geometries have been summarised and correlated in a number of works.[17,18,19]

2.2. Kinematically Developed, Thermally Developing Steady Flows in Parallel-sided Ducts

More realistic models of most melt flows are obtained by allowing for temperature variations and the temperature dependence of viscosity. In practice, however, ducts and dies are seldom long enough to allow full thermal development, and the case of interest is the thermally developing flow. Until recently only two-dimensional problems of this type had been studied in which, for example, the downstream evolution of radial velocity and temperature profiles were followed in cylindrical pipes or annuli. Slit dies have also been treated in two dimensions, based on the assumption of effectively infinite width. Almost all of this work has used the lubrication approximation (often implicitly), so that the velocity profile is taken as the fully developed one consistent with the local radial temperature profile. If this is known then the velocity profile can be obtained by quadrature as described for developed isothermal flows except that iteration may now be required to obtain the local axial pressure gradient.

The main problem is the calculation of the evolving temperature profile. The usual approach in the pipe flow problem, for example, is to neglect conduction in the axial direction, and, consistent with the kinematically developed flow assumption, to drop the radial convection term. A parabolic problem then results. For the axisymmetric constant property case, for example,

$$\rho C v_z \frac{\partial T}{\partial z} = \frac{1}{r}\frac{\partial}{\partial r}\left(rk\frac{\partial T}{\partial r}\right) \tag{3}$$

In some works, terms representing viscous heat generation, and/or temperature changes resulting from density changes with pressure, are added to eqn (3). Other variations in the physical models used involve assumptions concerning the constitutive equation for viscosity, the temperature and pressure dependence of physical properties and the boundary conditions applied at the duct wall. Table 2 summarises most work in this category. Regrettably, few publications provide adequate detail of their numerical techniques or demonstrate convergence of the solutions. Of the various works, one may single out that of Gee and Lyon[20] as an early publication employing a quite sophisticated physical model. Some of the most ambitious work is contained in the series of papers by Wilkinson and co-workers[24,25,27,30,31]. A substantial amount of data on Nusselt numbers in tube flow is presented. Unfortunately the 'dimensionless' temperatures used are unit dependent. Additionally, the presentation of heat transfer data for flows with significant viscous heating in terms of Nusselt numbers is unsatisfactory. Viscous heating can cause the mean flow temperature to equal the tube wall temperature, whilst the wall heat flux remains non-zero. The Nusselt number then goes to infinity. Several works include the cooling effect accompanying expansion as pressure falls along the duct, using equilibrium thermodynamic relationships. The physics of this phenomenon, however, is complex and probably involves a delayed response by the melt to pressure changes.

TABLE 2

NUMERICAL ANALYSES OF KINEMATICALLY DEVELOPED, THERMALLY DEVELOPING FLOWS

Reference	Duct geometry	Flow model	Incompressible flow	Viscous heating	Expansion cooling	Thermal properties	Duct wall thermal boundary conditions	Numerical method	Comments
20 (1957)	Cylindrical pipe	$\mu(\dot{\gamma}, T)$	Yes	Yes	Yes	$k(T)$ $C_p(T)$	$T = T_w$	Little detail given. Steady state solutions obtained by direct iteration or by following time transient	Good agreement with a few experimental results
21 (1965)	Cylindrical pipe	$\mu(T, P)$	Yes	Yes	No	Constant	$T = T_w(z)$ or $\frac{\partial T}{\partial r} = 0$	Exponential coordinate transformation to concentrate mesh near wall. Fully implicit backwards difference for $\partial T/\partial z$, with lagging velocity profile	Good agreement with experimental results in which temperature rise due to viscous heating of up to 35°C in a capillary caused 70% reduction in μ
22 (1969)	Cylindrical pipe	$\mu(\dot{\gamma}, T)$ power-law	Yes	Yes	Yes	$k(T)$ $C_p(T)$ $\rho(T, P)$	$T = T_w$	Little detail given	Moderate agreement with some experiments
23 (1971)	Cylindrical pipe	$\mu(\dot{\gamma}, T)$ in tabular form	Yes	Yes	Yes	Constant	$T = T_w$	Little detail given	Good agreement with a limited amount of experimental data
24 (1973)	Cylindrical pipe	$\mu(\dot{\gamma}, T)$ power-law	Yes	Yes	No	Constant	$T = T_w$	Energy equation stepped downstream using Crank–Nicholson with lagging velocity profile	Substantial number of computer results given for Nusselt number
25 (1973)	Cylindrical pipe	$\mu(\dot{\gamma}, T)$ with finite yield stress (Bingham plastic)	Yes	Yes	No	Constant	$T = T_w$	Energy equation stepped downstream using Crank–Nicholson with lagging velocity profile	Substantial number of computed results given for Nusselt number

26 (1974)	Cylindrical pipe, annulus and slit die	$\mu(\dot{\gamma}, T)$ power-law or polynomial	Yes	Yes	No	Constant	$T = T_w$	Little detail given	A few example results shown
27 (1974)	Cylindrical pipe	$\mu(\dot{\gamma}, T)$ power-law	Yes	Yes	No	Constant	Specified constant q_w	As Ref. 24	Substantial number of computed results given
28 (1975)	Cylindrical pipe, annulus and slit	$\mu(\dot{\gamma}, T)$ power-law	Yes	Yes	No	Constant or $\left.\frac{\partial T}{\partial r}\right\|_w = 0$	$T = T_w$	Energy equation stepped downstream using iterative implicit finite-difference scheme	Computed results presented graphically for $n = 0{\cdot}4$
29 (1975)	Cylindrical pipe	$\mu(\dot{\gamma})$ Prandtl–Eyring	Yes	Yes	No	Constant	$T = T_w$ or $\left.\frac{\partial T}{\partial r}\right\|_w = 0$	No details	Various examples of computed temperature profiles shown
30 (1978)	Cylindrical pipe	$\mu(\dot{\gamma}, T, P)$	Yes	Yes	Yes	Constant	$T = T_w$	As Ref. 24	Good agreement with experimental temperature measurements, but temperature changes only 10°C, and viscous heating slight $Na \leqslant 0{\cdot}06$
31 (1980)	Cylindrical pipe, annulus, conical flow	$\mu(\dot{\gamma}, T, P)$	Yes	Yes	Yes	Constant	$T = T_w$	As Ref. 24	Large overlap of content with reference 29. Also includes brief treatment of kinematically developing conical flow at $Re = 500$
32 (1980)	Cylindrical pipe	$\mu(\dot{\gamma}, T)$ power-law	Yes	Yes	No	Constant	$T = T_w(z)$ or $q = q_w(z)$ both linear in z, or constant	No details given	Shows computed result for pressure drop graphically and in tabular form

Reverting to a simpler approach, Parnaby *et al.*[33] have described a lumped parameter method for treatment of thermally developing flows. This amounts to the use of analytic solutions for developed isothermal non-Newtonian flow in parallel-wall channels, with viscosity taken at a mean temperature. Viscous heating rates based on these solutions, together with empirical heat transfer coefficients, are used to calculate the change in mean temperature in the flow direction. Tapered ducts are treated as series of parallel-sided ones. The method is illustrated with applications to extrusion dies and extruder channels.

2.3. Kinematically and Thermally Developing Two-dimensional Steady Flows in Ducts

Jarzebski and Wilkinson[34] have treated non-isothermal developing flow of a power-law fluid in a converging slot. A cylindrical polar coordinate system was used with the principal flow in the negative radial direction. The momentum equation is simplified only by dropping second radial derivatives of the radial velocity, inertia terms being retained. Similarly, in the energy equation, the radial conduction term is dropped, resulting in a system involving only first derivatives in r. The equations are coupled through temperature dependence of viscosity. Solutions are obtained by marching from the inlet, by a Crank–Nicholson method, linearised using parameters at the upstream end of the radial step. Certain numerical difficulties are mentioned but not elaborated upon. Results for Reynolds numbers up to 540 are presented, with some discussion of non-isothermal behaviour and velocity profile development. A correlation for Nusselt number is tentatively presented based on computations for the Newtonian case.

The problem of isothermal, shear-thinning flow into an axisymmetric sudden contraction has been studied in a thorough and detailed paper by Kim-E, Brown and Armstrong.[35] A finite element method was used, with careful mesh adjustment to achieve converged solutions. Momentum equations were retained in their full form, and the main objective of the work was to disentangle the effects of fluid inertia and shear-thinning upon the velocity field in the contraction. For polymer processing applications, the most important conclusions relate to the influence upon the size of the vortices formed upstream of the contraction. In melt flows, however, these are also strongly influenced by fluid elasticity. A similar problem is treated by Mitsoulis, Vlachopoulos and Mirza,[36] Mitsoulis and Vlachopoulos,[37] and by

Durst and Loy.[38] The axisymmetric thermal entry region in pipes has been extensively studied by Pittman and Nakazawa[8] and Rashid.[39] In contrast to previous works on pipe flow, the momentum conservation equations were retained in their full form. Similarly the energy equation retained all first and second derivative terms, and included viscous heat generation. A temperature-dependent power-law model was assumed for viscosity. Solutions were obtained for various pipe wall temperatures corresponding to heating or cooling and for a wide range of Reynolds, Nahme, Graetz and Péclet numbers. Extensive data on dimensionless pressure drops, bulk temperature and detailed velocity and temperature fields were presented. The calculations were carried out using the finite element method with a penalty function enforcement of incompressibility. Retention of the first derivative terms required a stabilisation technique in some high Péclet or Reynolds number flows; the streamline upwinding method was used. Similar techniques have been applied to the analysis of a pipe die.[39]

2.4. Wire-coating Dies

The first numerical analysis of a wire coating die seems to be that by Caswell and Tanner.[40] Creeping, isothermal, power-law flow was assumed in the axisymmetric die. Solutions were obtained using a finite element Galerkin method with the two velocity components v_r and v_z and pressure as unknowns, with quadratic interpolation for velocities and linear interpolation for pressure in triangular elements. In dies of various designs solutions displayed negative pressures near the impact point, and melt recirculation. The analysis did not include free surface flow of the extrudate. This problem was coupled with melt flow within the die in an analysis by Viriyayuthakorn and Deboo,[41] which again assumed creeping isothermal shear-thinning flow and used a similar finite element formulation. The location of the free surfaces and the contact point of the drawn-down jacket are found iteratively in a scheme which is described only very briefly, but is claimed to converge faster and more smoothly than the free surface location method used previously by Nickell *et al.*[42] in die swell studies. However, it should be noted that the cable jacketting analysis does not include melt elasticity, which will in practice influence draw-down significantly.

A numerical analysis of velocity and temperature fields in wire-coating co-extrusion dies has been described by Basu.[43] The approach is essentially that used in most work on thermally developing flow in parallel-walled channels, described above (Section 2.2). Locally developed velocity profiles are assumed, and found by quadrature, incorporating a shear-thinning temperature-dependent flow model. The energy equation, including viscous heating, is simplified to parabolic form and stepped in the flow direction, with the velocity field lagging. Results are shown for a HDPE/PS (high density polyethylene/polystyrene) system at wire speeds of up to 1000 cm/s, at which speed viscous heat generation produced temperature rises of up to 70°C.

A finite element analysis of high-speed wire coating has been carried out by Pittman and Nakazawa[44] at wire velocities of up to 3000 m/min. Under these conditions shear rates in the immersion zone of the die are of the order of 10^7/s, and intense shear heating produces substantial temperature rises. The flow and energy equations are thus very strongly coupled; however, converged solutions were obtained successfully giving details of velocity, temperature and pressure fields in the die. Shear stress on the wire was integrated to give the wire tension developed in the die. Le Nir briefly mentions a finite difference analysis of isothermal flow in a coating die, in a paper describing the development of computerised control for wire insulating lines.[45]

2.5. Flow Distribution in Dies and Systems of Ducts

The design of dies or systems of ducts to produce a required flow distribution is a problem of great practical importance in polymer processing. For example, uniformity of flow from the extrusion slot in sheet and film dies is required, and flow rates in the various branches of a runner–sprue network in injection moulding must be such as to give satisfactory mould filling. These problems are essentially three-dimensional but may be tackled approximately using various methods. Booy[46] describes an approach in which components of the various flow paths through the equipment are assigned flow resistances, based on analytic solutions for flow through simple geometries such as pipes, slits, etc. By inspection of the network of resistances, a system of simultaneous equations is set up representing flow conservation and pressure consistency. For constant viscosity flows the system is linear in the flow-rates; for the non-Newtonian case the flow resistances depend on the flow-rates in each part of the network, and the solution is obtained iteratively. Tadmor, Broyer and Gutfinger[47] have applied the so-called flow analysis network (FAN) method to the problem of flow distribution in a cross-head die. The annular channel is unwrapped onto a plane surface, and the flow regarded locally as developed isothermal flow between infinite parallel plates. A potential equation then governs the pressure field and this is solved by a finite difference method on a rectangular grid. Flows are obtained as gradients of the pressure potential. The method can be regarded as based on a generalised Hele Shaw flow assumption, closely related to the methods used for cavity filling simulation in injection moulding (see Section 5.3). Fenner and Nadiri[48] used essentially the same approach in a further analysis of a cross-head die to determine correct deflector design, but applied the finite element method to solve the pressure and flow fields. This has the advantage over the FAN method of more conveniently representing the curvilinear shape of the deflector.

2.6. Three-Dimensional Analysis of Flow in Dies and Ducts

In recent years, developments in computing power have begun to make possible the extension of analyses to three-dimensional flows, i.e. three velocity components and temperature, all dependent on three position coordinates. A group at IKV, Aachen, have developed finite element programmes using a penalty function formulation with the 20 node isoparametric element. The formulation is for creeping flow, but otherwise retains all terms in the incompressible flow equations. Initially, developing flows in star-cross-section dies and a transition zone were analysed, assuming isothermal Newtonian flow.[49,50] Later, the analysis was extended to include temperature fields and shear rate- and temperature-dependent viscosity.[2,51]

For ducts where the sides are parallel, or approximately parallel, computing demands can be reduced substantially by parabolising the energy and momentum equations (dropping second derivatives in the flow direction) as has been described for two-dimensional problems in the preceding sections. This approach was followed by Lawal and Mujumdar[52] in a detailed simulation of developing flow and heat transfer in ducts of arbitrary cross-section. A mesh was set up on the duct cross-section using an orthogonal body-fitted coordinate system and the parabolised conservation equations stepped downstream using a finite difference method. Flow behaviour was modelled using a temperature-dependent power-law constitutive equation; buoyancy and free convection were included using a quadratic temperature dependence of density, and thermal properties were allowed to vary linearly with temperature.

Computed results were tested successfully against pre-existing analytic and experimental data for developing flow and heat transfer in square-section ducts, and the code was then used to generate new data for regular pentagonal ducts. The techniques used in this work provide powerful means for simulation of the three-dimensional flow and heat transfer problem in parallel or approximately parallel sided ducts and dies.

3. SINGLE-SCREW EXTRUSION

3.1. Metering Zone Melt Flow

3.1.1. One Dimensional Analyses for Operating Characteristics and Melt Temperatures

Developments in the analysis of single-screw plasticating extruders up to 1976 have been excellently reviewed by Fenner.[53] A further review[3] includes a relatively small amount of additional work up to about 1980.

Metering zone melt flow has received a large amount of attention. A frequent assumption has been that the channel is sufficiently wide and shallow for cross-channel variations of velocity and temperature to be ignored, and also for the helical channel to be 'unwrapped' and treated as prismatic, using Cartesian coordinates. Together with the assumption of developed conditions in the down-channel direction, the early analyses were therefore one-dimensional with dependence on only the channel-depth coordinate y, and absence of the v_y velocity component (see Fig. 1). Following the linear analytic analyses of the 1950s, a number of papers appeared in the early 1960s in which melt viscosity was allowed first to be shear/rate-dependent, and later, also

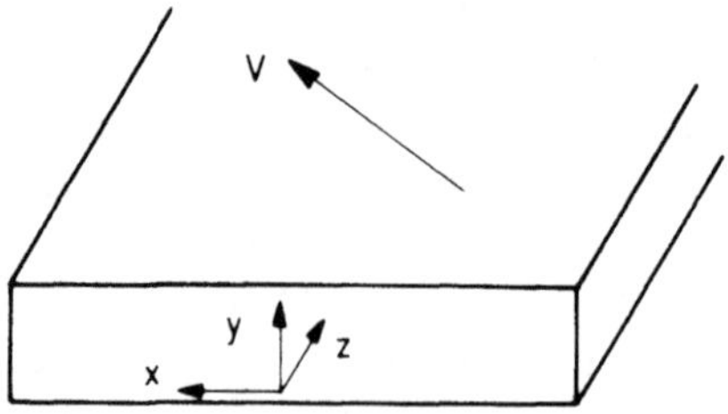

FIG. 1. Coordinate system in the 'unwrapped' representation of an extruder channel. V represents the relative motion of the barrel and screw.

temperature-dependent, leading to the need for numerical solutions. Notably examples of this class of analyses are the works of Griffith[54] and Zamodits and Pearson.[55]

Griffith[54] included the cross-channel velocity component v_x, allowing for its contribution to the viscous heat generation and shear-dependence of viscosity. He recognised that at the high Péclet numbers which obtained, streamlines of the cross-channel circulation coincided closely with isotherms, and utilised this in the thermal analysis. The coupled ordinary differential equations for momentum and energy conservation were solved iteratively using a Runge–Kutta method. Solutions took two to three hours on a Burroughs 205.

Zamodits'[55,56] analysis differed from the foregoing principally in using an insulated thermal boundary condition on the screw, in place of the isothermal screw and barrel condition used by Griffith. Fenner[53] also implemented this model using an alternative finite difference calculation scheme, which was solved ten times faster.

3.1.2. Two-dimensional Analyses for Operating Characteristics and Melt Temperatures

In the late 1960s analyses were extended to allow dependence on two position coordinates. In the work of Yates[57,59] these were the channel depth (y) and down-channel (z) directions, leading to a model for thermal development in the down-channel direction. The wide–shallow channel assumption was retained and the model allowed for heat transport in the down-channel direction by convection, together with y-direction conduction and viscous heat generation. The energy equation thus produced an initial value parabolic problem, coupled with the ordinary differential equations representing quasi-developed (creeping) flow at each z position. Regular perturbation expansions were made for the unknowns in terms of Nahme number, which measures the coupling of flow and energy equations due to the effects of viscous heating upon viscosity. The momentum and energy equations were thus uncoupled at each order of approximation, and solutions were obtained stepping down the channel using the Crank–Nicholson implicit method. The Yates model was also implemented by Fenner,[60] and used in the design of large hot melt extruders.

Yates established that his solution scheme was convergent for cases where the Nahme number is less than 1·5. Viscous heating in extrusion is often such as to give large values, and attempts to solve the parabolic energy equation by a direct numerical approach, stepping down-channel with a finite difference method, run into difficulties when back pressure in the extruder causes some 'down-channel' velocities to become negative. Elbirli and Lindt[61] have proposed a solution scheme to overcome this problem using a Lagrangian reference frame, i.e. a coordinate system which floats along with the material. Solutions were obtained for Nahme numbers up to 100 for Newtonian flow, but increasingly non-Newtonian behaviour reduced the range of convergence.

Martin[62] treated the alternative two-dimensional problem by considering flow to be thermally fully developed in the down-channel direction, and studied velocity and temperature fields on the channel cross-section. The inclusion of thermal convection as well as conduction produced a mixed hyperbolic–elliptic problem for the temperature field, which at the prevailing high Péclet numbers required stabilisation by a form of up-wind differencing in the finite difference formulation. A small number of solutions were produced, which nevertheless demonstrated the importance of cross-channel thermal convection. Additional isothermal Newtonian studies including leakage flow over the flights showed how pressure-driven back-flow through the clearance can dramatically alter the streamline pattern in the channel, with significant implications for heat transfer. Martin remarks that the programme was expensive to run and could not at that time be employed on a routine basis.

At about the same time Dyer[63] was studying the equivalent problem using a cylindrical polar representation of the channel. Fenner[58] later found faults in Dyer's programme, and re-worked it to yield solutions for deep, highly curved channels. Difficulties still remained, however, in treating the thermal convection terms and only isothermal solutions were obtained. With a 20×20 mesh a Newtonian solution took about 20 min on an IBM 7094. This level of mesh subdivision was not considered adequate for non-Newtonian cases.

The 1970s saw the introduction of finite element methods in the analysis of melt flow. Palit and Fenner[64,65] demonstrated the ability of the method to deal with channel cross-sections other than the simple rectangular ones to which previous finite difference analyses had been restricted. This work was isothermal, but Palit[66–68] produced a small number of non-isothermal solutions on a coarse, regular mesh for the type of problem studied by Martin. These were limited by the lack at that time of a satisfactory finite element version of up-wind differencing for treating convection–diffusion problems with significant first order derivative terms (high Péclet number problems).

The basis for a finite-element treatment using a helical coordinate system to represent the true channel geometry was set up by Nebrensky,[69] Pittman and Smith. Developed isothermal solutions were published by Hami and Pittman[70] and Choo, Hami and Pittman[71] and compared quite successfully with experimental results for a deep channel screw.

3.1.3. Pressure and Heat Transfer Effects Associated with the Clearance; Influence of Channel Geometry on Heat Transfer

Other than briefly in Martin's work, effects associated with leakage through the flight clearance had so far been treated only analytically. In 1970 Winter[72] showed, from a finite difference analysis of thermally developing flow in the clearance, regarded as flow between infinite parallel plates, that significant pressure variations could occur in the flow direction. These effects must be considered together with the journal-bearing pressure generation which occurs when the flight land and barrel surfaces are not parallel, in determining whether resultant pressure forces may lead to a breakdown of melt lubrication in the clearance.

A study of the effect that leakage flow and clearance geometry may have upon overall heat transfer in the extruder is contained in a paper by Pittman and Rashid.[73] Solutions were obtained for melt circulation and temperature fields over the channel cross-section using finite element methods, and included a detailed representation of the flight clearance. The model—which allowed for shear rate and temperature dependent viscosity, thermal convection and conduction, and viscous heat generation—assumed thermally fully developed flow, but omitted the down-channel velocity component. It is thus somewhat idealised, but suggested some interesting effects. At normal flight clearances of about 0·1 mm, thermal boundary conditions assigned on the flight land had little influence on temperatures attained within the channel as a result of viscous heating. However, at larger clearances corresponding to a worn screw or barrel, temperatures within the channel were found to rise or fall significantly depending upon whether (respectively) an insulated condition or an isothermal temperature equal to that of the barrel was assigned on the flight land. The effects are related to viscous heating in the clearance and the height of the 'thermal barrier' formed by the hot layer of melt left behind by the flight. In the same work the geometrical flexibility of the finite element methods was further exploited in an investigation of the effect of channel geometry upon developed melt temperatures arising from viscous heating.

The use of mathematical models of melt flow within a computerised optimisation process for screw design has been described by Helmy and Parnaby.[74] The requirement for repeated solutions in the steady-state hill-climbing optimisation led inevitably to the use of simple analytic models.

3.2. Solids Conveying

In traditionally designed screws the solids conveying zone often contributes little to pressure generation: it occupies only a short part of the screw, and has been regarded as of less importance than the other zones. Together with the formidable difficulty of analysing flow and stress in the granular material, this has resulted in a relatively small number of publications. However, the more recent development of very effective entry zones, where pressure rises rapidly, must revive interest in modelling solids conveying.

Numerical techniques were introduced to this area when Tadmor and Broyer[75] extended the analytic analysis of Darnell and Mol,[76] to take account of temperature rise in the solid plug. Heat is generated at the slipping plastic–metal interface at a rate governed by the local coefficient of friction, which depends on temperature and pressure at the interface. Part of this heat was considered to be lost through the barrel at a rate governed by a constant heat transfer coefficient, the remainder being conducted into the plug, regarded as a semi-infinite region. A transient heat conduction problem, corresponding to the progress of the plug down the channel, was solved using a forwards difference in time and central difference in space. In a continuation of this work Kacir and Tadmor[77] introduced the term 'delay zone' to describe the transition from the solids conveying to the melting zones: in this zone a melt film builds up at the barrel surface to a thickness sufficient to be swept off by the flight. Their model for this process is largely analytic, but leads to a non-linear equation for melt film thickness which was solved by a Runge–Kutta method.

A more detailed study of stress in the solid plug was carried out by Lovegrove and Williams.[78] The Darnell and Mol model predicts pressures in the solids conveying zone proportional to the pressure at the feed pocket. This conflicts with experience, which shows that extruder operation is not dependent on the head of solids in the hopper. The Lovegrove–Williams analysis introduced gravity forces as the mechanism whereby pressure build up is initiated. A one-dimensional analysis led to analytic results, but a two-dimensional version accounting for cross-channel stress variations led to a set of hyperbolic equations which were solved by the method of characteristics.

3.3. Melting

The melting zone is of great importance not only because of the function indicated by its name, but also because significant pressure generation often occurs there. It has therefore been the subject of a substantial amount of work. The complexity of the process, which combines most of the difficulties of solids conveying and melt flow analyses, together with the phase change phenomenon, has led to a piecemeal approach in which flow and heat transfer in various melt films and the solid bed are analysed separately and then coupled into an overall model. Although the various elements of the analysis may be analytic, the complete model is generally of sufficient complexity to require implementation in a computer program. The evolution of melting models from Tadmor's[79] original fully analytic approach of 1966 through to the sophisticated numerical and analytic models of the late seventies is traced in detail by Fenner.[53] Virtually all of these are based upon the melting mechanism described by Maddock[80] (see Fig. 2), although Lindt[81] has modelled an alternative mechanism observed in larger diameter (90 mm) screws. The use of numerical solutions as components of the models occurred early on, when Tadmor, Duvdevani and Klein[82] allowed non-Newtonian, non-isothermal flow in the melt film adjacent to the barrel. Subsequently numerical solutions were used for the other films. In the more elaborate models two-dimensional analyses are used, allowing for cross-channel varia-

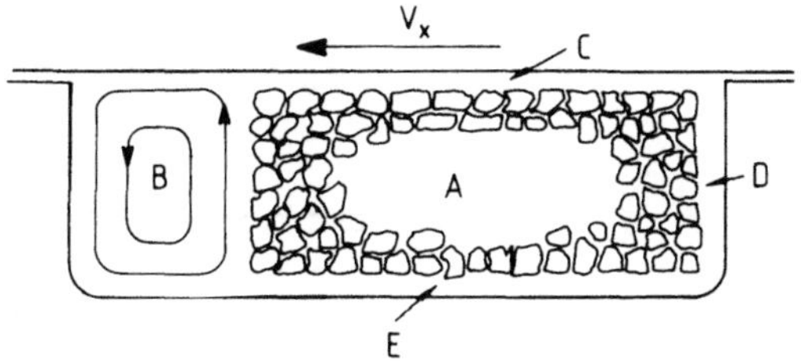

FIG. 2. Diagrammatic representation of melting in an extruder channel as described by Maddock. V_x represents the relative cross-channel motion of the barrel surface. Material melted from the solid bed *A* is swept across the channel in the melt film *C* into the melt pool *B*. Melt films also exist at the screw surfaces *E* and *D*.

TABLE 3

SUMMARY OF MELTING MODELS IN SINGLE-SCREW EXTRUSION

Reference	Melting mechanism		Solid bed analysis		Melt flow analysis					
	Location of solid bed[a]	Molten film formation[b]	Mechanical behaviour[c]	Heat transfer mechanism[f]	Fluid[g]		Kinematics[h]		Dynamics[i]	
					Film	Pool	Lower film	Pool	Film	Pool
Tadmor[79]	TF	B	R	CD	N	—	—	—	—	—
Tadmor *et al.*[82]	TF	B	R	CD	PL(T)	—	—	—	—	—
Chung[86]	TF	B	R	CD	PL(T)	—	—	—	—	—
Hinrichs and Lilleleht[87]	TF	B	R	R	PL(T)	—	1D	—	—	—
Donovan[88]	TF	B	R[d]	CD, CV	PL(T)	—	—	—	—	—
Mondvai *et al.*[89]	TF	B, S	R	CD	N	—	1D	—	—	—
Edmondson and Fenner[90,91]	TF	B, S	D	CD, CV	PL(T)	N	1D	2D	—	P
Lindt[92]	CTR	B, S	R	CD	PL(T)	NA[j]	1D	NA	P	NA
Shapiro *et al.*[93] and Halmos *et al.*[94]	TF	B, A	D	MX	PL(T)	N	1D	2D	—	P
Fukasi *et al.*[95]	TF	B	R[e]	CD, CV	PL(T)	PL(T)	—	1D	—	P
Elbirli *et al.*[96]	TF	B, S	R/D	CD, CV	PL(T)	PL(T)	1D	2D	P	P
Elbirli *et al.*[97]	TF	B, S	R/D	CD, CV	PL(T)	PL(T)	2D	2D	P	P
Lindt and Elbirli[98]	TF	B, S	R/D	CD, CV	PL(T)	PL(T)	2D[k]	2D	P[k]	P

Developed from Ref. 78.

[a] TF, bed at trailing flight; CTR, bed in channel centre.

[b] B, film assumed to exist at barrel surface; S, film assumed to exist at screw surface.

[c] R, perfectly rigid; D, freely deformable.

[d] Uniform acceleration allowed for.

[e] Rigid in constant depth channel, otherwise acceleration proportional to taper.

[f] CD, radial conduction; CV, down-channel convection; MX, 'mixing-cup' mean (radial) temperature.

[g] N, Newtonian; PL(T), temperature-dependent power-law or similar.

[h] 1D, one-dimensional flow; 2D, two-dimensional flow.

[i] P, pressure gradients included in flow equations.

[j] NA, not applicable.

[k] Flow equations in films adjacent to screw and barrel extended to include down-channel flow components, and to allow circulation of melt between these two films.

tions. Heat transfer into the solid bed has also been treated numerically. A useful recent summary of existing models is given by Lindt,[83] together with an assessment of their success and the importance of their various features. A summary of the main features of published melting models is shown in Table 3. Dynamic melting models for reciprocating-screw injection-moulding machines have also been described.[84,85]

3.4. Overall Models of Plasticating Extrusion

The first account of the coupling of models for the various functional zones of the plasticating extruder, into an overall computerised model, was published by Klein *et al.*[99–102], working for Western Electric, New Jersey, in 1966. For solids conveying, a modified form of the Darnell and Mol[76] approach was used; for melting, the model proposed by Tadmor, Duvdevani and Klein;[82] and for melt pumping, analytic non-Newtonian parallel-plate results for drag and pressure flow were superposed and empirical corrections applied. Generally successful comparison with experiment was described.

Since that time many other plastics companies have developed models of varying degrees of sophistication for their own use, and some have been made available commercially. However, there seems to have been little description or discussion of these in the literature. Klein[103] mentions a commercially available package, EXTRUD, and uses it to simulate the effect of screw wear on the performance of plasticating extruders.

A fully predictive six-section model has recently been developed by Agur and Vlachopoulos.[104] Pressure distribution in the hopper is first analysed to give the inlet pressure required to initiate pressure build up according to the Darnell and Mol[76] model which is used for conveying. Tadmor's[79] model is used for melting, and Yates's[59] for melt pumping. Flow through the breaker plate, adapter and die are also treated using a numerical, developing flow method, and die swell is estimated using computed shear stress together with the theory by Tanner. A limited comparison with experiment is given.

A dynamic model of a plasticating extruder has been described by Tadmor *et al.*[105]

4. TWIN-SCREW EXTRUSION

Very few numerical analyses of twin-screw extrusion have appeared. Denson and Hwang[106] used linear triangle finite elements to calculate down-channel flow rates in a self-wiping co-rotating twin-screw extruder. Isothermal, creeping, Newtonian flow was assumed and shape factors were obtained by comparing drag and pressure flows separately to the results of simple one-dimensional, parallel-plate extruder theory. In the study of heat transfer in recirculating channel flow by Pittman and Rashid,[73] the dish-shaped channels such as occur in this type of extruder were included.

A firm basis for the analysis of counter-rotating, intermeshing, twin screws has been provided by Janssen.[107] Prediction of the operating characteristic requires the analysis of a number of leakage flows. These were treated separately using Newtonian isothermal flow assumptions and the (analytic) results coupled in an iterative computer programme. Lai-Fook[108] extended Janssen's work by allowing non-Newtonian (power-law) flow. Finite difference solutions of the type used by Dyer[63] and Fenner[58] were used for flow and pressure fields in the C-shaped chamber, and a finite element method for leakage through the tetrahedron gap. The approach here followed that used earlier by Fenner and Nadiri[48] in the analysis of cross-head die flow. Riedler[109] has also used finite elements to analyse flow in the C-chamber.

5. INJECTION MOULDING

The commercial importance of the injection moulding process has provided the incentive for a substantial amount of work on computer methods for runner design, simulation of the injection, packing and cooling stages and for mechanical aspects of mould design. Mould cooling is treated separately in Section 6. In the following paragraphs the various approaches to cavity filling analysis are first described. These all rely on the assumption that the part thickness is small relative to its other dimensions, and that this thickness varies only gradually along the flow path (see Fig. 3). When the flow can be regarded as parallel, as in a fan-gated rectangular mould or radial, as in a centre-gated disc, two-space-dimensional analyses are possible. Generally these have used numerical solutions of the governing equations for momentum and energy conservation, but initially a method is outlined which uses asymptotically valid analytic solutions. Two dimensional numerical analyses are then described, followed by three-dimensional methods applicable to more complex cavity shapes. Some runner-system design procedures are next reviewed, and finally some commercially available mould design packages are outlined.

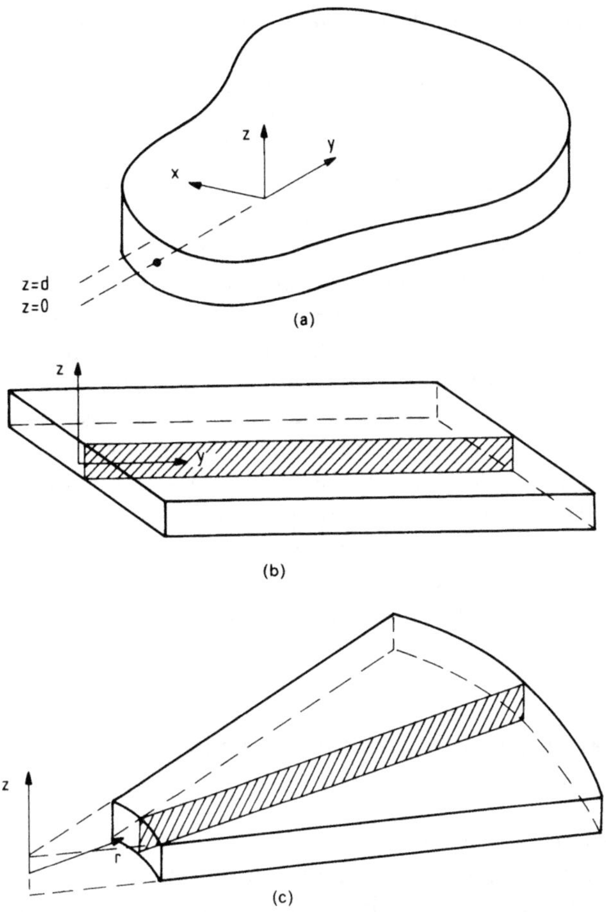

FIG. 3. Geometry and coordinate systems used in modelling of injection mould cavities. (a) A thin cavity of arbitrary shape. (b) A rectangular cavity. For parallel flow, analysis is in the z–y plane, shown hatched. (c) A disc cavity. For radial flow, analysis is in the r–z plane, shown hatched.

Computer simulation of injection moulding has been reviewed by Kamal and Lafleur,[110] who concentrate more on the physics of the process than the mathematical techniques. A more recent article by Richardson,[3] however, pays careful attention to details of the numerical techniques. Grmela[111] and Kamal and Tan[112] have also reviewed mathematical models of injection moulding.

5.1. Computerised Two-dimensional Analysis Using Selected Analytic Solutions

Richardson, Pearson and Pearson[113] have proposed a scheme in which the distribution (sprue–runner) system and mould cavity is approximated by simple geometrical units, namely a circular pipe, rectangular channel or axisymmetric disc. Six analytic solutions are listed for flow and heat transfer, which are asymptotically valid for various combinations of Graetz, Nahme and Brinkman numbers. These are all based on the lubrication approximation, and neglect thermal conduction in the flow direction. A computer program was written to put the scheme into operation; interaction with the user is required to adjust flow in order to satisfy pressure drop requirements. The procedure is described as having been useful in design of mould systems for rubber, but no quantitative examples or comparison are given.

Kuo[114] used a similar approach, but based on simpler analytic solutions. The idea of approximating actual mould flows by simpler two-space-dimensional representations, such as parallel or radial flow, has been called the coupled flow path method. Its implementation with numerical solutions of the flow and heat transfer equations is discussed below, in Section 5.4.

5.2. Rectangular or Disc Moulds: Two-Dimensional Analyses in the z–y or z–r Plane

5.2.1. Summary of the General Approach

A typical calculation scheme applied to parallel flow in a rectangular mould is first outlined. When the width of the flow path in the x-direction is constant, and the thickness in the z-direction varies only slowly with y (see Fig. 3), the lubrication approximation allows the momentum conservation equation to be reduced to

$$\frac{\partial P}{\partial y} = \frac{\partial}{\partial z}\left(\mu \frac{\partial v_y}{\partial z}\right) \tag{4}$$

Boundary conditions are

$$v_y = 0, \qquad z = d \tag{5}$$

$$\frac{\partial v_y}{\partial z} = 0, \qquad z = 0 \tag{6}$$

where d is half the mould thickness. Integrating once at constant y,

$$\frac{\partial P}{\partial y} z = \mu \frac{\partial v_y}{\partial z} \tag{7}$$

If μ is expressed by, for example, the temperature-dependent power-law

$$\mu = \mu_0(T)\left(\frac{\partial v_y}{\partial z}\right)^{n-1} \tag{8}$$

then eqn (4) can be substituted from eqn (8) and integrated to give

$$\frac{v_y}{\left(\frac{\partial P}{\partial y}\right)^{1/n}} = \int_z^d \left(\frac{z'}{\mu_0}\right)^{1/n} \mathrm{d}z' = \phi'(z) \tag{9a}$$

In the non-isothermal situation μ_0 will be a function of z and the integral can be obtained by quadrature if the temperature profile at the relevant y position is known.

Continuity requires

$$Q = \int_0^d v_y \, \mathrm{d}z \tag{10}$$

where Q is the instantaneous volumetric flow rate and, from eqn (9a),

$$Q = -\left(\frac{\partial P}{\partial y}\right)^{1/n} \int_0^d \phi'(z) \, \mathrm{d}z \tag{10a}$$

Again the integral is obtained by quadrature and hence, for a given Q, the local pressure gradient $\partial P/\partial y$ is obtained.

When viscosity is given by a relationship which cannot be solved for $\partial v_y/\partial z$ we write, instead of eqn (9a),

$$\frac{v_y}{\partial P/\partial y} = \int_z^d \frac{z'}{\mu} \mathrm{d}z = \phi(z) \tag{9b}$$

and an estimate of the shear-rate profile as well as the temperature profile is required for evaluation of the integral. Then, as before by continuity,

$$\frac{\partial P}{\partial y} \int_0^d \phi \, \mathrm{d}z = Q \tag{11}$$

and an estimate of $\partial P/\partial y$ is obtained. Using this, return to eqn (9b) and obtain v_y, giving a revised shear-rate profile. Re-evaluate the integral in eqn (9b) and iterate to convergence.

In the energy conservation equation, conduction in the flow direction is neglected and, consistent with the incompressible lubrication flow, we have

$$\rho C\left(\frac{\partial T}{\partial t}+v_y\frac{\partial T}{\partial y}\right)=k\frac{\partial^2 T}{\partial z^2}+\mu\left(\frac{\partial v_y}{\partial z}\right)^2 \tag{12}$$

with initial condition

$$t=0, \qquad y=0, \qquad T=T_{\text{in}}(z) \tag{13}$$

where T_{in} is a known inlet temperature; and boundary conditions

$$\begin{aligned} t\geqslant 0, \qquad z=0, \qquad &\frac{\partial T}{\partial t}=0 \\ z=d, \qquad &T=T_{\text{w}}(y,t) \end{aligned} \tag{14}$$

where T_{w} is a given wall temperature. Alternatively a heat flow condition may be used, expressed in terms of a heat transfer coefficient h and the coolant temperature T_{c}

$$t\geqslant 0, \qquad z=d, \qquad -k\frac{\partial T}{\partial z}=h(T-T_{\text{c}}) \tag{15}$$

Equation (12) can be stepped forwards in time using various finite difference methods. In most works a simple, forwards difference is used for the time derivative, giving an explicit method in terms of the velocity and temperature field at the start of the time step.

The solution procedure in outline is then as follows. For a known temperature field the velocity profile at each y position is evaluated as described above, based on a given flow-rate. In the early stages of filling, this will be the maximum delivery rate of the ram. The pressure gradient is integrated along the flow path to get the total pressure drop over the mould. If this is found to exceed a set limit then the flow-rate is reduced iteratively until the limiting pressure drop is achieved. The resulting velocities, shear rates and local viscosities are then used in the energy equation to advance the temperature at each mesh point over one time step. The flow front positon is also moved forwards, and the mesh extended appropriately in the flow direction. Temperatures have to be assigned to these new mesh points and, for example, the mean temperature immediately up-stream may be used, as an approximation to the effects of the 'fountain flow' at the melt front, which carries material from the core to the mould walls. With the new temperature field, the velocity solution is now repeated. When initial estimates for shear rate profiles across the mould are required (in eqn (9b)), these can be taken from the previous time step. As filling proceeds the melt front generally progresses less fast, and unequal mesh spacing in the y-direction results. Unequal spacing across the z-direction may also be used to provide better resolution of steep gradients near the mould walls.

5.2.2. Detailed Implementations

The various works in the literature either simplify or elaborate upon the physical model outlined here, and also differ in the details of their numerical methods.

The earliest numerical simulation of mould filling seems to be that by Harry and Parrott.[115] They considered that the 'simulation axis moves with the melt front' and omitted the convection term $v_y(\partial T/\partial y)$ from eqn (12). The problem then becomes dependent on z and t only: it is visualised that conditions near the melt front are being simulated. Pressure gradient was taken simply as delivery pressure divided by filled length, and viscosity was represented by a temperature-dependent power-law expression. In a limited comparison with experiment, a short shot was successfully predicted. Berger and Gogos[116] modelled the filling of a centre-gated disc mould, including thermal convection in the flow direction and following the approach suggested by Pearson.[5b] The momentum equation, however, was simplified by neglecting hoop stresses, and viscosity was correspondingly expressed by a power-law dependence on $\partial v_r/\partial z$ only. Hoop stress was included, though, in the viscous heat term of the energy equation, and this equation was stepped forward using the Crank–Nicholson method. When the temperature near the wall fell below the solidification point, the nominal mould half thickness d was reduced. Further work on disc moulds by Kenig and Kamal is available.[117–119]

Stevenson[120] later used essentially the same approach as Berger and Gogos, though this earlier work was not referred to. Design charts for injection pressure were produced[121] in terms of the four dimensionless parameters: power-law index, Nahme number, reduced time, and reduced melt inlet–mould wall temperature difference. Guidelines were provided for application of the charts to injection of semi-crystalline materials, and less-simple mould geometries. In a frequently cited work, Lord and Williams[122] analysed rectangular mould filling, using eqns (4)–(15) above. As with previous works, density and thermal conductivity of the melt were taken as contant; however, specific heat varied with temperature. Latent heat is not mentioned though, and the effect of solidification was achieved simply by melt viscosity near the mould wall becoming very high. Viscosity was expressed either using the temperature-dependent WLF equation or by a logarithmic polynomial. The energy equation was treated using central difference for the conduction term, upstream (backwards) difference for convection and forward difference for the time derivative. Iterations over the time step were also tried, but it was concluded that the procedure was not required. The simulation was successfully used to indicate the solution to a degradation problem, and predicted injection pressures were in good agreement with experimental measurement.

In further work based on the Lord and Williams approach, Titomanlio[123] established a criterion, by means of numerical experiments, for negligible viscous heating effect during mould filling, in which case the filling process is characterised by three dimensionless groups. A means is proposed for combining these to give a master curve for a normalised injection pressure. Thienel and Menges,[124] Wang,[125] and Echenagucia, Fricke and Hassler[126] have also used methods similar to those of Williams and Lord.[122] Titomanlio, Acierno and La Mantia[127] have extended the Lord and Williams model to the packing stage by allowing density to vary with temperature and pressure.

Isayev and Hariharan[128] have modelled shrinkage of injection moulded parts using the equation of state for specific volume, which was assumed to respond instantaneously to pressure and temperature changes, together with a two-dimensional simulation of mould filling and a one-dimensional simulation of subsequent melt cooling.

The lubrication assumption applied in the foregoing works precludes treatment of the 'fountain flow' at the melt front; however, this has been simulated for the isothermal, Newtonian case by Bigg[129] using the marker and cell method. Flow into a rectangular and an L-shaped cavity was considered.

In none of the works so far described have latent heat effects and crystallisation been taken into account. Recently Kamal and Lafleur[130] have linked a non-isothermal crystallisation kinetics model to the heat transfer problem in mould filling, using the methods described in Section 6. The flow problem was solved by a version of the marker and cell method, using the White–Metzner modification of the Maxwell model as the constitutive equation for melt flow. No detail of this implementation is provided. Birefringence of the amorphous phase is calculated from computed normal stresses, using the stress optical law, and thence, knowing crystallinity and orientation functions, the Young's modulus of the material is predicted. Results were within about 30–50% of experimental values.

5.3. Thin Cavities of More Complex Shape: Analyses Based on Hele–Shaw Flow

5.3.1. The Hele–Shaw Approximation

Extending the analysis to include the two velocitiy components v_x, v_y (Fig. 3), the lubrication approximation leads to

$$\frac{\partial P}{\partial x} = \frac{\partial}{\partial z}\left(\mu \frac{\partial v_y}{\partial z}\right) \tag{16}$$

$$\frac{\partial P}{\partial y} = \frac{\partial}{\partial z}\left(\mu \frac{\partial v_y}{\partial z}\right) \tag{17}$$

with boundary conditions

$$\begin{aligned} v_x = v_y = 0, \qquad & z = d \\ \frac{\partial v_x}{\partial z} = \frac{\partial v_y}{\partial z} = 0, \qquad & z = 0 \end{aligned} \tag{18}$$

Integrating eqns (16) and (17)

$$v_x = \frac{\partial P}{\partial x}\phi, \qquad v_y = \frac{\partial P}{\partial y}\phi \tag{19}$$

where ϕ is as in eqn (9b). In expressing the shear dependence of viscosity one needs

$$\dot{\gamma} = \left[\left(\frac{\partial v_y}{\partial z}\right)^2 + \left(\frac{\partial v_x}{\partial z}\right)^2\right]^{1/2} \tag{20}$$

and from eqn (9)

$$\dot{\gamma} = \left[\left(\frac{\partial P}{\partial x}\right)^2 + \left(\frac{\partial P}{\partial y}\right)^2\right]^{1/2} \frac{z}{\mu} \tag{21}$$

For incompressible flow

$$\frac{\partial v_x}{\partial x} + \frac{\partial v_y}{\partial y} = 0 \tag{22}$$

and hence from eqn (19)

$$\frac{\partial^2 P}{\partial x^2} + \frac{\partial^2 P}{\partial y^2} + \frac{\partial P}{\partial x}\frac{\partial \phi}{\partial x} + \frac{\partial P}{\partial y}\frac{\partial \phi}{\partial y} = 0 \tag{23}$$

If it is assumed that ϕ varies only slowly with x and y when the Laplace equation for pressure results,

$$\frac{\partial^2 P}{\partial x^2} + \frac{\partial^2 P}{\partial y^2} = 0 \tag{24}$$

A stream function ψ can be defined:

$$v_x = -\frac{\partial \psi}{\partial y}\phi G(t), \qquad v_y = \frac{\partial \psi}{\partial x}\phi G(t) \tag{25}$$

where $G(t)$ is a factor accounting for the time dependence of flow. From Eqns (16) and (17), and again neglecting gradients of ϕ,

$$\frac{\partial^2 \psi}{\partial x^2} + \frac{\partial^2 \psi}{\partial y^2} = 0 \tag{26}$$

Under the assumptions mentioned, flow in the cavity is thus a Hele–Shaw flow, governed by the potential eqns (24) and (26), and amenable to solution by conformal mapping techniques. It should be noted that it is not possible to apply the no-slip condition at the edges of the cavity. Streamlines and isobars are orthogonal over the x–y domain, and since the cavity edges coincide with a streamline, the isobars are perpendicular to the impermeable edges.

The energy equation is treated by neglecting conduction in the x and y direction

$$\rho C\left(\frac{\partial T}{\partial t} + v_y\frac{\partial T}{\partial y} + v_x\frac{\partial T}{\partial x}\right) = k\frac{\partial^2 T}{\partial z^2} + \mu\dot{\gamma}^2 \tag{27}$$

Then, provided that the inlet melt temperature is constant, isotherms in an x–y plane are parallel to streamlines. Again, note that the neglect of second derivatives with respect to x and y means that boundary conditions cannot be applied at the mould edges, or melt front.

5.3.2. Conformal Mapping Methods

The conformal mapping method has been used by Richardson,[131] Kuo and Kamal,[132,133] and Ryan and Chung.[134] Richardson used analytic methods to examine unbounded isothermal Newtonian Hele–Shaw flows. Kuo and Kamal and Ryan and Chung modelled symmetrical filling of an end-gated rectangular mould with shear rate- and temperature-dependent viscosity. It was necessary to solve the energy equation only on the centre-line, and temperatures elsewhere were then obtained by mapping. The first authors[132] dropped the convection term and were thereby able to use a semi-analytical solution. The work was also extended to the packing stage.[133] Ryan and Chung[134] retained convection, and stepped the temperature forward in time using an explicit finite difference scheme. Results from these two analyses were in quite close agreement, and compared well with experimental observations.

5.3.3. Direct Numerical Methods

Although the conformal mapping technique can be extended to cavities with curved walls by the choice of an appropriate coordinate system, the treatment of less simply shaped cavities and those containing inserts requires numerical solution of the potential equations. Broyer, Gutfinger and Tadmore[135] applied the flow analysis network method (see also Ref. 47) for isothermal non-Newtonian filling of thin cavities. Although described as a finite element method, this is, in effect, a finite difference formulation on a rectangular grid. Special techniques are described for dealing with grid squares crossed by the cavity edges, or melt front. Simulations of the filling of complex moulds, including weld-line location, compared well with experiment; however, experimental conditions were such as to minimise thermal effects. The most elaborate of the Hele–Shaw models is that developed by Hieber and Shen[136,137] in the injection moulding project at Cornell University, NY. The governing equations are eqns (16)–(21) above, with continuity written in terms of mean flow velocities over the gap

$$\frac{\partial}{\partial x}(d\bar{v}_x) + \frac{\partial}{\partial y}(d\bar{v}_y) = 0 \tag{28}$$

The pressure field equation becomes

$$\frac{\partial}{\partial x}\left(S\frac{\partial P}{\partial x}\right) + \frac{\partial}{\partial y}\left(S\frac{\partial P}{\partial y}\right) = 0 \tag{29}$$

where

$$S = \int_0^d \frac{z^2}{\mu} \mathrm{d}z \qquad (30)$$

A temperature-dependent power-law expression is used for μ. Note that the x–y dependence of S is not neglected here, and the formulation can thus be described as a generalised, rather than a classical Hele–Shaw flow. Following earlier work using the marker and cell method,[137] current work uses finite elements to solve for the pressure field, with straight-sided triangular elements over the filled portion of the x–y domain, and quadratic interpolation for pressure. The resulting equations are non-linear because of the dependence of μ in eqn (30) upon shear rate $\dot{\gamma}$, and temperature. They are accordingly solved by over-relaxation with periodic updating of the local value of S. The iterative solution scheme makes it easier (as compared with conventional elimination methods) to handle the irregular node and element numbering pattern which develops as the flow front advances.

The form of the energy equation is, as before, eqn (27). Temperatures as a function of z and t are obtained by a finite difference solution at the x–y points corresponding to the vertices of the finite element triangles, and then interpolated linearly with respect to x and y. A forward difference was used for the time derivative, giving an explicit scheme. Central differences were used for the conduction term, but for the convection term a form of 'up-winding' was found necessary for stability. The terms $v_x(\partial T/\partial x)$ and $v_y(\partial T/\partial y)$ were evaluated at the centroids of the triangles surrounding the vertex node in question. Values were averaged in proportion to the angles subtended at the node, but omitting the contributions from elements lying on the downstream side. A further complication was the need for pressure gradients at the vertices to evaluate the viscous heating term. The gradients are discontinuous across element boundaries, and values were averaged over all elements surrounding the node, in proportion to angle subtended. Thermal properties were taken constant and latent heat effects were not included. The solution scheme, briefly, was then as follows. For a known temperature field eqn (29) is solved iteratively, accounting for the shear dependence of S. The mean velocities at nodes on the flow front are found from

$$\bar{v}_x = \frac{\partial P}{\partial x}\frac{S}{d} \qquad \bar{v}_y = \frac{\partial P}{\partial y}\frac{S}{d} \qquad (31)$$

The melt front is then advanced over the chosen time-step, and the finite element mesh extended with additional elements to cover the newly filled area. The temperatures assigned at the new nodes are not commented upon, and it is not clear whether the mesh extension is carried out automatically or manually. To avoid oscillatory melt fronts, it was found preferable to use a predictor–corrector scheme in which the flow field is re-solved at the end of the time-step, and the front then re-positioned using the mean velocity over the step. If a portion of the new front lies outside the cavity edge, the mass crossing the edge is re-positioned manually within the domain. Required computing times were not quoted.

In comparisons with experiments, generally good agreement was found for front position, pressure traces and weld-line location in cavities of uniform[136] or gradually varying[138,139] thickness and including inserts.

5.4. The Coupled-Flow-Path Method

Analyses of the type just described require substantial amounts of computing, and are difficult to make entirely automatic as regards repositioning of the flow front. A simpler approach, which Hieber[140] has called the coupled-flow-path method, approximates the filling of a complex (thin) cavity using analyses which are only two-space-dimensional. The cavity is broken down into regions in which the flow can be regarded, approximately, as either parallel flow between flat plates, $v_y(y, z, t)$, or radial flow, $v_r(r, z, t)$. The flow and energy equations are then solved as a series of coupled two-dimensional problems. The precise way in which the filling is approximated by these simpler flows depends on the judgement of the analyst.

Figure 4 shows two possible ways of representing a two-gated rectangular mould. In this instance the position of the weld line must be estimated *a priori*, based on flow-rates through the two runners. To get the best estimate of filling times, it is important that the representation of the mould should have the correct volume and it can be seen in Fig. 4 how this is achieved. When flow branches, as on encountering an insert, iteration is required to find the flow-rate into each branch—the requirement being, of course, equal pressure drop from the branch point to the flow front in both arms.

Hieber[140] describes applications of the coupled-flow-path method to a two-gated mould such as that shown in Fig. 4, and found that the way the mould was subdivided was not critical. He also applied it to a

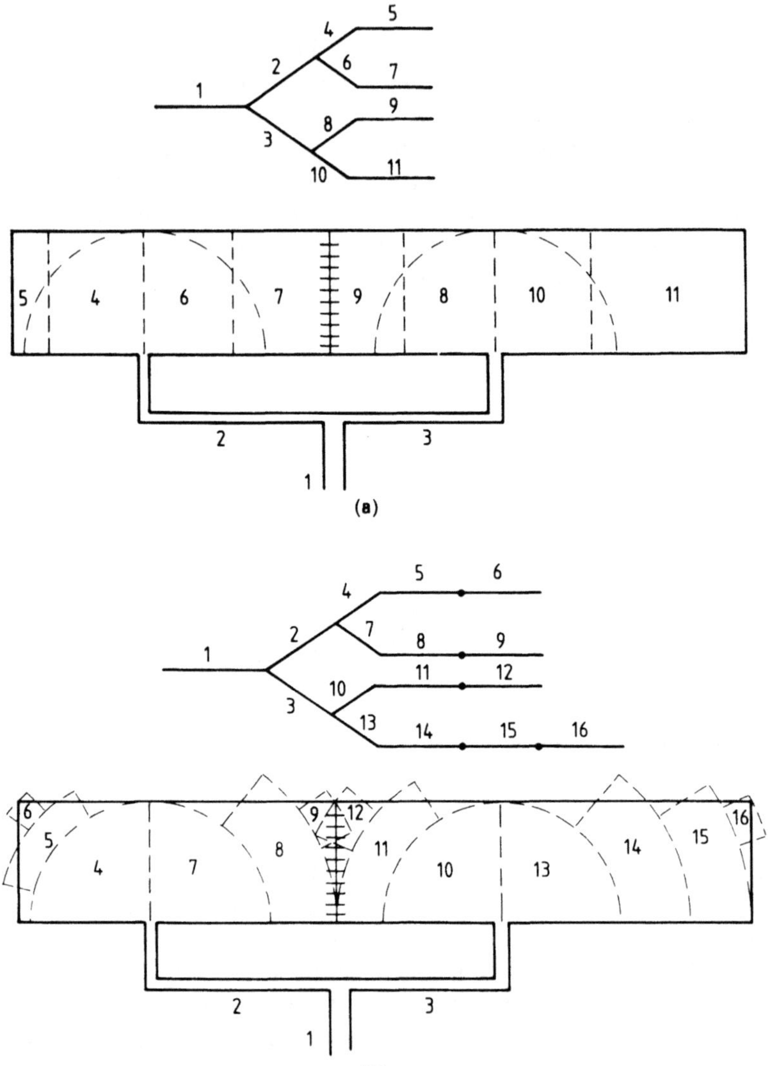

FIG. 4. Coupled flow path method: representation of flows in a twin-gated rectangular mould by a number of radial or parallel flows. The desired weld line position is initially fixed as indicated, ++++. Two possible representations are: (a) From the gates, flow expands radially to the indicated arc. Mould filling is completed by parallel flows. The coupling of flow paths is indicated in the tree diagrams. (b) Mould filling is represented throughout by a series of radial flows.

more complicated automobile door trim panel moulding, and in both cases obtained good agreement with the more elaborate Hele–Shaw analysis[136] described previously. No details of the two-dimensional numerical solution techniques were given.

The coupled-flow-path method had earlier been applied in the original version of the commercial Moldflow simulation package and in a number of other commercial packages.

5.5. Nozzle, Sprue and Runner Systems

The transient flow and temperature fields in an injection moulder nozzle have been analysed by Nunn and Fenner,[141] using the axisymmetric versions of eqns (4)–(9a), and eqn (12). The energy equation was solved by an implicit finite difference scheme. Results are given for a typical nozzle, where the Graetz number is 3×10^3, the Péclet number is 2×10^4 and the Nahme number is 16. The mean temperature rise was found to be about 10°C, though local values were as high as 100°C. Flow and temperature fields became effectively steady after a few milliseconds.

Start-up flow in a circular duct was also analysed by Williams and Lord[142] as a preliminary to their work on runner design. It was found that conditions at the exit from a 6 in × 0·4 in diameter sprue carrying PVC at 4 in^3/s were essentially steady after 0·5 s. On this basis a steady-state analysis was proposed[142] for sprue and runner design. Again the axisymmetric version of eqns (4)–(7), eqn (9b) and eqns (10)–(12) were used. The temperature boundary condition at the duct wall was expressed in terms of a heat transfer coefficient, simulating penetration of heat into the metal. The solution proceeded by stepping the energy equation downstream using an implicit formula near the wall to avoid undue limitation of the step size in this low velocity region, and an explicit method in the core. Iteration over the z-step was carried out to find the downstream velocity profile required in the implicit formula. Latent heat effects were not included and 'solidification' occurred when the melt temperature near the wall was such that viscosity was high enough virtually to prevent flow. Results were compared with experiments on PVC and ABS in capillaries and runners, and pressure drops agreed within about 15%. The analysis was successfully used to balance runners in a six cavity mould.

As a further part of the Cornell project mentioned above, Wang, Wang and Hieber[143] have described an iterative computer program for runner-system design. This is based on isothermal power-law flow in circular runners—non-circular runners are represented by equivalent circular ones. Relationships between runner lengths and diameters are set up such that melt reaches all gates simultaneously. Then gates are sized such that the cavities all fill at the same time. The resulting design forms a first approximation, which can be improved using the coupled-flow-path method,[140] based on steady non-isothermal solutions in cylindrical ducts. This work is backed up by a number of numerical and experimental studies of elastic effects in abrupt expansions and contractions,[144,145] in short tubes,[146] at junctions,[147] and in transient flows.[148] Non-isothermal viscous flow in tubes of non-circular section has also been studied[149] (see also Section 2.1).

A scheme for runner-system design using asymptotically valid analytic solutions is included in the work of Richardson, Pearson and Pearson[113] (see Section 5.1).

5.6. Commercial Packages for Mould Design

Injection mould design is the area where commercial computer-aided design packages are most firmly established in the plastics industry. Nevertheless their use is still quite limited. A recent summary of presently available packages is provided by Naitove,[150,151] and other essays on CAD for injection moulders are also of interest.[152–155] Various types of program–data bank packages are available; aspects covered include mould cooling and design of cooling systems; cavity filling simulation and runner-system design; and part selection and mechanical design.

Mould cooling analysis will be discussed separately in Section 6. Of the cavity and runner design packages, the best known is Moldflow, introduced into Britain in 1980. Details of the calculation methods used are not available in the open literature, but the general approach can be outlined. The design principles implemented within the package are described in the Moldflow Philosophy,[156] and some hints on the calculation methods are provided by Austin,[157] who makes the point that the temperature field in cavity filling rapidly reaches steady state, except near the flow front, implying the use of steady-state solutions in at least some Moldflow analyses.

There were recently about 40 programs available from Moldflow, some obsolescent, but retained for users familiar with them. Typically six or seven programs might be used in an analysis. This could start with the programme MCHECK to give a rapid assessment of filling behaviour. The input data includes flow path length, thickness, plastic type, mould and melt inlet temperatures. Output provides, for a range of filling times, injection pressure, shear stress at the gate and flow front, temperature at the front, and coooling time to the softening point. From this an initial assessment of plastic type and moulding conditions will be made.

A more detailed analysis of the filling process is next provided by programme MF. The cavity is represented by a layflat, which is divided into a number of submoulds, by choosing acceptable positions for weld-lines. Each submould is treated by the coupled-flow-path method, representing it as a series of parallel or radial flow regions. Clearly, each submould must fill at the same instant, and simulations are run to determine the injection pressure required at each gate to achieve this. At this stage iterative adjustment of weld-line position may be required. Next, submoulds are replaced by nominal cavities having equivalent flow resistances, and the runner system designed using the coupled-flow-path method based on flow in circular ducts, so that the pressure drops over every path, plus the required drops over the cavities, are equal. This can be done for conventional or heated runners. Moldflow[156] assert that gates provide poor flow control, and prefer to control flows to the cavities by using appropriate runner diameters. Further detail on the filling process is provided by the programme MP, which gives a progressive account of temperature, shear-rate, clamp-force, etc. during the injection.

The Rubber and Plastics Research Asssociation, UK, has carried out an assessment of Moldflow, and a report[158] is available to members. Assessments are also available in the literature. Cox, Mentzer and Custer[159] compared simulations with measurements for a relatively simple bar moulding with polypropylene; predicted injection pressures were 3–5 MPa low, and with a glass-filled nylon about 4 MPa high, but these discrepancies were scarcely significant relative to experimental accuracy. Mohnberg and Bach[160] made comparisons for a spiral mould. Injection pressures agreed within 5–25%, though the larger discrepancy was probably due to inadequacies of some of their flow property data.

More recently Moldflow have introduced a finite element simulation. Only a sketchy description of how this works is available.[157,161] The three-dimensional cavity is represented by a mesh of triangular elements. A flow field is assumed for the instant of filling and the corresponding pressure field calculated. Isobars are displayed as an approximation to successive flow front positions. The flow field is adjusted iteratively until the 'true' filling pattern is obtained; temperature distribution is also obtained.

Among other packages, Buckinski[162] describes briefly a design package including a coupled-flow-path simulation and also computer aids for mechanical design of the mould. Singh[163] mentions a runner design mould-filling system, without giving any detail.

One of the largest programmes of work on development of CAD systems for moulds is in progress at the Institut für Kunststoffverarbeitung, Aachen. The objective approaches the creation of an expert system for mould design. It includes computer-aided plastic material selection[164]—using a property data bank; economic assessment including optimisation of number of mould cavities;[165] cavity filling simulation and runner design;[124,166,167] mould cooling system design;[166] stress–strain analysis for plastic parts and moulds;[164] and computer-aided part selection and mechanical design.[165] An overview of the IKV objectives in this area is provided by Menges.[154] The foregoing references provide little detail of the various calculation procedures, for which the reader is referred to numerous reports and dissertations of the IKV. In 1983 some 20 companies were collaborating with IKV in testing parts of the system.

A substantial programme on CAD applied to moulding is in progress at Cornell University, Ithaca, New York, in collaboration with a number of companies, and component parts of this have been mentioned above. Outlines of the work have been given by Wang.[155]

Further computer packages are referred to in a recent book,[168] which broadly surveys computer-aided engineering for injection moulding.

6. TRANSIENT COOLING WITH SOLIDIFICATION

The cooling phase of injection-moulding or blow-moulding commonly occupies 60%–80% of the cycle time, and to avoid wasting production capacity it is obviously important to ensure that it continues for no longer than is strictly necessary. A substantial amount of work has therefore been devoted to analysing the transient cooling problem with phase change, in order to predict the minimum time at which an article may safely be ejected from the mould, and to assess the effectiveness of coolant circulation. For blow moulding, the analyses can also indicate the benefits to be had from various internal forced cooling systems. An economic evaluation and description of a number of these is provided by Ryder.[169] Additionally, details of temperature history in the polymer allow inferences or predictions about crystallinity in the moulded article, thermal stresses and susceptibility to warping and shrinkage.

Due to the low thermal diffusivity of plastics, most of the resistance to heat transfer lies within the plastic itself, and it is difficult to see how the approach of Singh and Wang,[170] which appears to take account only of thermal resistances in the metal and to the circulating coolant, can be effective. These resistances must be coupled with a solution of the differential equation for transient conduction in the plastic, including phase change.

6.1. Options for the Mathematical Modelling of Cooling with Phase Change

In a wider context, the cooling and freezing problem has generated an extensive literature. In its classical form it is known as the Stefan problem, stated in one dimension with a discontinuity in the heat flux at a sharp solid–liquid interface. For example, if material is solid (subscript 1) in $x < X(t)$ and liquid (subscript 2) in $x > X(t)$ then

$$T_1 = T_2 = T_f, \quad x = X(t) \tag{32a}$$

$$k_2 \frac{\partial T}{\partial x}\bigg|_{X(t)+} - k_1 \frac{\partial T}{\partial x}\bigg|_{X(t)-} = -\rho_1 L \frac{\mathrm{d}X(t)}{\mathrm{d}t} \tag{32b}$$

where $\rho_1 L$ is the latent heat release on freezing unit volume of the material, and T_f is the freezing temperature. Heat conduction in both phases is governed by

$$\rho_i C_i \frac{\partial T}{\partial t} = k_i \frac{\partial^2 T}{\partial x^2} \qquad i = 1, 2 \tag{33}$$

The analytic solution to this constant property problem in a semi-infinite region with constant initial and boundary conditions is well known[171a] and shows that

$$X(t) \propto t^{1/2} \tag{34}$$

For less simple boundary conditions and cases where properties are temperature dependent, numerical solution techniques must be applied. For a review, see Ref. 172. Treatment of the problem in the form corresponding to that just described involves the difficulty of tracking the interface $X(t)$ and including the appropriate latent heat release. An alternative formulation is in terms of enthalpy, H.

$$\rho_i \frac{\partial H}{\partial t} = \frac{\partial}{\partial x}\left(k_i \frac{\partial T}{\partial x}\right) \tag{35}$$

where

$$H = \int_0^T C_1 \, \mathrm{d}T' \qquad T < T_\mathrm{f} \tag{36}$$

$$H = \int_0^{T_\mathrm{f}} C_1 \, \mathrm{d}T' + L + \int_{T_\mathrm{f}}^T C_2 \, \mathrm{d}T', \qquad T > T_\mathrm{f} \tag{37}$$

The advantages of this approach are that the interface no longer has to be tracked explicitly and it is not necessary to consider solid and liquid regions separately. It is also easier to generalise to more than one space-dimension. The mathematical basis of this weak solution method is examined by Atthey.[173] However, difficulties remain in the numerical solution: spurious plateaux in the cooling curves are likely to occur. The problem is eased if the phase change is allowed to occur over a temperature interval $T_\mathrm{f} \pm \varepsilon$

$$H = \int_0^T C_1 \, \mathrm{d}t \qquad T < T_\mathrm{f} - \varepsilon \tag{38}$$

$$H = \int_0^{T_\mathrm{f}-\varepsilon} C_1 \, \mathrm{d}t + \lambda^* + \int_{T_\mathrm{f}+\varepsilon}^T C_2 \, \mathrm{d}t \qquad T > T_\mathrm{f} + \varepsilon \tag{39}$$

Here λ^* is not a true latent heat since it also includes sensible heat of the liquid and solid over the solidification range, and some additional information must be provided on the rate of heat release in this range. Weak solution methods are reviewed by Voller *et al.*[174] This last formulation corresponds with the physics of the problem most often of interest in polymer processing, namely the solidification of a partially crystalline material over a finite temperature interval, and this is usually sufficiently wide to avoid the spurious cooling plateaux noted by Voller and others. In this application the required information on λ^* can strictly be obtained only by taking into account the kinetics and degree of crystallisation in the material. However, as an approximation, the problem can be formulated in terms of temperature and an apparent specific heat, solving the variable property conduction equation over the whole region containing both melt and solid

$$\frac{\partial}{\partial t}(\rho C T) = \nabla \, . \, \nabla(kT)$$

The specific heat will contain a peak corresponding to latent heat effects, and its form is likely to depend on cooling rate and pressure through their influence on the crystallisation process.

6.2. Analyses Based on the Use of Apparent Specific Heat

In the simplest implementations, the dependence of apparent specific heat on temperature and pressure is neglected. On this basis finite difference solutions in one space dimension have been obtained and applied to polymer processing by Menges *et al.*;[175] Dietz;[176] Singh;[177] Kamal and Kalyon;[178] Gutfinger *et al.*[179] and Ryan and Dutta;[180] using central differences in space and first order explicit time-stepping. Edwards *et al.*[181,182] have used the implicit Crank–Nicholson method, and show comparisons with experimental measurements of surface temperature, including the reheating after ejection of the moulding to ambient air. Dietz[183] modelled cooling and phase change under varying pressure, including pressure-dependence of properties and the expansion cooling effect, using data obtained previously,[184] and compared results with his experimental measurements. Dietz used a first order forward difference in time giving an explicit scheme, and in common with the other works mentioned immediately above, the equations were linearised using physical properties at the start of the time-step. A more accurate treatment, for a given time-step size, can be obtained using the Lees explicit three-level time-stepping scheme, with properties evaluated at the central level; this type was employed by Bonancina *et al.*[185] in their one-space-dimensional finite difference analysis. Pham[186] also used this method, with a procedure for evaluation of effective specific heat, which it was claimed provided the advantages of an enthalpy formulation, within an explicit temperature stepping scheme.

Two-space-dimensional finite difference analyses applied to polymer processing are less common, though the work of Edwards *et al.*[181,182] has been extended to two dimensions.[187] Atthey[173] has described a two-dimensional scheme based on an enthalpy formulation, including an internal heat generation term, and applied it to spot welding of metals. Voller and Cross[174] also used an enthalpy formulation and described a special averaging technique for enthalpy, designed to avoid spurious plateaux in the cooling curve. These ideas were carried further by Tacke.[188]

The geometrical flexibility of finite elment methods, and their ability to interpolate variable physical properties within elements, make them particularly suitable for many practical two-dimensional problems. Comini *et al.*[189] describe a scheme based on the apparent specific heat formulation, but in cases where phase change occurs over a small temperature interval, they prefer to interpolate enthalpy rather than specific heat within elements. Various ways of differentiating enthalpy to get integration point values of specific heat are discussed.[189,190] Time-stepping is by the Lees 3-level scheme as used previously by Bonancina *et al.*[185] Automatic time-step adjustment is included to economise on computing as equilibrium is approached. Lewis *et al.*[191] have recently summarised this work and go on to use it in a sophisticated elastoviscoplastic analysis for thermal stresses. The work is directed principally towards metallurgical applications, but the techniques have wide applicability.

For problems involving mixed media, with widely differing thermal diffusivities, substantial reductions in computation time can be obtained using mixed implicit–explicit time stepping. Samonds, Morgan and Lewis[192] implemented this approach with spatial discretisation by finite elements in an analysis of cooling and solidification of metal within a sand form (mould).

Alternating Direction Implicit Methods have also been used to increase computational efficiency in finite element methods.[193]

In the polymer processing context, Pittman and Nakazawa[194,195] have described a two-dimensional finite element analysis, which uses a linearised Crank–Nicholson time-stepping. In an example, the cooling and solidification of the complicated L-ring section of a blow-moulded barrel was modelled, including conduction to the mould and heat loss to internal blown air, governed by a forced convection heat transfer coefficient.

In none of the work so far described has any account been taken of the influence on apparent specific heat of pressure or cooling rate. Kenig and Kamal[196] approached this problem by shifting their specific heat–temperature curve with respect to temperature, to account for the influence of pressure and cooling rate on the melting point. Thermal conductivity and density were taken as functions of temperature and pressure in a one-dimensional finite difference simulation, which included conduction in the mould metal. In experiments, the melt was cooled in an instrumented cylinder from a known initial temperature and pressure. Predicted temperature and pressure transients were close to the observed ones.

6.3. Analyses Including Crystallisation Kinetics

A limitation of the foregoing approach is that it takes no account of the temperature history of the plastic, upon which its degree of crystallinity depends. Additionally, initial cooling rates in the surface layers of mouldings are often much greater than can be achieved in the differential scanning calorimetry used to obtain apparent specific heat traces. The correct approach, of course, is to model crystallisation kinetics in parallel with the thermal problem. However, the complexity of the physics and the difficulty of obtaining the necessary rate parameters make the approach difficult to implement. Early work was carried out by Sifleet *et al.*[197] The one-dimensional constant property heat conduction equation was written with a heat source term g representing latent heat release upon solidification

$$\rho C \frac{\partial T}{\partial t} = k \frac{\mathrm{d}^2 T}{\mathrm{d}x^2} + g \tag{40}$$

and g was related to crystallisation rate at the given temperature through the Avrami equation for relative crystallinity, θ

$$1 - \theta = e^{-Kt^m} \tag{41}$$

where K is a kinetic rate constant and m depends on the crystallisation and nucleation mechanism. Values were obtained as a function of temperature from isothermal crystallisation studies on a chlorinated polyether. Mention was made of an induction time required before crystallisation starts, and data on this were also obtained as a function of temperature. The crystallisation model was coupled with the conduction problem, which was solved using the Crank–Nicholson method. In contrast to all other works in this area, simulated cooling curves were not monotonic, but showed reheating during phase change of up to 5°C. Degrees of crystallinity were predicted to be significantly different for surface and internal layers in typical simulations, but no experimental comparisons were made.

More recently Kamal and Chu[198] have studied the isothermal crystallisation kinetics of a high density polyethylene and then used the data to model non-isothermal crystallisation, in a similar way to Sifleet *et al.*,[197] using an extension of the Avrami equation based on the idea that crystallisation under transient temperature conditions can be regarded as a sequence of isothermal steps. A delay or induction time was again mentioned, but in this case as a problem of instrument response, rather than the response of the polymer, as seems to have been assumed by Sifleet *et al.* Results of the non-isothermal crystallisation model were compared with experimental values for relative crystallinity as a function of time obtained from DSC at various scanning speeds. Agreement was within 15%.

In a continuation of the work Kamal and Lafleur[130] coupled the crystallisation model with a one-dimensional finite difference heat conduction analysis and compared predictions with results obtained from solidification experiments on a disk of HDPE (high density polyethylene). Degrees of crystallinity varied from about 45% at the wall to 55% in the core, and experimental and predicted values agreed within about 3% relative crystallinity. The crystallinity model was also coupled with an analysis of transient flow and heat transfer in mould filling.[130] Results for degree of crystallinity in a moulded sample were similar to those obtained in the static cooling experiment, which was to be expected in view of the short injection time in the experiment.

In addition to the foregoing analyses, which concentrate on conditions within the polymer, computer-aided methods for design of the heat transfer and hydraulic aspects of coolant circuits are available. Tseng[200] describes software to assist in fixing bubbler and baffle details. The commercially available package Moldcool[201] contains programs dealing with layout and specification of coolant circuits for injection moulding. Other programs use finite difference and finite element analyses to provide mould temperature profiles.[202]

7. MELT THERMOCOUPLE PERFORMANCE

Reliable measurement of temperature is fundamental to the correct operation of processing equipment. Measurement errors of electrical origin should be easy to eliminate, but those due to thermal interactions between the melt, the transducer and the machine are more complex. These are inherently difficult to investigate experimentally, and mathematical modelling of the relevant flow and heat transfer phenomena provides useful information.

Measurement errors in melt thermocouples have been analysed using finite element methods by Pittman and Mahmoudzadeh.[203,304] Two types of thermocouple were considered. For the first, where the junction is located flush with an inner surface, e.g. of an extruder barrel, it was concluded that in typical designs the registered temperature was very close to that of the machine wall, and that a ceramic guard ring round the thermocouple junction was ineffective in insulating it from the wall. For the second type, the upstream-pointing parallel-to-flow design, errors can arise due to additional viscous heat generation in the perturbed flow field around the probe, and by conduction along the probe to or from the duct walls. These errors were found generally to be negligible in typical cases.

Van Dam and Janeschitz-Kriegl[205] have applied finite difference methods to the solution of temperatures in the parallel-to-flow type of probe. The mathematical model is based on a number of simplifications, including neglect of viscous heating and the assumption of fully developed annular flow in contact with the probe. The results confirmed the usefulness of analytic solutions at low and high values of a Graetz number based on probe dimensions, and with the help of the numerical results these solutions were blended to give a master curve for convenient estimation of probe temperature errors due to conduction through the stem.

8. CALENDERING

It is debatable whether calendering should be included in the present class of internal flows, but when elastic effects are neglected it has more in common with them than with the free-surface-dominated flows which are omitted here. In fact, there has been relatively little work on numerical analysis of calendering in the polymer processing context. Mathematical modelling of the process has been reviewed in a recent article by Vlachopoulos and Mitsoulis.[206]

Early work modelled the process assuming inelastic flow and applying the lubrication approximation. Kiparissides and Vlachopoulos[207] used finite elements to solve for the flow field; viscous heating and the resulting temperature field were simulated separately using finite difference methods.[208] More recently, the lubrication approximation has been dropped and finite element methods used in analyses which include simulation of the circulation patterns in the melt bank.[209,210]

Agassant and Espy[209] computed circulation patterns in the bank assuming two-dimensional Newtonian isothermal creeping flow, with no slip at the roll surfaces. The position of the bank free surfaces was fixed, based on experimental observations, and the velocity field was obtained using finite elements with a stream function–vorticity formulation. Streamline patterns agreed qualitatively with the striations observable on sectioning a bank of rigid PVC. A much more complete analysis has been carried out by Mitsoulis, Vlachopoulos and Mirza.[210] Again, the full form of the momentum equations for creeping two-dimensional flow were used, but these were coupled with the two-dimensional energy equation retained in the full form, including conduction and convection heat transfer and viscous heat generation. A temperature-dependent power-law model for melt flow was used, and slip of the melt at the roll surface was allowed for. The position of the bank free surface, and the separation point at the calender exit were found iteratively. A primitive variable finite element method was used, and the streamline upwinding method was applied in treating the energy equation, to suppress temperature oscillations at high mesh Péclet numbers.

Results from this two-dimensional analysis for Newtonian, isothermal cases agreed well with earlier results based on the lubrication approximation, for pressure distribution, roll-separating force, torque, etc. However, substantial discrepancies occurred for the more realistic, non-isothermal cases. Comparison of simulations with experimental results for calendering of PVC showed generally good agreement, and it was suggested that discrepancies were due to the fact that flow is not truly two-dimensional; spread of material in the cross-machine direction occurs. Three-dimensional analysis of spread in rolling has been undertaken, in the metal-forming context, by Li and Kobayashi.[2]

Mitsoulis *et al.*[210] included melt elasticity in some of their computations using the Criminale–Ericksen–Filbey constitutive equation, but the range of convergence of the calculation was limited to Deborah numbers less than one, under which conditions elasticity had little effect upon the results. These convergence difficulties have also been met by Thomson and Berman[2] in a two-dimensional simulation of aluminium rolling, and they remain an unsolved problem in all numerical treatments of visco-elastic flow.

9. EVALUATION OF QUANTITIES RELATED TO MATERIAL EXPERIENCE

Previous sections have described analyses leading to evaluation of flow velocities, temperatures, pressures, stresses, etc., and whilst these are of great value in studying processing operations, they leave unanswered a set of important questions related to the *effects* that processing has had on the polymeric material. We are concerned here with the accumulated experience of a material particle as it passes through the process, leading to the development of practically important properties in the product. Pittman and Nakazawa[195,211] have identified a class of these 'history-dependent' quantities, which include the degree of laminar mixing, orientation, degradation, etc. The development of these is governed by equations having the general form

$$\frac{\mathrm{D}U}{\mathrm{D}t}=\frac{\partial U}{\partial t}+\mathbf{v}\,.\,\nabla U=F(U,\mathbf{v},T\cdots) \tag{42}$$

that is, the convected (material) derivative of the required quantity, U, is equated to a source function, F, which may depend on local values of U, the velocity field, temperature etc. The form of eqn (42) required for the calculation of residence time, and various deformation tensors used in the analysis of mixing and orientation, are given. These first order hyperbolic equations were discretised by finite element methods, and solutions are described for steady and time-transient cases. Applications of this approach, together with other work, are outlined in the following paragraphs.

9.1. Laminar Mixing

Two types of mixing may be distinguished: distributive mixing refers to the homogenisation process, whereby, for example, a drop of dye is distributed throughout a liquid; dispersive mixing refers to the break up of more or less rigid agglomerates of particles, within a fluid or plastic matrix.

In polymer melts, molecular diffusion is slow, and distributive mixing occurs predominantly by inhomogeneities being smeared or stretched out into striations or filaments, which eventually become so fine as to be individually indistinguishable. Analysis of this process has generally been based on the assumption that the components of the mixture have identical flow properties, and that it deforms as a homogeneous continuum. As striations of the minor component becomes finer, their surface area increases, and this area increase has been used as a measure of the extent of mixing.

To apply this measure in analysing the performance of mixing devices it is, of course, necessary first to know the flow field in the device, and then to work out the surface area increase of the blobs of material passing through it. Until recently the relationship between material surface area increase and flow field had been given explicitly for only two simple cases: simple shear flow and pure extensional flow. The result for shear flow has been used to assess the mixing effectiveness of melt flow in extruder channels.[212–215].

The universal assumption in these works has been that only the magnitude of the shearing is of importance, not its sense. However, it seems possible that on some streamlines of the flow, deformation occurring near the barrel is reversed as material circulates to the base of the channel, as indicated in Fig. 5. The need for a more general and complete approach to laminar mixing analysis is apparent, and is provided by classical continuum mechanics results for the stretching of material line surface, and volume elements in terms of the deformation tensors of the flow field. These results have been applied only recently.[195,211,216–218] Ottino and co-workers[216,217] gave a rigorous discussion of laminar mixing and derived a number of measures of mixing efficiency. Application to analysis of mixing in an extruder channel has been made, based on a simple analytic description of the flow field.[219] Application of the continuum mechanics results to more complicated flow fields requires numerical evaluation of the deformation tensors. Pittman and Nakazawa[195,211] and Pittman[218] have described a finite element procedure involving integration of forms of eqn (42) above, where the dependent variable U represents a deformation tensor. Pittman[218] also argues that mixing analysis based on calculation of material surface area increase is less effective than

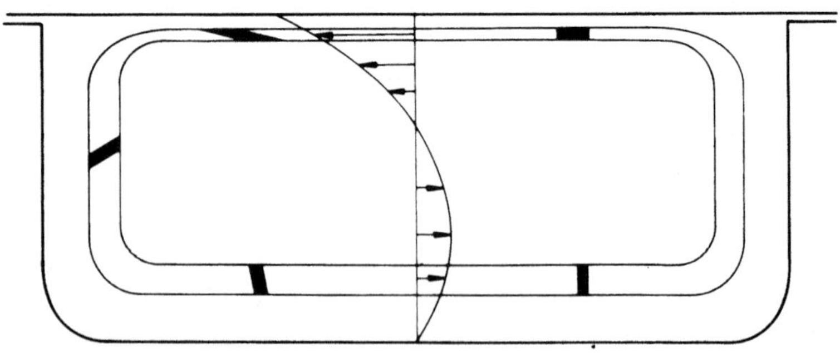

FIG. 5. Shearing deformation of an element of melt as it circulates on the extruder channel cross-section shown qualitatively. Stretching, which occurs near the barrel surface, tends to be reversed near the screw root.

one based on consideration of the deformation of material volume elements. The use of principal stretch ratios is proposed in place of area ratios.

In contrast to these methods it is possible, using numerical techniques, to track directly the position of an interface between two components of a mixture. For this, the marker and cell method is particularly suitable, and has been applied to mixing in recirculating channel flow by Bigg and Middleman[220] and Erwin and Mokhtarian.[221] For computational reasons, though, it seems unlikely that the large degrees of deformation required in practical mixing processes can be simulated, and only selected parts of the interface can be tracked.

9.2. Fibre Orientation

In the processing of resins containing short fibre reinforcement, flow can give rise to complex patterns of preferential orientation, which have important consequences for the mechanical properties of the product. Prediction of orientation distributions resulting from given flows is thus of considerable importance.

Givler, Crochet and Pipes[222] approached this problem using the equations governing low Reynolds number rotation of isolated particles suspended in an unbounded Newtonian flow (Jeffery's equations). The ordinary differential equations for orientation angles were integrated, stepping along selected streamlines and using the local velocity gradients. To build up orientation distributions it was necessary to consider a number of initial fibre orientations on each streamline. The method was applied to predict orientation distribution in a moulded bar,[223] using a finite element solution to obtain the flow field. Reasonably good agreement was obtained with experiment.

Folgar and Tucker[224] also took Jeffery's equations as their starting point, but worked in terms of an orientation distribution function rather than the orientation of individual fibres. An equation is developed for the evolution of the distribution function which modifies the Jeffery theory by addition of a term allowing for random re-orientation due to fibre interactions. This is analogous to diffusivity in rotary Brownian motion, and in this case allows for rotation of fibres away from the direction of flow induced alignment at a rate proportional to the gradient of the orientation distribution function. The proportionality constant (orientation diffusivity) is itself linearly related to the local shear rate, which governs the frequency of collisions. The theory is not able to provide the coefficient in this relationship. It was determined experimentally by fitting the theory to steady-state orientation distributions produced in Couette flow, and by appropriate choice of the interaction coefficient it was possible to fit the distributions rather well. The theory was applied to fibre orientation prediction in compression moulding.[225,226] Flow patterns in the mould were simulated using finite element solutions of Hele–Shaw equations (see Section 5.3). The orientation distribution was then calculated by integrating the partial differential evolution equation along selected pathlines. Experimental orientations were obtained from radiographs. The required interaction coefficient was found to depend on moulding conditions, but once determined allowed the theory to be used predictively.

An alternative starting point for fibre orientation prediction is to regard the suspension as a homogeneous continuum and assume that fibres rotate like material line elements of fluid. This was the approach used by Modlen,[227] who showed that for an initially random distribution in three dimensions, the probability of orientation in a given direction after flow is proportional to the cube of the radius vector of the strain ellipsoid. On this basis, orientation distributions were computed resulting from simple deformations, and compared successfully with experimental results.

Pittman and Nakazawa[8,195,211] extended these ideas, pointing out that the assumption that fibres move like material elements can be justified from the Jeffery equations for slender particles, except that fibres, because of their finite thickness, can rotate across streamlines. However, rotation in this region is very slow and can be neglected to a good approximation. Calculation of orientation distributions then amounts to finding the magnitude and direction of the principal axes of the strain ellipsoid (the principal stretch ratios) throughout the flow field. These are available from the eigenvalues and eigenvectors of the Cauchy or Finger deformation tensors, which, in turn, are found by integration of the appropriate form of eqn (42). This gives the field of principal stretches, providing orientation distributions at all points, without the need to track deformation along selected streamlines. In addition, the complete distribution is immediately available, without the need to consider several individual initial fibre positions. The calculations were illustrated by finite element results based on expanding radial flow between parallel disks.[211] Predicted alignment patterns were qualitatively similar to what is usually observed in centre-gate disc moulds: in the core, extensional deformation produces predominantly circumferential alignment, whilst in the skin region, shear results in predominantly radial alignment.

9.3. Residence Time and Polymer Degradation

Residence time distributions in processing equipment are of practical interest for a number of reasons. They can provide an indication of the extent of mixing in the flow direction, and are related to the occurrence of degradation.

Pinto and Tadmor[213] have calculated residence time distribution functions for melt flow in single-screw extruders, using one-dimensional analytic Newtonian flow solutions for the flow field. Ingen Housz[228] used the same approach to obtain residence times, and then used an analytic (fully developed) solution for temperature to obtain the thermal experience of melt particles. Based on an Arrhenius rate equation for thermal degradation, a method was proposed for calculation of a reduced residence time, from which the extent of degradation may be assessed. The analysis was extended to pipe extrusion, using a finite difference solution for the temperature field in the pipe. Morette and Gogos[229] included viscosity changes resulting from thermal degradation in their finite difference solutions for flow of PVC in capillaries. Residence times of material particles were logged, and when these exceeded an induction time, viscosity was altered according to an experimentally determined function.

Thompson, Pittman and Zienkiewicz[230] have proposed a method for calculation of residence time fields of arbitrary flows. This is by integration of the appropriate form of eqn (42), which for residence time, R, is simply

$$\frac{\mathrm{D}R}{\mathrm{D}t} = 1 \tag{43}$$

Finite element solutions obtained in this way have been shown by Thompson *et al.*[230] and Pittman and Nakazawa.[195] A residence time field is perhaps a less familiar idea than a residence time distribution. It provides a map showing, for each material particle in the flow field, the time which has elapsed since its entry. High residence time zones can be identified, together with the volume of material held up in them. A residence time distribution, being a flow-rate-weighted distribution of residence times at the exit, provides no information on the amount or location of material held up in 'dead-zones'. Pittman and Nakazawa[195] propose an extension of the approach based on integration of eqn (42) to calculate the extent of degradation. This requires definition of an appropriate measure of degradation (to replace U in eqn (42)) and the formulation of the source function F expressing the rate of degradation as a function of local temperature, deformation, concentration of stabiliser, etc.

10. SOME ILLUSTRATIVE FINITE ELEMENT SOLUTIONS

In this section are shown, briefly, examples of results obtained in recent analyses using the finite element techniques which have been described previously.[8,44,195] Full accounts will be published elsewhere.

10.1. Chill Rolling

It is required to find the cooling rate and temperature field in a polymer strip as it passes between chilled rollers. The problem is illustrated in Fig. 6. Spread is neglected and a two-dimensional analysis applied. Boundary conditions are treated as follow: v_y is set equal to zero on AB and CD, the free surfaces; on FE, the line of symmetry; and on AF and DE, to produce parallel flow at entry and exit of the solution domain. The default, or natural, boundary condition ensures that the gradient of v_x perpendicular to these surfaces is zero. Thus we have parallel plug flow entering at AF and leaving at DE. The velocity of these flows is determined in the solution, based on a specified circumferential velocity on the roll surface, BC, where the no-slip condition is assumed.

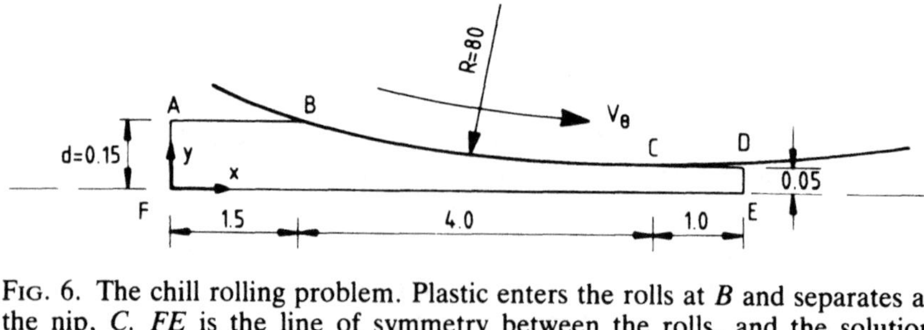

FIG. 6. The chill rolling problem. Plastic enters the rolls at B and separates at the nip, C. FE is the line of symmetry between the rolls, and the solution domain is $ABCDEF$. Note that the vertical scale here is expanded times five. Dimensions in mm.

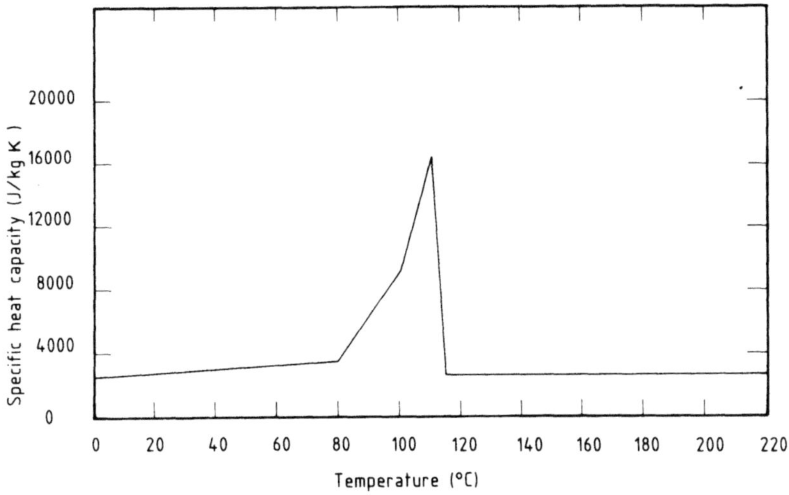

FIG. 7. Representation of temperature dependence of apparent specific heat in the chill rolling problem, by five linear segments.

Melt enters AF at a temperature of 453 K, and the roll surface is at 298 K.

Specific heat of the melt as a function of temperature is represented by five linear segments, as shown in Fig. 7. Phase change is included in the modelling by means of the peak in specific heat, which includes latent heat of solidification. Thermal conductivity is taken as 0·34 W/m K. Melt flow properties are modelled using a temperature-dependent power-law relationship

$$\mu = \mu_0(\dot{\gamma})^{n-1} \exp[-b(T - T_0)]$$

where $\mu_0 = 1{\cdot}88 \times 10^4\,\mathrm{N\,s/m^2}$, $b = 0{\cdot}006/\mathrm{K}$, $n = 0{\cdot}46$ and $T_0 = 463\,\mathrm{K}$. Velocity vectors for the case where roll speed is 0·3 rpm are shown in Fig. 8, and temperature contours in Fig. 9. A Péclet number for the problem can be defined (see Fig. 6)

$$Pe = \frac{\rho C d V_\theta}{k}$$

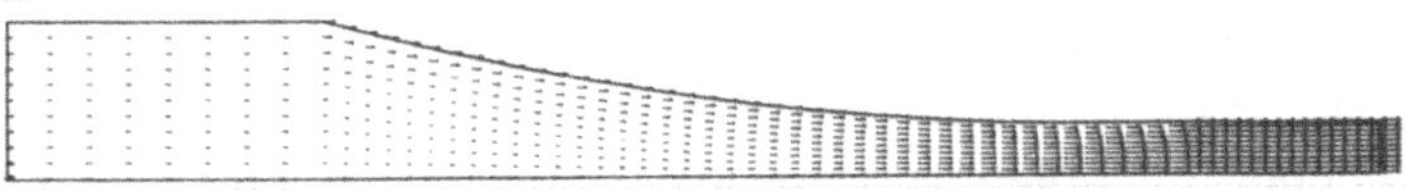

FIG. 8. Velocity vectors in chill rolling. Roll speed 0·3 rpm.

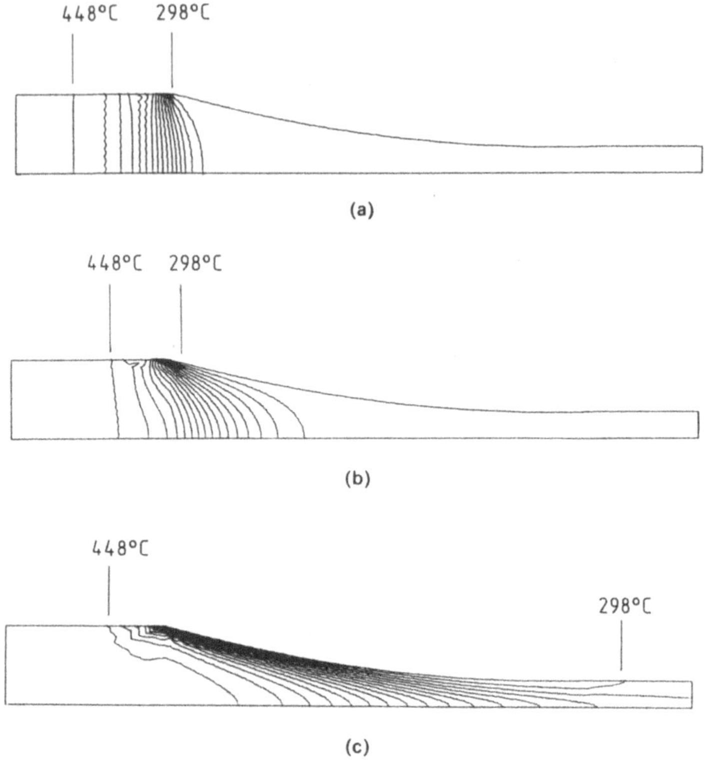

FIG. 9. Temperature contours in chill rolling. Contour interval 10 K. (a) $Pe = 2{\cdot}4$; (b) $Pe = 24$; (c) $Pe = 240$.

For the present case, its value is 2·4, indicating rapid cooling by conduction over a short flow distance, as evident in Fig. 9(a). Temperature contours for $Pe = 24$ and $Pe = 240$ are shown in Figs 9(b) and 9(c).

10.2. Blow-mould Cooling

An analysis of heat transfer is carried out including conduction through the polymer layer, with convective heat transfer to internal blown air, and conduction through the mould metal, with convective heat transfer to circulating coolant in the channels. The objective is to obtain details of cooling patterns in the polymer, and their dependence on location of the coolant channels, mould metal thermal conductivity

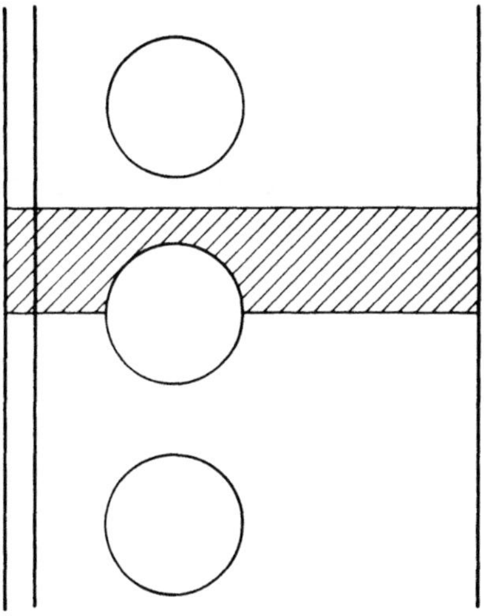

FIG. 10. The cooling phase of blow-moulding. The diagram shows the cross-section of the melt layer in contact with a mould wall containing an array of circular coolant channels. The domain of the solution is shown hatched.

and rate of heat removal into the coolant and internal air. The problem is illustrated in Fig. 10. Figure 11 shows the finite element mesh $A'BHI'$ represents the melt layer in contact with the mould, $BCDE'F'G'H$. $G'F'E'$ represents half the cross-section of a coolant channel; lines ABC and $IHGED$ are lines of symmetry. No heat is conducted across these, or across CD, and this is achieved using the

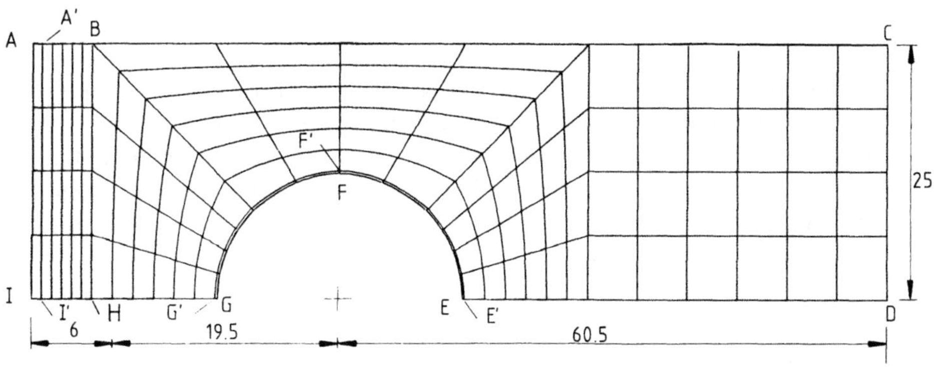

FIG. 11. The finite element mesh used in the blow-mould cooling simulation. Dimensions in mm.

default (natural) boundary condition for temperature. Heat transfer to the coolant at 283 K across $G'F'E'$ is governed by a heat transfer coefficient $h = 8000\ \mathrm{W/m^2\,K}$. For transfer across $A'I'$ to the internal blown air at 283 K, the coefficient $h = 40\ \mathrm{W/m^2\,K}$.

In the present analysis, these convection boundary conditions are handled in a novel way. The conventional approach in finite elements involves evaluation of a line integral along the transfer surface, and this is computationally cumbersome. More convenient, especially in the present codes, which are designed to handle multiple materials, is to attach to the transfer surfaces layers of virtual elements $AA'I'I$ and $G'F'E'EFG$ with appropriate properties, and use the simpler Dirichlet boundary condition of 283 K on the virtual surfaces AI and GFE. Heat loss from the material surface is then governed by conduction through the virtual elements. If the heat flux in these is normal to the surface, and if their temperature profile is the analytic steady-state one, then the virtual conduction flux can be equated to the required surface heat flux for a flat surface as follows

$$h(T_{\mathrm{surface}} - T_{\mathrm{coolant}}) = \frac{k_{\mathrm{v}}}{d_{\mathrm{v}}}(T_{\mathrm{surface}} - T_{\mathrm{coolant}})$$

Thus the virtual conductivity must be chosen

$$k_{\mathrm{v}} = d_{\mathrm{v}} h$$

where d_{v} is the virtual element thickness.

For a curved surface, the required result is

$$k_{\mathrm{v}} = d_{\mathrm{v}} \frac{r_{\mathrm{s}}}{r_{\mathrm{lm}}} h$$

where r_{s} is the local radius of curvature of the transfer surface, and r_{lm} is the log mean radius of the transfer and virtual surfaces. A sufficiently good approximation to a virtual heat flux normal to the material surface can be achieved by making the virtual elements thin. The steady-state temperature profile can be obtained by making the virtual specific heat zero.

A further feature of the analysis is the treatment of the initial conditions. At the moment when melt contacts the mould a temperature discontinuity exists across the interface. This cannot be represented numerically, and to assign an arbitrary initial interfacial temperature and temperature gradients may give false results during

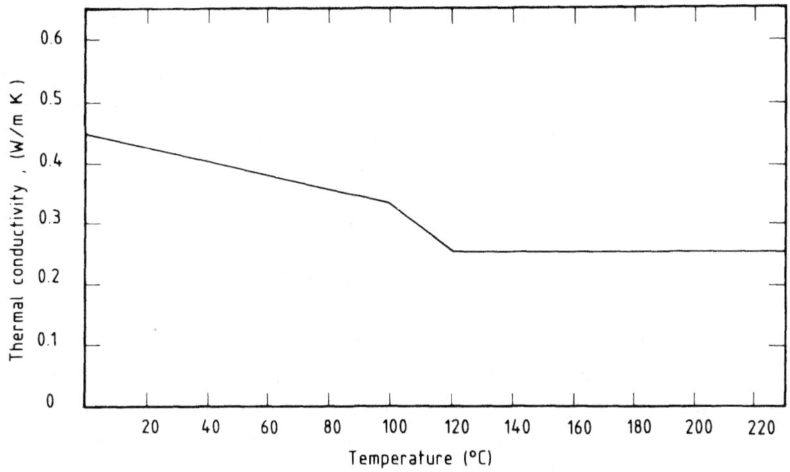

FIG. 12. Representation of the temperature dependence of thermal conductivity by three linear segments, in the mould cooling problem.

the important, rapid, early stages of cooling. To avoid this difficulty, the initial moments of cooling are treated using an analytic solution for temperature profiles in the region of a plane interface between two semi-infinite solids.[171b] The numerical analysis in the present case was started after 2 s, using the analytic solution as initial condition. A further advantage of this procedure is that one can then ensure that mesh subdivision is sufficiently fine to represent the short time (2 s) gradients adequately. Failure to do this results in oscillatory solutions.

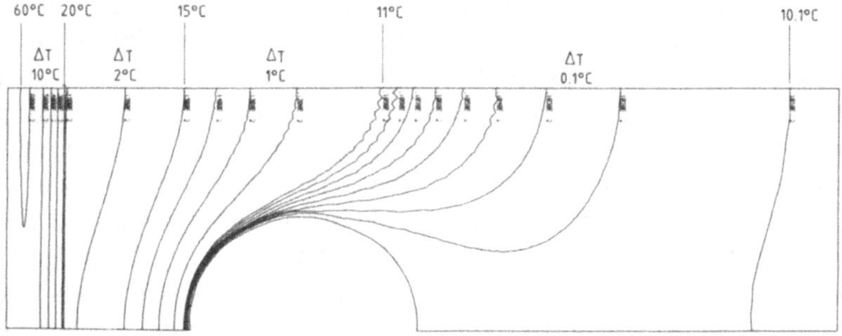

FIG. 13. Temperature contours in the mould cooling simulation after 100 s cooling time. Contour interval in the various regions is indicated on the diagram.

A further analytic result was used to provide the 2 s profile near the melt–air interface.[171c]

Thermal properties of the melt are shown in Figs 7 and 12 and again latent heat is handled using a peak in specific heat. The mould metal conductivity was 20 W/m K and ρC was $3{\cdot}6 \times 10^6$ J/m^3 K. Temperature contours obtained after cooling for 100 s are shown in Fig. 13.

11. CONCLUSION

In this concluding section, some suggestions are offered of ways and means for advancing the application of numerical analyses to polymer processing operations. First, as regards the numerical techniques to be used, it seems clear that finite element methods should nowadays be the choice for spatial discretisation, though finite differences remain the preferred method for time-stepping. An important advantage of finite element methods is their geometrical flexibility. Solutions can be made computationally efficient by concentrating mesh refinement in 'difficult' regions of the solution domain. Curvilinear boundaries are easily fitted, and the geometrical specifications and boundary conditions form input data to the program. Thus the same program can readily be used to solve problems on an unlimited range of domain geometries. This is exactly what the designer needs—to explore, through numerical simulations, the consequences of varying the geometry and lay-out of an item of processing equipment. There are also important advantages in finite element methods connected with the convenience with which different types of boundary conditions are handled, and with the treatment of composite structures involving materials with differing physical properties. The flexibility of finite element methods leads to the development of program packages, where the central solution codes are supported by pre- and post-processors designed to facilitate data preparation and input, and display of the solution results. Automatic mesh generation codes reduce the labour of preparing geometrical and boundary condition data; digitising facilities can be used to derive meshes from designer's sketches or engineering drawings. Graphics programs can display results as velocity vector plots, contours of temperature, isobars, etc.

Development of packages such as these requires many years of work and experience; however, in many instances the greater part of this work has already been carried out by academic research groups, and it is unnecessary for companies interested in applying numerical methods to undertake developments themselves.

Secondly, as regards use of such techniques in the polymer processing area, it is clear that a level of expertise both in the numerical analysis and the physics and engineering aspects of the topic are required. The practical problems are complex, both mathematically and physically, and it is highly dangerous to regard sophisticated numerical schemes as black boxes producing infallible results. Only large companies are likely to have the personnel and computer hardware to support and use the methods in-house. Smaller companies will obtain access to the techniques through consultancy. Indeed, one can identify as a determining factor in the wider use of computer-aided analysis and design, the need for an increased number of organisations offering the combination of engineering and computational expertise mentioned above on a consultancy or contract basis. These organisations and the industry as a whole would then reap significant economic benefits.

REFERENCES

1. Pittman, J. F. T., Wood, R. D., Alexander, J. M. and Zienkiewicz, O. C. (Eds), *Numerical Methods in Industrial Forming Processes,* Proc. Int. Conf., Swansea, Pineridge Press, Swansea, UK, 1982.
2. Pittman, J. F. T., Wood, R. D., Zienkiewicz, O. C. and Alexander, J. M. (Eds), *Numerical Analysis of Forming Processes,* J. Wiley & Sons, Chichester, Sussex, 1984.
3. Pearson, J. R. A. and Richardson, S. M. (Eds), *Computational Analysis of Polymer Processing,* Applied Science Publishers, London, 1983.
4. White, J. L., 'Dynamics, heat transfer and rheological aspects of melt spinning: a critical review', *Poly. Engng Rev.,* **1,** 1981, 297–362.
5. (a) Pearson, J. R. A., *Mechanical Principles of Polymer Melt Processing,* Pergamon Press, Oxford, 1966, pp. 60–3.
5. (b) *Ibid.,* pp. 128–34.
6. Crochet, M. J., Davies, A. R. and Walters, K., *Numerical Simulation of Non-Newtonian Flow,* Elsevier Science Publishers, Amsterdam, 1984.
7. Various authors, Special issue on numerical simulation of viscoelastic flows, *J. Non-Newtonian Fl. Mech.,* **16** (1–2), 1984.
8. Pittman, J. F. T. and Nakazawa, S., 'Analysis of melt flow and heat transfer using finite elements', *Polymer Engng Rev.* **4,** 1984, 143–75.

9. Schechter, R. S., 'On steady flow of a non-Newtonian fluid in cylinder ducts', *A.I.Ch.E. J.*, **7,** 1961, 445–48.
10. Middleman, S., 'Flow of power law fluids in rectangular ducts', *Trans. Soc. Rheol.*, **9,** 1965, 83–93.
11. Wheeler, J. A. and Wissler, E. H., 'The friction factor–Reynolds number relation for the steady flow of pseudoplastic fluids through rectangular ducts', *A.I.Ch.E. J.*, **11,** 1965, 207–16.
12. McEachern, D. W., 'Axial laminar flow of a non-Newtonian fluid in an annulus', *A.I.Ch.E. J.*, **12,** March 1966, 328–32.
13. Nebrensky, J. and Ulbrecht, J., 'Non-Newtonian flow in annular ducts', *Collection of Czech Chem. Commun.*, **33,** 1968, 363–75.
14. Mitsuishi, N., Kitayama, Y. and Aoyogi, Y., 'Non-Newtonian flow in non-circular ducts', *Int. Chem. Engng,* **8,** 1968, 168–74.
15. Mitsuishi, N. and Aoyagi, Y., 'Non-Newtonian flow in non-circular ducts', *Chem. Engng Sci.*, **24,** 1969, 309–319.
16. Röthemeyer, F., 'Das Druckströmungsfeld in zylindrischen Fleisskanalen mit beliebigem Querschnitt', *Rheol. Acta,* **9,** 1970, 259–63.
17. Kozicki, W., Chou, C. H. and Tiu, C., 'Non-Newtonian flow in ducts of arbitrary cross-sectional shape', *Chem. Eng. Sci.*, **21,** 1966, 665–79.
18. Miller, C., 'Predicting non-Newtonian flow behaviour in ducts of unusual cross-section', *Ind. Eng. Chem. Fundam.*, **11,** 1972, 524–8.
19. Hanks, R. W., 'On the prediction of non-Newtonian flow behaviour in ducts of non-circular cross-section', *Ind. Eng. Chem. Fundam.*, **13,** 1974, 62–6.
20. Gee, R. E. and Lyon, J. B., 'Non-isothermal flow of viscous non-Newtonian fluids', *Ind. Engng Chem.*, **49,** 1957, 956–60.
21. Gerrard, J. E., Steidler, J. K. and Appeldoorn, J. K., 'Viscous heating in capillaries', *I. & E. C. Fund.*, **4,** 1965, 332–40.
22. Forsyth, T. H. and Murphy, N. F., 'Temperature profiles of molten flowing polymers in a heat exchanger', *A.I.Ch.E. J.*, **15,** 1969, 758–63.
23. Kim, H. T. and Collins, E. A., 'Temperature profiles for polymer melt in tube flow, Part II, Conduction and shear heating corrections'. *Poly. Engng and Sci.*, **11,** 1971, 83–92.
24. Forrest, G. and Wilkinson, W. L., 'Laminar heat transfer to power-law fluids in tubes with constant wall temperature', *Trans. I. Chem. Engnrs,* **51,** 1973, 331–8.
25. Forrest, G. and Wilkinson, W. L., 'Laminar heat transfer to temperature-dependent Bingham fluids in tubes', *Int. J. Heat and Mass Transf.*, **16,** 1973, 2377–91.
26. Losson, J. M., 'A mathematical model is used in predicting die performance', *Japan Plastics,* Jan. 1974, 22–8.
27. Forrest, G. and Wilkinson, W. L., 'Laminar heat transfer to power-law fluids in tubes with constant wall heat flux', *Trans. I. Chem. Engnrs,* **52,** 1974, 10–16.
28. Winter, H. H., 'Temperature fields in extruder dies with circular, annular or slit cross-sections', *Poly. Engng and Sci.*, **15,** 1975, 84–9.
29. Mennig, G., 'Zur Berechnung der Temperaturverteilung bei der Düsenströmung hochviskoser Massen', *Kunststoffe,* **65,** 1975, 693–5.

30. HULATT, M. and WILKINSON, W. L., 'Heat Transfer to molten polymers flowing through tubes', *Poly. Engng and Sci.*, **18,** 1978, 1148–54.
31. HULLAT, M., JARZEBSKI, A. B. and WILKINSON, W. L., 'Heat transfer to flowing polymer melts', *J. Non-Newtonian Fluid Mech.*, **7,** 1980, 345–57.
32. NOWAK, Z., GRYGLASZEWSKI, P. and STRACHARSKA-TARGOSZ, J., 'Pressure drop in non-isothermal laminar flow of power-law fluids in tubes with various thermal conditions', *Warme- und Stoffübertragung,* **14,** 1980, 181–288.
33. PARNABY, J., HASSAN, G. A., HELMY, H. A. A. and ALI, A., 'Design of plastics processing machinery using lumped parameter methods', *Plastics and Rubber Processing and Applications,* **1,** 1981, 303–15.
34. JARZEBSKI, A. B. and WILKINSON, W. L., 'Non-isothermal developing flow of a power-law fluid in a converging slit', *J. Non-Newtonian Fluid Mech.,* **12,** 1983, 1–11.
35. KIM-E, M. E., BROWN, R. A. and ARMSTRONG, R. C., 'The roles of inertia and shear thinning in flow of an inelastic liquid through an axisymmetric sudden contraction', *J. Non-Newtonian Fluid Mech.,* **13,** 1983, 341–63.
36. MITSOULIS, E., VLACHOPOULOS, J. and MIRZA, F. A.'Numerical simulation of entry and exit flows in slit dies', *Poly. Engng and Sci.*, **24,** 1984 707–15.
37. MITSOULIS, E. and VLACHOPOULOS, J., 'Effect of Reynolds number in laminar flow through a sudden planar contraction', *A.I.Ch.E. J.,* **31,** 1985, 1736–9.
38. DURST, F. and LOY, T., 'Investigation of laminar flow in a pipe with sudden contraction of the cross section area', *Computers and Fluids,* **13,** 1985, 15–36.
39. RASHID, K., '*Finite Element Analysis of Non-Newtonian Flow and Heat Transfer*', PhD Thesis, University College, Swansea, UK, 1983.
40. CASWELL, B. and TANNER, R. I., 'Wirecoating die design using finite element methods', *Poly. Engnr. and Sci.*, **18,** 1973, 416–21.
41. VIRIYAYUTHAKORN, M. and DEBOO, R. V., 'A finite element model for cable jacketing simulation', *ANTEC* (Soc. Plastics Engnrs US), 1983, 178–82.
42. NICKELL, R. E., TANNER, R. I. and CASWELL, B., 'The solution of viscous incompressible jet and free surface flows using finite element methods', *J. Fluid. Mech.*, **65,** 1974, 189–206.
43. BASU, S., 'A theoretical analysis of non-isothermal flow in wire-coating co-extrusion dies', *Poly. Engng and Sci.*, **21,** 1981, 1128–38.
44. NAKAZAWA, S., PITTMAN, J. F. T. and ZIENKIEWICZ, O. C., 'Numerical solution of flow and heat transfer in polymer melts', in *Finite Elements in Fluids,* R. H. Gallagher, D. H. Norrie, J. T. Oden and O. C. Zienkiewicz, Eds, J. Wiley & Sons, Chichester, Sussex, 1982, pp. 251–303.
45. LE NIR, V. L., 'Computerization of wire insulating lines—An industrial approach'. *Polymer Engng and Sci.,* **24,** 1984, 633–44.
46. BOOY, M. L., 'A network flow analysis of extrusion dies and other flow systems', *Poly. Engng and Sci.*, **22,** 1982, 432–7.

47. Tadmor, Z., Broyer, E. and Gutfinger, C., 'Flow analysis network (FAN)—a method for solving flow problems in polymer processing'. *Poly. Engng and Sci.*, **14,** 1974, 660–5.
48. Fenner, R. T. and Nadiri, F., 'Finite element analysis of polymer melt flow in cable-covering crossheads', *Poly. Engng and Sci.*, **19,** 1979, 203–9.
49. Menges, G., Cordes, H. and Masberg, U., 'Strömungsfelder in komplexen Fliesskanalen', *Kunststoffe,* **70,** 1980, 138–43.
50. Masberg, U., 'Three-dimensional calculation of velocity fields in complex flow channels', *Practical Rheology in Polymer Processing* (Proc. PRI Conf. Loughborough, UK), 1980, pp. 3.1–3.9.
51. Wortberg, J., Beiss, K., Dierkes, A., Masberg, U., Franzkoch, B. and Michaeli, W., 'Wege zum Berechnen von Extrusionswerkzeugen'. *IKV Colloquium 10,* Aachen, W. Germany, 1980, pp. 267–96.
52. Lawal, A. and Mujumdar, A. S., 'Laminar flow and heat transfer in power-law fluids flowing in arbitrary cross-sectional ducts'. *Numerical Heat Transfer,* **8,** 1985, 217–44.
53. Fenner, R. T., 'Developments in the analysis of steady screw extrusion of polymers'. *Polymer,* **18,** 1977, 617–35.
54. Griffith, R. M., 'Fully developed flow in screw extruders', *I. & E. C. Fund.,* **1**, 1962, 180–7.
55. Zamodits, H. J. and Pearson, J. R. A., 'Flow of polymer melts in extruders. Part I. The effect of the transverse flow and of a superposed, steady temperature profile'. *Trans. Soc. Rheol.,* **13,** 1969, 357–85.
56. Zamodits, H. J., PhD Thesis, Cambridge University, UK, 1964.
57. Martin, B., Pearson, J. R. A. and Yates, B., *Report No.* 5, Polymer Processing Research Centre, Department of Chemical Engineering, Cambridge University, UK, Oct. 1969.
58. Fenner, R. T., *Extruder Screw Design,* Iliffe Books for the Plastics Institute, London, 1970.
59. Yates, B., PhD Thesis, Cambridge University, UK, 1968.
60. Fenner, R. T., 'The design of large hot melt extruders', *Polymer,* **16,** 1975, 298–304.
61. Elbirli, B. and Lindt, J. T., 'A note on the numerical treatment of the thermally developing flow in screw extruders', *Poly. Engng and Sci.,* **24,** 1984, 482–7.
62. Martin, B., PhD Thesis, Cambridge University, UK, 1969.
63. Dyer, D. F., 'A numerical solution for the single screw extrusion of a polymer melt', *A.I.Ch.E. J.,* **15,** 1969, 823–8.
64. Palit, K. and Fenner, R. T., 'Finite element analysis of slow non-Newtonian flow', *A.I.Ch.E. J.,* **18,** 1972, 628–33.
65. Palit, K. and Fenner, R. T., 'Finite element analysis of slow non-Newtonian flows', *A.I.Ch.E. J.,* **18,** 1972, 1163–70.
66. Palit, K., PhD Thesis, Imperial College, London University, UK, 1972.
67. Palit, K., *Polymer Processing Research Report No. 2.,* Dept of Mech. Engng, Imperial College, London, March 1974.
68. Palit, K., 'Analysis of melt flow in polymer processing using finite

elements', *Proc. Int. Conf., Science and Technology of Polymer Processing,* N. P. Suh and N-H. Sung, Eds, MIT Press, Cambridge, Massachusetts, 1977, pp. 357–79.

69. NEBRENSKY, J., SMITH, J. M. and PITTMAN, J. F. T., 'Flow and heat transfer in extruders: a variational analysis applied in helical co-ordinates', *Poly. Engng and Sci.*, **13,** 1973, 209–15.
70. HAMI, M. L. and PITTMAN, J. F. T., 'Finite element solutions for flow in a single screw extruder, including curvature effects', *Poly. Engng and Sci.,* **20,** 1980, 339–48.
71. CHOO, K. P., HAMI, M. L. and PITTMAN, J. F. T., 'Deep channel operating characteristics of a single-screw extruder: finite element predictions and experimental results for isothermal non-Newtonian flow', *Poly. Engng and Sci.,* **21,** 1981, 100–4.
72. WINTER, H. H., 'Temperature induced pressure gradient in the clearance between screw flight and barrel in a single screw extruder', *Poly. Engng and Sci.,* **20,** 1980, 406–12.
73. PITTMAN, J. F. T. and RASHID, K., 'Heat transfer in recirculating extruder channel flow', *J. Polymer Engng,* **5,** 1985, 1–48.
74. HELMY, H. A. A. and PARNABY, J., 'Computer aided optimal melt screw design', *Poly. Engng and Sci.,* **16,** 1976, 437–49.
75. TADMOR, Z. and BROYER, E., Solids conveying in screw extruders. Part III. Non-isothermal model', *Poly. Engng and Sci.,* **12,** 1972, 378–86.
76. DARNELL, W. H. and MOL, E. A. J., 'Solids conveying in extruders', *S.P.E. J.,* April 1956, 20–9.
77. KACIR, L. and TADMOR, Z., 'Solids conveying in screw extruders. Part II. The delay zone', *Poly. Engng and Sci.,* **12,** 1972, 387–95.
78. LOVEGROVE, J. G. A. and WILLIAMS, J. G., 'Solids conveying in a single screw extruder; the role of gravity forces', *J. Mech. Engng Sci.,* **15,** 1973, 114–22.
79. TADMOR, Z., 'Fundamentals of plasticating extrusion', *Poly. Engng and Sci.,* July 1966, 185–90.
80. MADDOCK, B. H., 'A visual analysis of flow and mixing in extruder screws'. *S.P.E. J.,* May 1959, 383–9.
81. LINDT, J. T., 'A dynamic melting model for a single-screw extruder', *Poly. Engng and Sci.,* **21,** 1981, 1162–6.
82. TADMOR, Z., DUVDEVANI, I. J. and KLEIN, I., 'Melting in plasticating extruders. Theory and experiment', *Poly. Engng and Sci.,* July 1967, 2–21.
83. LINDT, J. T., Mathematical modelling of melting of polymers in single-screw extruders: a critical review', *ANTEC* (Soc. Plastics Engnrs US), 1984, 73–6.
84. LIPSHITZ, S. D., LAVIE, R. and TADMOR, Z., 'A melting model for reciprocating screw injection-moulding machines', *Poly. Engng and Sci.,* **14,** 1974, 553–9.
85. DONOVAN, R. C., 'The plasticating process in injection moulding', *Poly. Engng and Sci.,* **14,** 1974, 101–11.
86. CHUNG, C. I., 'A new theory for single screw extrusion', *Modern Plastics,* **45,** 1968, 110–15 and 178–98.

87. HINRICHS, D. R. and LILLELEHT, L. U., 'A modified melting model for plastifying extruders', *Poly. Engng and Sci.,* **10,** 1970, 268–73.
88. DONOVAN, R. C., 'A theoretical melting model for plasticating extruders', *Poly. Engng and Sci.,* **11,** 1971, 247–57.
89. MONDVAI, I., HALASZ, L. and MOLNAR, I., 'Extrusion von Thermoplasten II', *Plaste Kautschuk,* **20,** 1973, 630–8.
90. EDMONDSON, I. R. and FENNER, R. T., 'Melting of thermoplastics in single screw extruders', *Polymer,* **16,** 1975, 49.
91. COX, A. P. D. and FENNER, R. T., 'Melting performance in the single screw extrusion of thermoplastics', *Poly. Engng and Sci.,* **20,** 1980, 562–71.
92. LINDT, J. T., 'A dynamic melting model for single-screw extruders'. *Poly. Engng and Sci.,* **16,** 1976, 284–91.
93. SHAPIRO, J., HALMOS, A. L. and PEARSON, J. R. A., 'Melting in single screw extruders'. *Polymer,* **17,** 1976, 905–18.
94. HALMOS, A. L., PEARSON, J. R. A. and TROTTNOW, R., 'Melting in single screw extruders', *Polymer,* **19,** 1978, 1199–216.
95. FUKASI, H., KUNIO, T., SHINYA, S. and NOMOURA, A., 'A plasticating model for single-screw extruders', *Poly. Engng and Sci.,* **22,** 1982, 578–86.
96. ELBIRLI, B., LINDT, J. T., GOTTGETREU, S. R. and BABA, S. M., 'Mathematical modeling of melting of polymers in barrier-screw extruders', *Poly. Engng and Sci.* **23,** 1983, 86–94.
97. ELBIRLI, B., LINDT, J. T., GOTTGETREU, S. R. and BABA, S. M., 'Mathematical modeling of melting of polymers in a single-screw extruder'. *Poly. Engng and Sci.,* **24,** 1984, 988–99.
98. LINDT, J. T. and ELBIRLI, B., Effect of the cross-channel flow on the melting performance of single screw extruders, *Poly. Engng and Sci.,* **25,** 1985, 412–18.
99. KLEIN, I. and MARSHALL, D. I., 'Fundamentals of plasticating extrusion', *Poly. Engng and Sci.,* July 1966, 198–202.
100. KLEIN, I. and TADMOR, Z., 'The simulation of the plasticating screw extrusion process with a computer programmed theoretical model', *Poly. Engng and Sci.,* **9,** 1969, 11–21.
101. TADMOR, Z. and KLEIN, I., *Engineering Principles of Plasticating Extrusion,* Krieger, New York, 1978.
102. KLEIN, I. and MARSHALL, D. I. (Eds), *Computer Programs for Plastic Engineers,* Reinhold, New York, 1968.
103. KLEIN, I., 'Predicting the effect of screw wear on the performance of plasticating extruders', *Poly. Engng and Sci.,* **15,** 1975, 444–50.
104. AGUR, E. E. and VLACHOPOULOS, J., 'Numerical simulation of a single-screw plasticating extruder', *ANTEC* (Soc. Plastics Engnrs, US) 1982, 465–8.
105. TADMOR, Z., LIPSHITZ, S. D. and LAVIE, R., 'Dynamic model of a plasticating extruder', *Poly. Engng and Sci.,* **14,** 1974, 112–19.
106. DENSON, C. D. and HWANG Jr, J. R. 'The influence of axial pressure gradient on flow rate for Newtonian liquids in a self-wiping, co-rotating twin screw extruder', *Poly. Engng and Sci.,* **20,** 1980, 965–71.

107. JANSSEN, L. B. P. M., 'Twin screw extrusion', Elsevier Science Publishers, Amsterdam, 1978.
108. LAI-FOOK, R. A. L-T., PhD Thesis, Dept Mech. Engng, Imperial College, London University, UK, 1985.
109. RIEDLER, J., 'Quasi three-dimensional finite element solution for a twin-screw extruder', in *Numerical methods in laminar and turbulent flow, Conference Proc.*, C. Taylor, B. A. Schrefter, Eds, Pineridge Press, Swansea, UK, 1981, pp. 191–206.
110. KAMAL, M. R. and LAFLEUR, P. G., 'Computer simulation of injection moulding', *Poly. Engng and Sci.*, **22,** 1982, 1066–74.
111. GRMELA, M., 'Common structure compatibility and complementarity of mathematical models for injection moulding', *Poly. Engng and Sci.*, **24,** 1984, 673–85.
112. KAMAL, M. R., TAN, V. and RYAN, M. E., 'Injection molding: a critical profile', in *Proc. Int. Conf., Science and Technology of Polymer Processing,* N. P. Suh and N-H. Sung, Eds. MIT Press, Cambridge, Massachusetts, 1977, pp. 34–75.
113. (a) RICHARDSON, S. M., PEARSON, H. J. and PEARSON, J. R. A., 'Simulation of injection moulding', *Plastics and Rubber: Processing,* June 1980, 55–60.
113. (b) *Idem, Proc. Practical Rheology in Polymer Processing,* Loughborough, UK, 1980, pp. 2.1–2.13.
114. KUO, Y., 'Some applications of mathematical modeling to the injection moulding of complex production parts', in *Proc. Int. Conf., Science and Technology of Polymer Processing,* N. P. Suh and N-H. Sung, Eds, MIT Press, Cambridge, Massachusetts, 1977, pp. 580–604.
115. HARRY, D. H. and PARROTT, R. G., 'Numerical simulation of injection mold filling', *Poly. Engng and Sci.*, **10,** 1970, 209–14.
116. BERGER, J. L. and GOGOS, C. G., 'A numerical simulation of the cavity filling process with PVC in injection molding', *Poly. Engng and Sci.*, **13,** 1973, 102–12.
117. KENIG, S. and KAMAL, M. R., 'Cooling molded parts—a rigorous analysis', *S.P.E. J.*, **26,** July 1970, 50–7.
118. KAMAL, M. R. and KENIG, S., 'The injection molding of thermoplastics, Part I; theortical model', *Poly. Engng and Sci.*, **12,** 1972, 294–301.
119. KAMAL, M. R. and KENIG, S., 'The injection molding of thermoplastics, Part II. Experimental test of the model', *Poly. Engng and Sci.*, **12,** 1972, 302–8.
120. STEVENSON, J. F., 'A simplified method for analyzing mold filling dynamics. Part I: Theory', *Poly. Engng and Sci.*, **18,** 1978, 577–82.
121. STEVENSON, J. F., 'A simplified method for analyzing mold filling dynamics. Part II. Extensions and comparisons with experiment', *Poly. Engng and Sci.*, **19,** 1979, 849–57.
122. WILLIAMS, G. and LORD, H. A., 'Mold filling studies for the injection molding of thermoplastic materials. Part II. The transient flow of plastic materials in the cavities of injection-molding dies', *Poly. Engng and Sci.*, **15,** 1975, 569–82.

123. Titomanlio, G., 'Estimation of injection pressure during mold filling'. *Poly. Engng and Sci.*, **22,** 1982, 324–8.
124. Thienel, P. and Menges, G., 'Mathematical and experimental determination of temperature, velocity and pressure fields in flat molds during the filling process in injection molding of thermoplastic', *Poly. Engng and Sci.*, **18,** 1978, 314–20.
125. Wang, K. K., 'A system approach to injection molding process', *Polym.-Plast. Technol. Eng.*, **14,** 1980, 75–93.
126. Echenagucia, J., Fricke, A. L. and Hassler, J. C., 'Nodal positioning (NPA)—a numerical method for determining flow behaviour of thermoplastic materials', *Poly. Engng and Sci.*, **20,** 1980, 769–72.
127. Titomanlio, G., Acierno, D. and La Mantia, F. P., 'Modelling of the packing step in the injection molding of thermoplastic materials', *Ing. Chim. Ital.*, **16,** 1980, 110–12.
128. Isayev, A. I. and Hariharan, T., 'Volumetric effects in the injection moulding of polymers', *Poly. Engng and Sci.*, **25,** 1985, 271–8.
129. Bigg, D. M., 'Application of the marker and cell technique to polymer processing problems', *ANTEC,* 1975, 301–6.
130. Kamal, M. R. and Lafleur, P. G., 'A model for the prediction of microstructure distribution in thermoplastics injection molding'. *Poly. Engng and Sci.*, **24,** 1984, 692–7.
131. Richardson, S., 'Hele Shaw flows with a free boundary produced by the injection of fluid into a narrow channel'. *J. Fluid Mech.*, **56,** 1972, 609–18.
132. Kuo, Y. and Kamal, M. R., 'The fluid mechanics and heat transfer of injection mold filling of thermoplastic materials', *A.I.Ch.E. J.*, **22,** 1976, 661–9.
133. Kuo, Y. and Kamal, M. R., 'Flows of thermoplastics in the filling and packing stages of injection moulding', in *Proc. Int. Conf., Science and Technology of Polymer Processing,* N. P. Suh and N-H. Sung, Eds, MIT Press, Cambridge, Massachusetts, 1977, pp. 329–48.
134. Ryan, M. E. and Chung, T-S., 'Conformal mapping analysis of injection mold filling', *Poly. Engng and Sci.*, **20,** 1980, 642–51.
135. Broyer, E., Gutfinger, C. and Tadmor, Z., 'A theoretial model for the cavity filling process in injection molding', *Trans. Soc. Rheol.*, **19,** 1975, 423–44.
136. Hieber, C. A. and Shen, S. F., 'A finite element/finite difference simulation of the injection-molding process', *J. Non-Newtonian Fluid Mech.*, **7,** 1980, 1–32.
137. Wang, K. K., Shen, S. F., Hieber, C. A. and Stevenson, J. F., 'Cornell Injection moulding project' in *Proc. Int. Conf., Science and Technology of Polymer Processing,* N. P. Suh and N-H. Sung, Eds, MIT Press, Cambridge, Massachusetts. 1977, pp. 293–314.
138. Hieber, C. A., Socha, L. S., Shen, S. F., Wang, K. K. and Isayev, A. I., 'Experimentation and simulation for filling thin cavities of variable gap thickness', *ANTEC,* 1981, 759–63.
139. Hieber, C. A., Socha, L. S., Shen, S. F., Wang, K. K. and Isayev, A. I., 'Filling thin cavities of variable gap thickness: a numerical and experimental investigation', *Poly. Engng and Sci.*, **23,** 1983, 20–6.

140. HIEBER, C. A., 'Coupled-flow-path method for simulating the injection molding filling stage', *ANTEC,* 1982, 356–8.
141. NUNN, R. E. and FENNER, R. T., 'Flow and heat transfer in the nozzle of an injection moulding machine', *Poly. Engng and Sci.,* **17,** 1977, 811–18.
142. WILLIAMS, G. and LORD, H. A., 'Mold filling studies for the injection molding of thermoplastic materials. Part I. The flow of plastic material in hot- and cold-walled circular channels'. *Poly. Engng and Sci.,* **15,** 1975, 553–68.
143. WANG, V. W., WANG, K. K. and HIEBER, C. A., 'An iterative computer program for runner-system design in injection molding', *ANTEC,* 1983, 663–5.
144. UPADHYAY, R. K. and ISAYEV, A. I., 'Experimental and theoretical investigation of polymer flow in a channel with abrupt contraction or expansion', *ANTEC,* 1983, 714–16.
145. ISAYEV, A. I., UPADHYAY, R. K. and SHEN, S. F., 'Experimental and theoretical investigation of polymer flow in a converging and diverging channel', *ANTEC,* 1982, 298–301.
146. ISAYEV, A. I. and CHUNG, B., 'Flow of polymeric melts in short tubes', *ANTEC,* 1984, 433–8; *Poly. Engng and Sci.,* **25,** 1985, 264–70.
147. HIEBER, C. A., WANG, V. W., WANG, K. K. and CHUNG, B., 'The effect of processing conditions on the pressure losses in a runner-gate-cavity system', *ANTEC,* 1984, 769–71.
148. ISAYEV, I., 'Pressure and velocity jump in channel flow of polymers—modelling and experiment', *ANTEC,* 1983, 710–13.
149. HIEBER, C. A., UPADHYAY, R. K. and ISAYEV, A. I., 'Non-isothermal polymer flow in noncircular runners', *ANTEC,* 1983, 698–700.
150. NAITOVE, M. H., 'Practical questions and answers on computerized mold filling analysis', *Plastics Technology,* April 1984, 74–9.
151. NAITOVE, M. H., 'More questions and answers on computer mold filling analysis', *Plastics Technology,* May 1984, 75–8.
152. GATES, R. D., 'The changing relationship between designer and mold-maker due to CAD/CAM', *ANTEC,* 1983, 907–8.
153. COLLINS, S. H., CAD/CAM: its implications for the plastics industry', *ANTEC,* 1983, 904–6.
154. MENGES, G., 'Computer-aided engineering: the coming revolution', *Plastics Engng,* August 1983, 37–42.
155. WANG, K. K., 'A system approach to injection moulding', *Polym.–Plast. Technol. Eng.,* **14,** 1980, 75–93.
156. ANON., *The Moldflow Philosophy,* Moldflow (Europe) Ltd, Chislehurst, Kent, UK and Moldflow Australia Pty, Boronia, Victoria.
157. AUSTIN, C., *Computer Simulation of the Flow of Plastic into an Injection Mold,* Soc. Plast. Engnrs USA, June 1982.
158. WATKINSON, K., *Evaluation of Moldflow Computer Aided Injection Mould Tool Design,* Members' Report No. 81, RAPRA, Shawbury, Shrewsbury, UK, 1983.
159. COX, H. W., MENTZER, C. C. and CUSTER, R. C., 'The flow of thermoplastic melts: experimental and predicted', *ANTEC,* 1983, 694–7.
160. MOHNBERG, H. J. and BACH, K., 'Das Moldflow System in Vergleich zwischen Theorie und Praxis', *Plastverarbeiter,* **34,** 1983, 1087–91.

161. ANON., *Introducing the Moldflow IBM/PC XT System,* Moldflow (Europe) Ltd, Chislehurst, Kent, UK and Moldflow Australia Pty, Beronia, Victoria.
162. BUCKINSKI, A. M., 'An integrated CAD/CAM system for plastic injection molds', *ANTEC,* 1983, 909–11.
163. SINGH, K. J., 'Complete mold filling analysis of thermoplastics with a single computer run', *ANTEC,* 1984, 783–5.
164. MENGES, G. and HAACK, W., 'How to design a moulding part to suit particular plastic rules—the computer as an important aid', *ANTEC,* 1984, 954–8.
165. BANGERT, H. and MENGES, G., 'Systematic design of injection moulds and use of computers', *Polymer Engng Reviews.,* **1,** 1981, 113–68.
166. KEMPER, W., GEISBÜCH, P., SCHMIDT, L. and MENGES, G., 'Rheological layout of injection moulds', *ANTEC,* 1981, 767–70.
167. KEMPER, W., BANGERT, H., LICHIUS, U., PAAR, M., SCHLÜTER, H., SCHMIDT, L. and SCHMIDT, J., 'Kalkulations—und Konstruktionshilfen beim Entwurf von Spritzgiesswerkzeugen', *IKV Colloquium No. 11,* 1982, pp. 239–75.
168. BERNHARDT, E. C. (Ed.) *Computer Aided Engineering for Injection Moulding,* Hanser, for Soc. Plastics Engnrs, USA, 1983.
169. RYDER, L. B., 'Faster cooling for blow-molding', *Plastics Engng,* May 1975, 18–25.
170. SINGH, E. J. and WANG, H-P., 'Computer analysis to optimize mold cooling process', *ANTEC,* 1982, 330–7.
171. (a) CARSLAW, H. S. and JAEGER, J. C., *Conduction of Heat in Solids,* Clarendon Press, Oxford, UK, 1959, pp. 282–3.
171. (b) *Ibid.,* pp. 87–8.
171. (c) *Ibid.,* pp. 121–3.
172. OCKENDON, J. R. and HODGKINS, W. R. (Eds), *Moving boundary problems in heat flow and diffusion,* Oxford University Press, Oxford, 1975.
173. ATTHEY, D. R., 'A finite difference scheme for melting problems', *J. Inst. Maths. Applics.,* **13,** 1974, 353–66.
174. VOLLER, V. R., CROSS, M. and WALTON, P. G., 'Assessment of weak solution numerical techniques for solving Stefan problems', *Proc. Int. Conf. Numerical Methods in Thermal Problems.* R. W. Lewis and K. Morgan, Eds, Pineridge Press, Swansea, UK, 1979, p. 172.
175. MENGES, G., KULIK, M. and RHIEL, F., 'Entformungstemperatur und Kühlzeit bei der Herstellung extrusionsgeblasener Hohlkörper', *Plastverarbeiter,* **24,** 1973, 685–90.
176. DIETZ, W., 'Refroidissement des plastiques; methods de calcul', *Plastiques Modernes et Elastomers,* May 1975, 104–11.
177. SINGH, K. J., 'Computer software for plastic cooling analysis—a new approach', *ANTEC,* 1984, 962–4.
178. KAMAL, M. R. and KALYON, D., 'Heat transfer and microstructure in extrusion blow-molding', *Poly. Engng and Sci.,* **23,** 1983, 503–7.
179. GUTFINGER, C., BROYER, E. and TADMOR, Z., 'Melt solidification in polymer processing', *Poly. Engng and Sci.,* **15,** 1975, 515–24.

180. RYAN, M. E. and DUTTA, A., 'Mathematical modeling of the blow moulding process', *Poly. Engng and Sci.*, **22,** 1982, 1075–82.
181. EDWARDS, M. F., SUVANAPHEN, P. K. and WILKINSON, W. L., 'Heat transfer in blow moulding operations', *Poly. Engng and Sci.*, **19,** 1979, 910–16.
182. EDWARDS, M. F., GEORGHIADES, S. and SUVANAPHEN, P. K., 'A study of the cooling of blow-moulded objects', *Plastics and Rubber Processing and Applications,* **1,** 1981, 161–5.
183. DIETZ, W., 'A cooling time model for plastics processing operations', *Poly. Engng and Sci.*, **18,** 1978, 1030–6.
184. DIETZ, W., 'Die Wärme—und Temperaturleitfähigkeit von Kunststoffen', *Colloid and Polymer Sci.*, **255,** 1977, 755–72.
185. BONANCINA, C., COMINI, G., FASANO, A. and PRIMICERIO, M., 'Numerical solution of phase-change problems', *Int. J. Heat and Mass Transfer,* **16,** 1973, 1825–32.
186. PHAM, Q. T., 'A fast, unconditionally stable finite-difference scheme for heat conduction with phase change', *Int. J. Heat and Mass Transfer,* **28,** 1985, 2079–84.
187. EDWARDS, M. F., Private communication, 1984.
188. TACKE, K-H., 'Discretization of the explicit enthalpy method for planar phase change', *Int. J. for Num. Meth. in Engng,* **21,** 1985, 543–54.
189. COMINI, G., DEL GUIDICE, S., LEWIS, R. W. and ZIENKIEWICZ, O. C., 'Finite element solution of non-linear heat conduction problems with special reference to phase change', *Int. J. Num. Meth. Engng,* **8,** 1974, 613–24.
190. MORGAN, K., LEWIS, R. W. and ZIENKIEWICZ, O. C., 'An improved algorithm for heat conduction problems with phase change', *Int. J. Num. Meth. Engng,* **12,** 1978, 1191–5.
191. LEWIS, R. W., MORGAN, K. and ROBERTS, P. M., 'Determination of thermal stress in solidification problems', in *Numerical Analysis of Forming Processes,* J. F. T. Pittman, R. D. Wood, O. C. Zienkiewicz and J. M. Alexander, Eds, J. Wiley & Sons, Chichester, Sussex, 1984, pp. 405–31.
192. SAMONDS, M., MORGAN, K. and LEWIS, R. W., 'Finite element modelling of solidification in sand castings employing an implicit–explicit algorithm', *Appl. Math. Modelling,* **9,** 1985, 170–4.
193. LEWIS, R. W., MORGAN, K. and ROBERTS, P. M., 'Application of an alternating-direction finite element method to heat transfer problems involving phase change', *Numerical Heat Transfer,* **7,** 1984, 471–82.
194. PITTMAN, J. F. T. and NAKAZAWA, S., 'A finite element system for analysis of melt flow and heat transfer', in *Numerical Methods in Industrial Forming Processes,* Proc. Int. Conf., Swansea, J. F. T. Pittman, R. D. Wood, J. M. Alexander and O. C. Zienkiewicz, Eds, Pineridge Press, Swansea, UK, 1982, pp. 523–34.
195. PITTMAN, J. F. T. and NAKAZAWA, S., 'Finite element analysis of polymer processing operations', *Numerical Analysis of Forming Processes,* J. F. T. Pittman, R. D. Wood, O. C. Zienkiewicz and J. M. Alexander, Eds, J. Wiley & Sons, Chichester, Sussex, 1984, pp. 165–218.

196. Kenig, S. and Kamal, M. R., 'Temperature profiles in the injection molding of thermoplastics', *ANTEC,* 1982, 595–603.
197. Sifleet, W. L., Dinos, N. and Collier, J. R., 'Unsteady-state heat transfer in a crystallizing polymer', *Poly. Engng and Sci.,* **13,** 1973, 10–14.
198. Kamal, M. R. and Chu, E., 'Isothermal and nonisothermal crystallization of polyethylene', *Poly. Engng and Sci.,* **23,** 1983, 27–31.
199. Malkin, Ya. A., Beghisev, V. P., Keapin, I. A. and Andrianova, Z. S., 'General treatment of polymer crystallization kinetics—Part 2. The kinetics of nonisothermal crystallization', *Poly. Engng and Sci.,* **24,** 1984, 1402–8.
200. Tseng, M. M., 'Computer cooling analysis to select bubbler or baffle', *ANTEC,* 1984, 959–61.
201. Schauer, K. R., 'The MOLDCOOL computer-aided design program', *ANTEC,* 1983, 863–64.
202. Anon., 'Mold cooling software now for blow molding, and mold heating for thermosets', *Plastics Technology,* April 1984, 11.
203. Pittman, J. F. T. and Mahmoudzadeh, H., 'Performance of melt thermocouples. 1. The flush-mounted type', *Poly. Engng Rev.,* **3,** 1983, 63–74.
204. Pittman, J. F. T. and Mahmoudzadeh, H., 'Performance of melt thermocouples. 2. The upstream-pointing parallel-to-flow type', *Poly. Engng Rev.,* **3,** 1983, 75–107.
205. Van Dam, M. and Janeschitz-Kriegl, H., 'Temperature measurement and heat transfer in flowing polymer melts', *Int. J. Heat and Mass Transfer,* **28,** 1985, 395–406.
206. Vlachopoulos, J. and Mitsoulis, E., 'Fluid flow and heat transfer in calendering', To appear in *Advances in Transport Processes,* J. Wiley and Sons, New York.
207. Kiparissides, C. and Vlachopoulos, J., 'Finite element analysis of calendering', *Poly. Engng and Sci.,* **16,** 1976, 712–19.
208. Kiparissides, C. and Vlachopoulos, J., 'A study of viscous dissipation in the calendering of power-law fluids', *Poly. Engng and Sci.,* **18,** 1978, 210–14.
209. Agassant, J. F. and Espy, M., 'Theoretical and experimental study of the molten polymer flow in the calender bank', *Poly. Engng and Sci.,* **25,** 1985, 118–21.
210. Mitsoulis, E., Vlachopoulos, J. and Mirza, F. A., 'Calendering analysis without the lubrication approximation', *Poly. Engng and Sci.,* **25,** 1985, 6–18.
211. Pittman, J. F. T. and Nakazawa, S., 'Finite element calculation of history dependent quantities in polymer processing', in *Numerical Methods in Industrial Forming Processes,* Proc. Int. Conf., Swansea, J. F. T. Pittman, R. D. Wood, J. M. Alexander and O. C. Zienkiewicz, Eds, Pineridge Press, Swansea, UK, 1982, pp. 597--609.
212. Mohr, W. C., Saxton, R. L. and Jepson, C. H., 'Theory of mixing in single-screw extruder', *Ind. Engng Chem.,* **49,** 1957, 1857–62.
213. Pinto, G. and Tadmor, Z., 'Mixing and residence time distribution in melt screw extruders', *Poly. Engng and Sci.,* **10,** 1970, 279–88.

214. BIGG, D. M. and MIDDLEMAN, S., 'Mixing in a screw extruder. A model for residence time distribution and strain', *Ind. Engng Chem. Fundam.*, **13,** 1974, 66–71.
215. BIGG, D. M., 'On mixing in polymer flow systems', *Poly. Engng and Sci.*, **15,** 1975, 684–9.
216. OTTINO, J. M., RANZ, W. E. and MACOSKO, C. W., 'A framework for description of mechanical mixing of fluids', *A.I.Ch.E. J.*, **27,** 1981, 565–74.
217. OTTINO, J. M. and CHELLA, R., 'Laminar mixing of polymeric liquids; A brief review and recent theoretical developments', *Poly. Engng and Sci.*, **23,** 1983, 357–79.
218. PITTMAN, J. F. T., 'A new approach to the analysis of mixing in viscous liquids', *Proc. Multistream 84. IChE* (UK) Symposium Series No. **94,** 1984, 3.10–18.
219. CHELLA, R. and OTTINO, J. M., 'Fluid mechanics of mixing in a single-screw extruder', *Ind. Eng. Chem. Fundam.*, **24,** 1985, 170–80.
220. BIGG, D. and MIDDLEMAN, S., 'Laminar mixing of a pair of fluids in a rectangular cavity', *Ind. Eng. Chem. Fundam.*, **13,** 1974, 184–90.
221. ERWIN, L. and MOKHTARIAN, F., 'Analysis of mixing in modified single screw extruders', *Poly. Engng and Sci.*, **23,** 1983, 49–60.
222. GIVLER, R. C., CROCHET, M. J. and PIPES, R. B., 'Numerically predicted fiber orientation in dilute suspensions', in *Numerical Methods in Industrial Forming Processes,* Proc. Int. Conf., Swansea, J. F. T. Pittman, R. D. Wood, J. M. Alexander and O. C. Zienkiewicz, Eds, Pineridge Press, Swansea, UK, 1982, pp. 559–76.
223. GILLESPIE, J. W. JR, VANDERSCHUREN, J. A. and PIPES, R. B., 'Process induced fiber orientation. Numerical simulation with experimental verification', *Polymer Composites,* **6,** 1985, 82–6.
224. FOLGAR, F. and TUCKER, C. L. III, 'Orientation behaviour of fibers in concentrated suspensions', *J. of Reinforced Plastics and Composites,* **3,** 1984, 98–119.
225. JACKSON, W. C., FOLGAR, F. and TUCKER, C. L. III, 'Prediction and control of fiber orientation in molded parts', in *Polymer Blends and Composites in Multiphase Systems,* C. D. Han, Ed., Adv. Chem. Ser. (ACS) No 206, 1984, 279–99.
226. TUCKER, C. L. III. 'Compression molding of polymers and short fiber composites', in *Injection and Compression Molding Fundamentals,* A. I. Isayev, Ed., Marcel Dekker, New York, 1984.
227. MODLEN, G. F., 'Re-orientation of fibres during mechanical working', *J. Mat. Sci.*, **4,** 1969, 283–9.
228. INGEN HOUSZ, J. F., 'Thermal load in the processing of thermoplastic', *Poly. Engng and Sci.*, **17,** 1977, 828–31.
229. MORETTE, R. A. and GOGOS, G. G., 'Viscous dissipation in capillary flow of rigid PVC and PVC degradation', *Poly. Engng and Sci.*, **8,** 1968, 272–80.
230. THOMPSON, E. G., PITTMAN, J. F. T. and ZIENKIEWICZ, O. C., 'Some integration techniques for the analysis of viscoelastic flows', *Int. J. Num. Methods in Engng,* **3,** 1983, 165–77.

Chapter 7

SOLID PHASE COMPACTION OF POLYMERIC POWDERS

R. CRAWFORD

Department of Mechanical and Industrial Engineering, The Queen's University of Belfast, UK

1. INTRODUCTION

The technique of fabricating articles or preforms by the application of pressure to cold powdered materials is commonplace in a number of industries. In particular it is widely used with ceramics, pharmaceutical materials and metals.[1-10] It is not surprising therefore that, although there are a wide range of melt processing methods for plastics, there has in recent years been an interest in cold forming of plastic powders.

The initial attraction of solid phase compaction of polymeric materials is that it provides a method of manufacturing articles in the speciality plastics such as ultra-high molecular weight polyethylene, polytetrafluoroethylene or poly(vinylidene chloride). These plastics have unique physical properties but are severely limited in application because they do not lend themselves to shaping by the conventional melt processing methods such as injection moulding or extrusion. However, it is now being recognised that solid phase compaction need not be restricted to these 'difficult' materials and indeed, when it is used with the more common plastics, there are special advantages. For example, because it is more energy efficient, it is not limited to thin wall sections, it permits higher levels of fillers to be incorporated and can give faster cycle times since the relatively slow heating and cooling stages are removed from the cycle. There is also the exciting possibility of producing alloys of different materials—either polymer/metal combinations, to produce new materials with uniquely tailored properties, or polymer A/polymer B combinations to produce new polymeric compounds not previously considered feasible because of incompatible melt processing characteristics.

2. COMPACTION TECHNIQUES

It has been envisaged that cold compaction of polymeric powders can develop along two separate routes.[11] These are illustrated in Fig. 1.

2.1. Direct Compaction

The first alternative is to mould plastic articles or preforms by the direct compaction of cold powder. It is very likely that a plastic article

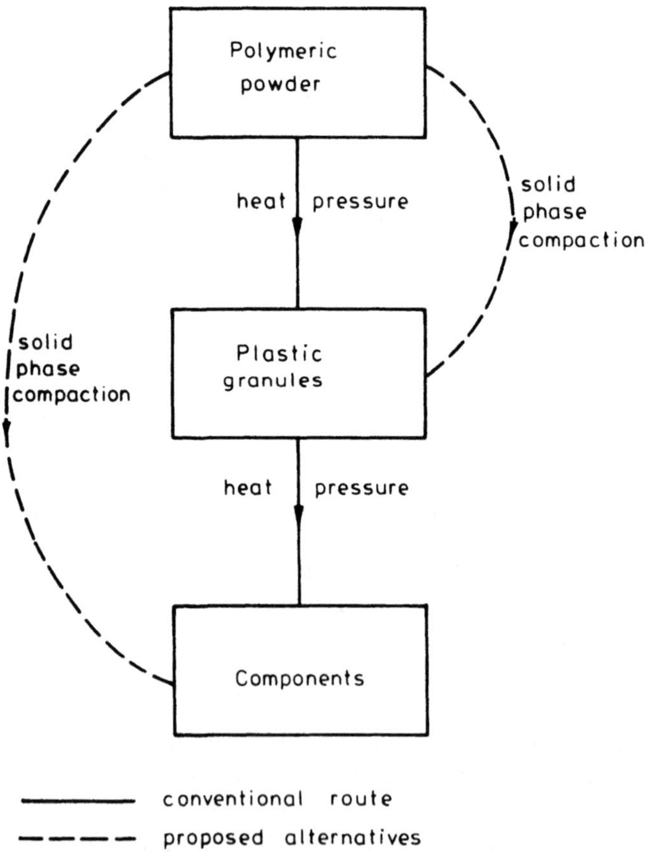

FIG. 1. Production of plastic components from polymeric powder.

produced in this way would also require a sintering operation although the 'green' compact of a number of polymeric powders has been found to be quite strong. However, even with a sintering stage the temperatures involved would be much less than those in melt processing, so among the advantages of the cold forming method would be energy savings and the elimination of stabilisers and processing aids.

2.2. Pelletting

The second alternative illustrated in Fig. 1 is to produce pellets by cold compaction, which could then be used in conventional processing equipment, thereby removing one of the heat/pressure stages in the production of the component. Unlike the previous method, it is not necessary to obtain very high strengths because the pellet only needs to be sufficiently strong to withstand handling. If the technical difficulties involved with the very high production rates can be overcome, this method of pelletising is potentially very attractive. In particular, it can alleviate problems of thermal degradation and the harmful effects of fumes from additives such as fire retardants. It is envisaged that pelletising by a cold compaction route would offer considerable savings in energy and capital cost of equipment. It is also likely that the properties of the end-product would be improved through the use of a better quality feedstock.

2.3. Early Work on Solid Phase Compaction

In 1972 Maeda[12] referred to cold forming of plastic powders in his work on cold working of thermoplastics. Then in 1974 Palmer and Rowe[13,14] reported on some studies on the porosity of articles produced by compressing poly(vinyl chloride) powders.

In 1975 Bigg[15] and Maeda and Matsuoka[16,17] described the results of the first serious attempts to use solid phase compaction as a moulding method. Bigg found that most semi-crystalline polymers and non gas-evolving, reactive polymers could be processed in this way but no success was obtained with glassy polymers. Maeda and Matsuoka successfully used polyethylene, polypropylene and, in contrast to the findings of Bigg, poly(vinyl chloride). In 1976 Jayaraman *et al.*[18] confirmed that a glassy polymer (polystyrene) could in fact be moulded successfully by cold compaction. In 1977 Bigg[19,20] followed up with the results of experiments to establish the effects of cold compaction variables and in this case eleven polymers, ranging from poly(amide–imide) to polysulphone, were studied. At this time Halldin and Kamel[21] also carried out an in-depth study of the compaction characteristics of ultra-high molecular weight polyethylene (UHMWPE). In 1979 Kulkami[22] provided a review of solid phase forming of plastics and subsequently Crawford *et al.*[23–29] presented detailed information covering all aspects of the cold compaction of a range of polymeric powders. It is interesting to note that at this time Dean[30] also reported work on the cold compaction of polymeric granules (as opposed to powder). In this case it was found that, provided flow could be ensured when the granules were under pressure, then useful articles could in fact be moulded in this way.

It is apparent therefore from the studies which have been made over the past 10 years that solid phase compaction of polymeric powders is a useful and viable method for manufacturing plastic articles.

To date the main methods of manufacture have involved either isostatic compaction of powders in flexible dies or pressing of powders in rigid metal dies using single-end or double-end compaction. In the first case the powder is placed in a rubber mould which is sealed and immersed in an oil bath. The oil is then pressurised so that the powder in the mould may be squeezed into shape. In the second case the powder is placed in a metal mould and a close-fitting piston comes down on top of the powder to press it into shape. These methods are already well established in ceramics, pharmaceuticals, agricultural and powder metallurgy industries.[1–3] However, in order that they can be used correctly and efficiently, for plastic powders it is essential to have an understanding of the effects of the compaction variables.

These include compaction pressure, compaction rate, powder characteristics, sintering conditions and die geometry. In the following sections the current information which is available will be reviewed for a wide range of plastics.

3. COMPACTION PRESSURE

3.1. Effect on Density and Strength

The main effect of pressure is to increase the density and the strength of the compact. Bigg,[20] Maeda,[16] Jayaraman[18] and Crawford[11] reported this for all their materials tested and typical results are shown in Figs 2–5. Referring to Fig. 5, it may be seen that for PVdC (poly(vinylidene chloride)) copolymer and PVC (poly(vinyl chloride))

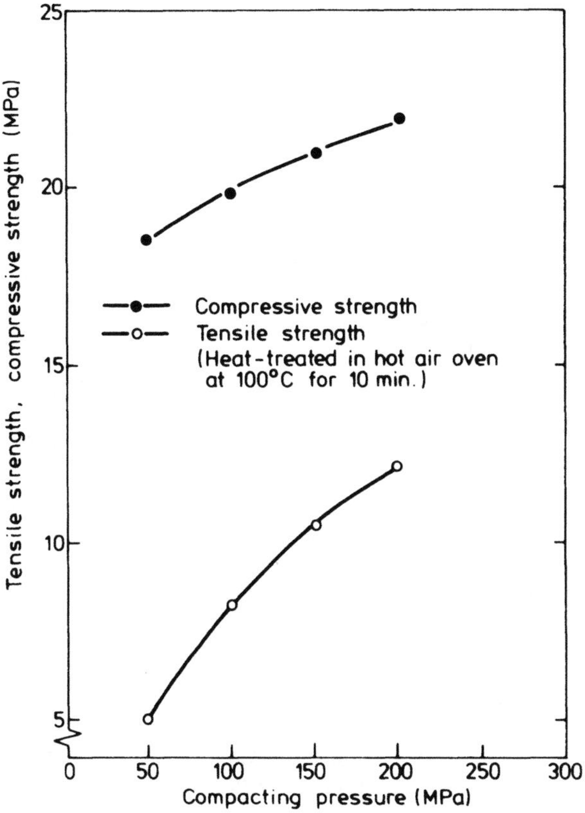

FIG. 2. Effects of compacting pressure on compressive and tensile strength of statically cold compacted PVC (Maeda[16]).

there is a terminal densification as the compaction pressure is increased. For PVdC copolymer this occurs at about 500 MPa, whereas in PVC, terminal densification occurs at 350 MPa. Bigg[20] found this value to be 138 MPa for PTFE (polytetrafluoroethylene) and 690 MPa for PPO (poly(phenylene oxide)). Halldin and Kamel[21] observed a value of 100 MPa for UHMWPE. The tensile strength of PVdC copolymer rises with an increase in density and reaches a maximum value of about 10 MPa as shown. The tensile strengths of the unsintered PVC products were very low, which made testing very difficult and so no strength values are shown.

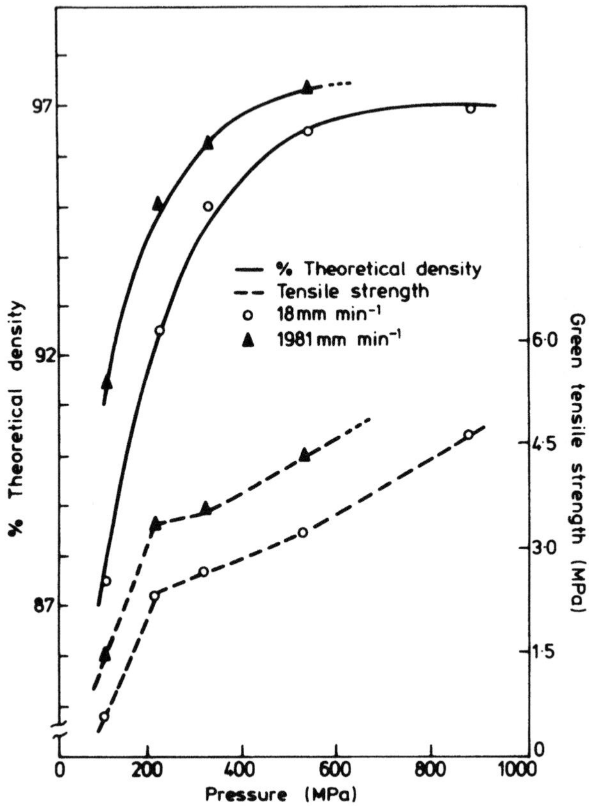

FIG. 3. Effect of pressure on density and tensile strength of polystyrene at two pressing speeds (Jayaraman *et al.*[18]).

In Fig. 6(a)–(c) the effect of pressure on the microstructure of PVdC compacts may be observed. These scanning electron micrographs substantiate the experimental observation regarding the general increase in tensile strength and density with pressure. In Fig. 6(a) mutual sliding and reorientating of the particles has taken place but discrete particles are very easily seen and there is little or no evidence of frictional heating. The porosity in the compacts seems quite high and the small tensile strength achieved is attributed to electrostatic forces. In Fig. 6(b) it may be seen that at a compaction pressure of 310 MPa, porosity has greatly decreased and the overall average size of particles is smaller. Figure 6(c) shows that at 650 MPa there is little

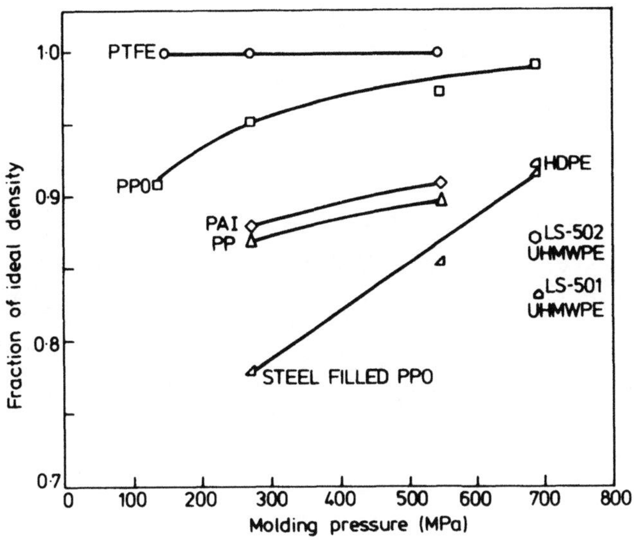

FIG. 4. Preform density versus moulding pressure for polymeric powders (Bigg *et al.*[20]).

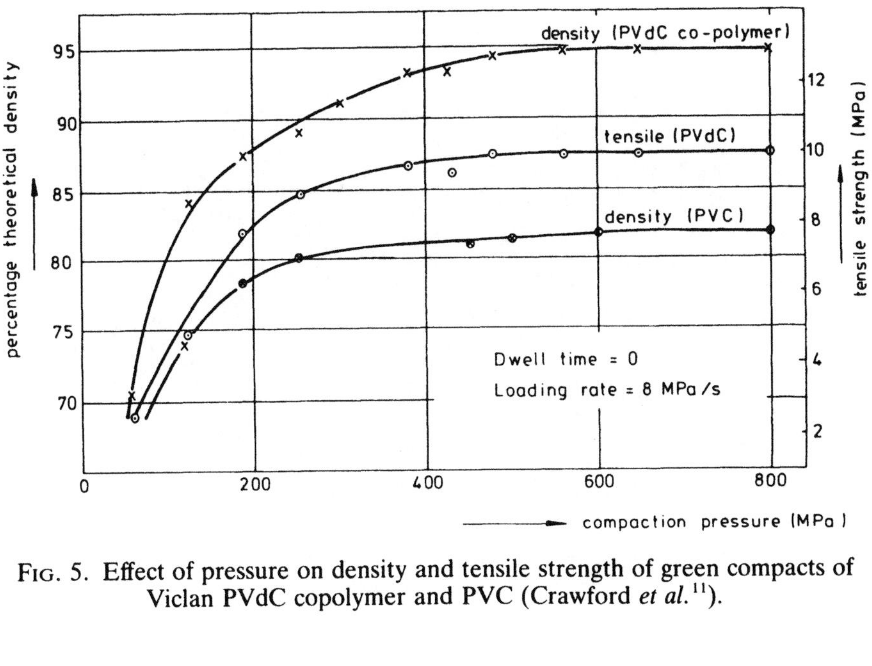

FIG. 5. Effect of pressure on density and tensile strength of green compacts of Viclan PVdC copolymer and PVC (Crawford *et al.*[11]).

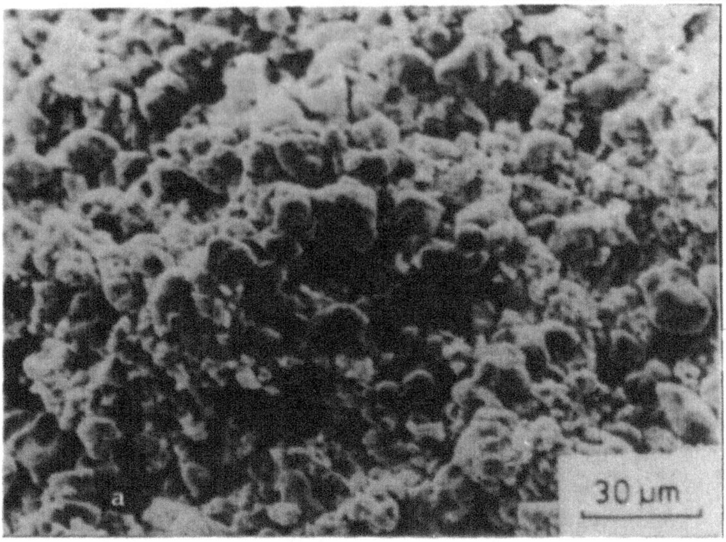

FIG. 6(a). PVdC copolymer compacted with a pressure of 50 MPa, loading rate 8 MPa/s and zero dwell time.

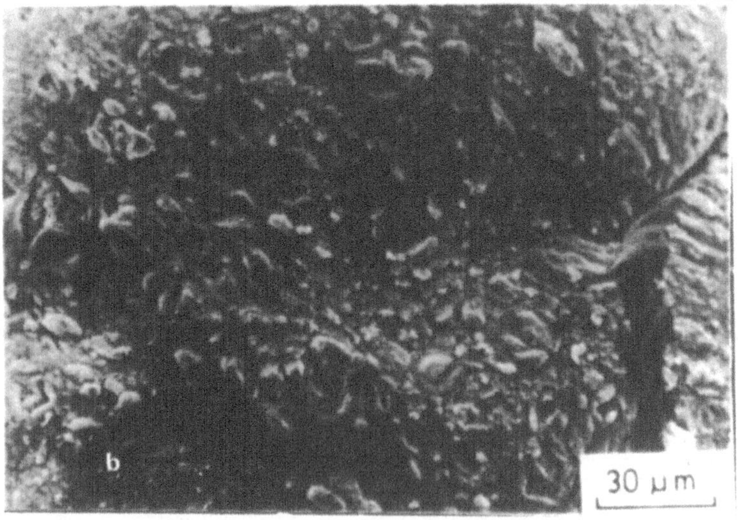

FIG. 6(b). PVdC copolymer compacted with a pressure of 310 MPa, loading rate 8 MPa/s and zero dwell time.

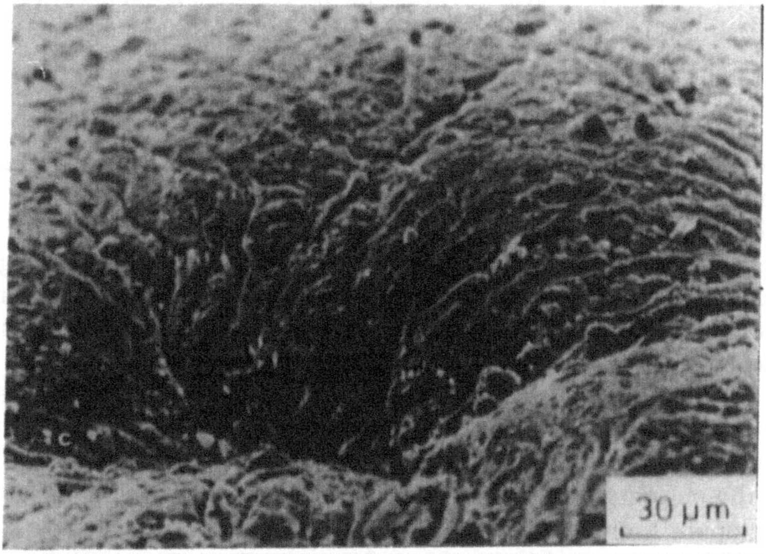

FIG. 6(c). PVdC copolymer compacted with a pressure of 650 MPa, loading rate 8 MPa/s and zero dwell time.

evidence from the microstructure of the material that it was ever of a particulate nature. It should be noted that the closer the preform density is to the density of the bulk polymer, the greater is the contact area for subsequent sintering. Palmer and Rowe[13,14] made quantitative measurements of the porosity of compacted PVC samples and observed that the porosity decreased rapidly at the lower pressures and more slowly at the high pressures, which agrees with the trends shown in Figs 1–5.

3.2. Pressure Losses

It is important to note that during compaction there are considerable frictional losses within the powder mass and between the powder and the die walls. This has a major influence on the quality of the product because it results in pressure gradients within the powder.

3.2.1. Measurement Techniques

There have been many direct and indirect methods for measuring the stresses within the powder mass and the pressure exerted on the die wall during compaction. Train[31] measured the pressure distribution within magnesium carbonate powder by embedding manganin gauges in the powder prior to compaction. McLeod and Marshall[32] measured the internal stress distribution in uranium dioxide powder using autoradiography whereas Takada *et al.*[33] utilised a photoelastic technique by compacting their powders (carbon black, carborundum glass and lactose) in an epoxy die. Seelig and Wulff[34] used a soft wire grid embedded in iron powder and the deformation of the grid was monitored by radiography. Duwez and Zwell,[35] in their tests with copper and stainless steel powders, concentrated on measuring the reactions at the die/powder interface using pressure transducers mounted in the wall of the die. Strijbos *et al.*[36] have also used this technique for their investigations with ferric oxide powder.

Crawford *et al.*[24,29] have used an instrumented die to measure the pressure at various points along the compaction punch and the die wall during single-end compaction of polymeric powders. One of the most useful outcomes of this work was the establishing of a relation between the applied force F_1 and the force F_2, transmitted to the bottom of the die. This takes the form

$$\frac{F_2}{F_1} = K^{h/D} \tag{1}$$

where K is a material constant, h is the height of the compact and D is the die diameter.

Figure 7 shows the variation in F_2 and compact height h for a constant rate of increase of F_1. These are average values across the compaction ram and die base and so measurements were also taken at specific points within the die. Typical results are shown in Figs 8 and 9. From the tests on PVdC it may be seen that at a compaction pressure of 200 MPa, the radial stress decreases linearly from a maximum at the top to almost zero at the bottom. On the lower punch the axial pressure is greatest at the centre and decreases in an exponential fashion towards zero at the outer edge. On the upper (moving) punch the pressure variation is the reverse of that observed on the lower punch, i.e. the axial pressure is lowest at the centre of the punch and then there is an exponential increase in pressure towards the outside edge. The powder in the top outside edges will therefore experience the highest axial and radial pressures.

3.2.2. Pressure Distribution

From the measurements taken on non-polymeric powders there has been some disagreement about the shape of the pressure distribution

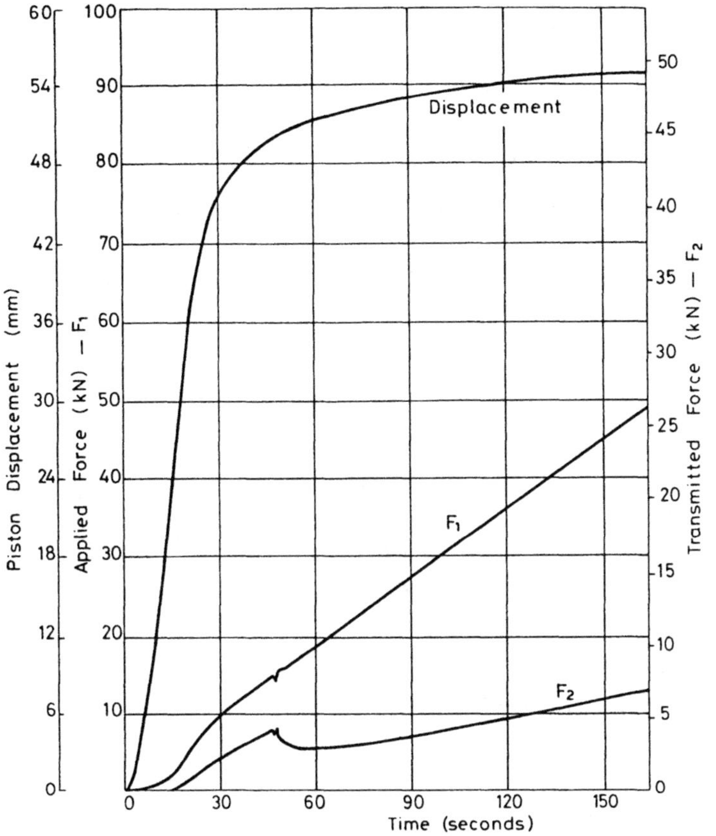

FIG. 7. Compaction characteristics for PVdC at a pressing rate of 1 MPa/s (Crawford *et al.*[27]).

on the walls of the die. Duwez and Zwell[35] suggested a linear relationship for copper and stainless steel powders, whereas Strijbos *et al.*[36] found that for ferric oxide powder the side wall pressure was constant for the top one-third to one-half of the die and then decreased linearly towards the bottom of the die. It may be seen in Fig. 8 that for PVdC powder both types of distribution are observed depending on the applied axial pressure. At the highest compaction pressure used (400 MPa), there is even evidence of the decrease of radial pressure close to the moving punch as reported by Takada *et al.*[33] This is probably a direct result of the friction between the compact and the upper punch (analogous to the barrelling effect

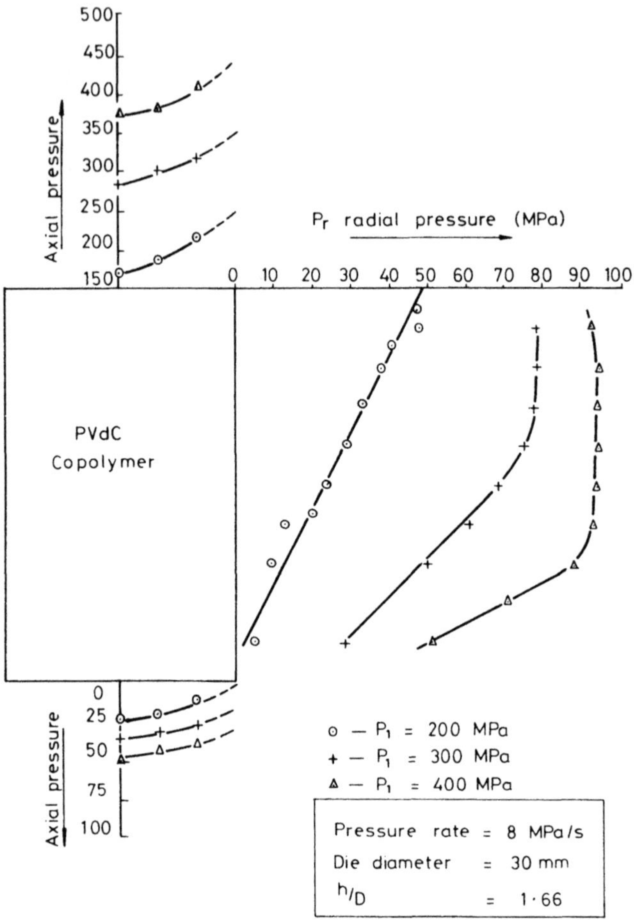

FIG. 8. Effect of compaction pressure on the boundary stress distributions for PVdC copolymer (Crawford *et al.*[24]).

observed when a cylindrical billet of metal is compressed between flat platens).

The reason for the change from a linear to a non-linear radial pressure profile is not immediately obvious. It is considered to be related to the sequence of events when the loose powder mass in the die is transformed into a solid object. At low pressures the successive stages involve reorganisation of particles and elastic deformation at contact points. At higher pressures the deformations become non-

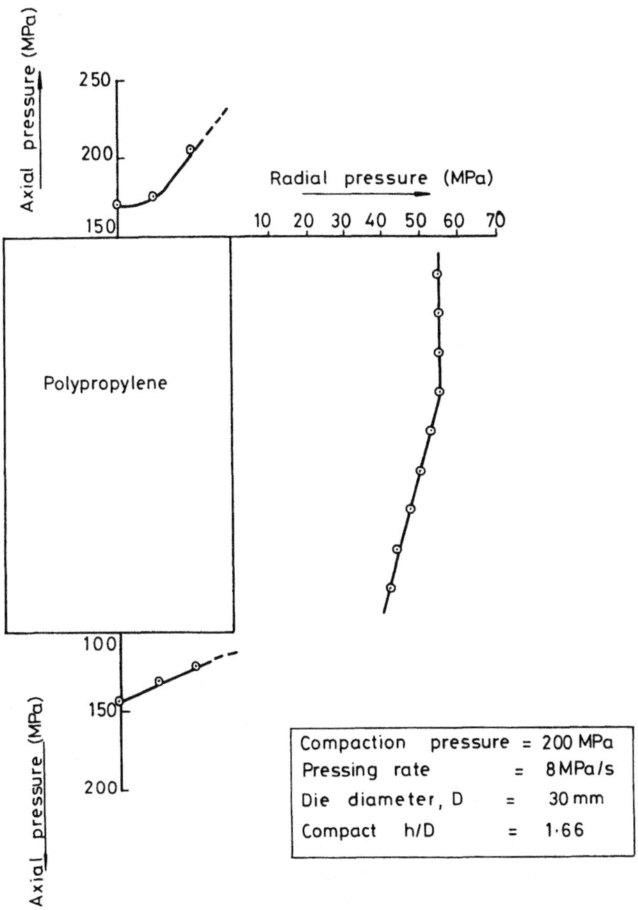

FIG. 9. The boundary stress distribution during compaction of polypropylene (Crawford *et al.*[24]).

recoverable and eventually, when bonding between the particles has occurred, the radial stresses will be due to the bulk compression of a solid object. It is interesting to note that the 'knee' in the radial pressure distributions occurs at the section of the compact above which the density has reached 95% of the theoretical density of the material. This has been confirmed in additional tests on PVdC at 350 and 450 MPa. It also explains why no 'knee' is observed at a compaction pressure of 200 MPa; although this pressure produces a solid compact, the maximum density which can be achieved is only about 92%. This line of reasoning suggests that the non-linear radial pressure distribution would only be observed in PVdC for pressures in excess of 265 MPa.

The results in Fig. 9 are interesting in that this particular type of polypropylene could not be compacted even at quite high applied pressures. In all cases when the axial pressure was released from the powder it could be poured from the die and scanning electron micrographs of the powder before and after compaction showed no changes in particle size or shape. The results in Fig. 9 suggest that even at the relatively low pressure of 200 MPa the material is undergoing sufficient densification to behave like a solid object even though there is no bonding between particles and their deformations are fully recoverable when the pressure is released.

3.2.3. Use of Hardness and Density Measurements

Measurements of hardness and density variations across polymeric compacts have provided valuable information to confirm the types of pressure variations shown in Fig. 8.[29] A typical plot of hardness variation across and along the centre of a 19 mm diameter cylindrical compact is shown in Fig. 10. It has been found[23] that there is a direct relationship between hardness and density of the compacts so this diagram also gives an indication of the density variations in a cold formed article. Of particular interest in this type of compaction (single-ended in a cylindrical die) is the fact that maximum densification occurs at the upper end where the compaction pressures are high. It is also significant that under all compaction conditions there is an area of hard (high density) material in the lower centre of the compact. This is particularly clear in Fig. 11 and has been explained in terms of the resolved components of axial and radial forces set up due to the resisting effect of the die walls. McLeod and Marshall[32] measured the density variations within ceramic compacts using autoradiography and explained the regions of high density in terms of the particle movement within the powder mass. In their model it is visualised that a central core of material is forced down into the unconsolidated powder below it. This continues until the reaction from the lower punch exceeds the die wall reaction, at which point the material begins to move outwards radially. It is suggested that it is this change of direction which causes the high density zone, due to the large shearing forces which are set up in this area.

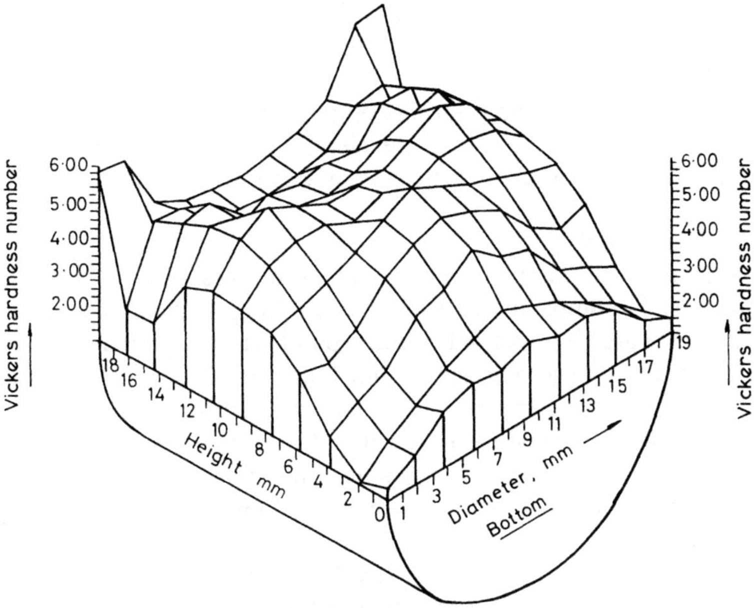

FIG. 10. Hardness distribution in PVdC compacts produced at 200 MPa ($D = 19$ mm) (Crawford *et al.*[24]).

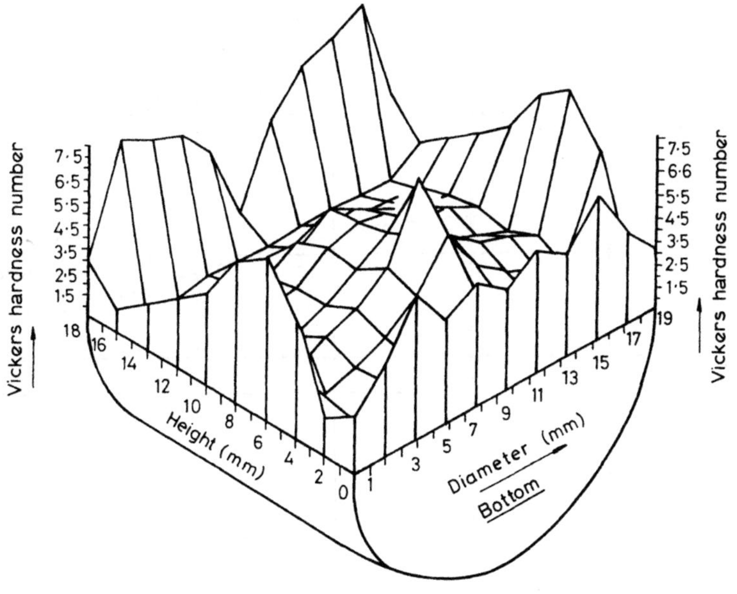

FIG. 11. Hardness distribution in PVdC compacts produced at 500 MPa ($h/D = 1$, $D = 19$ mm) (Crawford *et al.*[24]).

3.3. Die Tapering

The force losses occurring during compaction are very important from the design point of view because clearly it is desirable to have the structure of the compacted article as uniform as possible.

3.3.1. Taper Angle

Unfortunately in many cases the pressure loss can be very high, particularly in compacted articles that are long in the direction of the applied pressure. For example, from eqn (1) for a typical value of K of 0·3, when forming a cylinder 40 mm long and 20 mm diameter using single-ended compaction, the pressure at the bottom of the die is only 9% of the applied pressure at the top. It has been suggested that this problem might be alleviated in practical die designs by tapering the die section so that the loss in force would be compensated by a reduction in cross-sectional area.

Crawford and Sprevak[28] have explored the effect of compressing polymeric powders in tapered dies and their results are shown in Fig. 12. It was found that by using a tapered die it was possible to increase

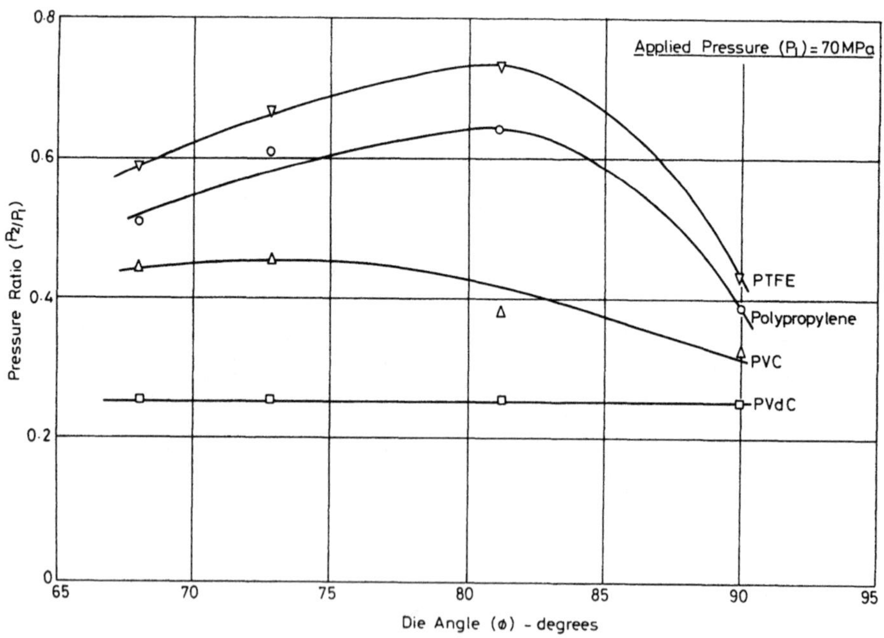

FIG. 12. Transmitted pressure ratio for four plastic powders in tapered dies (Crawford *et al.*[28]).

the transmitted pressure but the effect was not as large as would be expected. It was evident that there is a trade-off between increasing the taper angle in order to improve the area ratio but as a result of this, force is lost due to the increased resolved component at the die wall. For the polymeric powders tested the optimum taper angle was in the range 75–80°.

3.3.2. Force Loss Analysis

The force losses which occur during compaction in a tapered die may be analysed as follows:

Consider a section of tapered die in which the force at the top is F_t and the force at the bottom is F_b as illustrated in Fig. 13. From eqn (1) for a cylindrical die these forces would be related by

$$\frac{F_b}{F_t} = K^{h/D} \tag{2}$$

where K is a material constant associated with the friction properties of the die surface and the ratio of normal to axial forces at the die wall.

To a first approximation, this equation can be applied to the tapered die by considering it to be made up of n cylindrical slabs of height Δh where

$$\Delta h = h/n \tag{3}$$

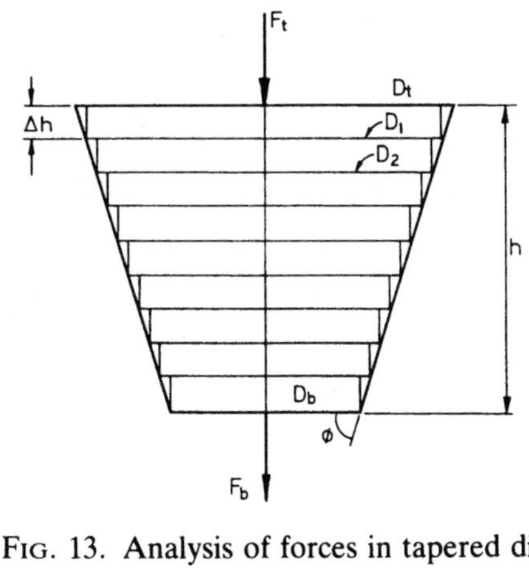

FIG. 13. Analysis of forces in tapered die.

For the first cylindrical section we have

$$D_1 = D_t - \frac{2\Delta h}{\tan\phi} \tag{4}$$

and for *i*th section

$$D_i = D_t - \frac{2i\Delta h}{\tan\phi} \tag{5}$$

It is assumed that eqn (2) is valid for each of the cylindrical sections. The force at the bottom of the section 1 is

$$F_1 = K^{\Delta h/D_1} \,.\, F_t\left(\frac{D_1}{D_t}\right)^2 \tag{6}$$

where the ratio $(D_1/D_2)^2$ is used to relate forces and pressures to the area at the bottom of the cylindrical section since this is the form of the model as illustrated in Fig. 13.

The force at the bottom of the tapered die section is then

$$F_b = K^{\Delta h/D_1 + \Delta h/D_2 + \cdots \Delta h/D_n} \,.\, F_t\left(\frac{D_b}{D_t}\right)^2 \tag{7}$$

which may be written as

$$F_b = K^{\gamma} \,.\, F_t\left(\frac{D_b}{D_t}\right)^2 \tag{8}$$

where

$$\gamma = \sum_{j=0}^{n}\left\{\frac{\Delta h}{D_t - \dfrac{2j\Delta h}{\tan\phi}}\right\}$$

Letting $n \rightarrow \infty$, then $\Delta h \rightarrow \mathrm{d}h$, the approximating die tends to the real die, and

$$\gamma \rightarrow \lim_{n\rightarrow\infty}\sum_{j=0}^{n}\left(\frac{\Delta h}{D_t - \dfrac{2j\Delta h}{\tan\phi}}\right) = \int_0^h \left(\frac{\mathrm{d}h}{D_t - \dfrac{2h}{\tan\phi}}\right)$$

that is

$$\gamma \rightarrow \frac{\tan\phi}{2}\int_{D_t}^{D_b}\left(\frac{\mathrm{d}u}{n}\right)$$

where

$$u = D_t - \frac{2h}{\tan \phi}$$

and

$$du = -\frac{2\,dh}{\tan \phi}$$

Thus

$$\gamma \rightarrow -\tan \phi \ln \left(\frac{D_t}{D_b}\right)$$

and

$$F_b = F_t K^{[1/2\tan\phi . \ln(D_t/D_b)]} . \left(\frac{D_b}{D_t}\right)^2 \tag{9}$$

or, in terms of pressures P_b and P_t,

$$\frac{P_b}{P_t} = K^{[1/2\tan\phi . \ln(D_t/D_b)]} \tag{10}$$

However, the parameter K should be a function of ϕ as the friction effect is related to the normal force acting on the walls of the cylinder. In the experiments reported earlier, it is possible to see the shape of this relationship. As expected $K(\phi) \rightarrow K$ as $\phi \rightarrow 90°$ and as a first approximation it would be expected that $K(\phi) \rightarrow 0$ as $\phi \rightarrow 0°$.

It would be convenient to relate $K(\phi)$ to K, the friction coefficient for the cylinder, and this may be done using an expression of the form

$$K(\phi) = K . f(\phi)$$

A function having the required properties is the logistic function

$$f(\phi) = \frac{\exp(a + b\phi)}{1 + \exp(a + b\phi)} \tag{11}$$

where a and b are constants.

It was found that eqn (10) gave predictions which agreed quite closely with the experimental results from the tapered dies.

4. COMPACTION RATE AND DWELL TIME

As well as the magnitude of the applied pressure, two other compaction variables are important as shown in Fig. 14. These are the compaction rate and the dwell time at maximum pressure.

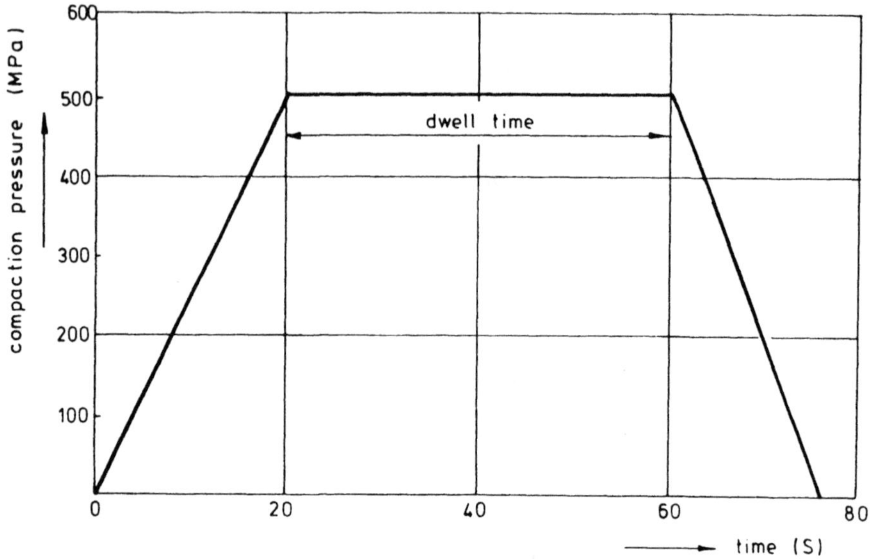

FIG. 14. Typical pressure–time profile obtained during static compaction.

4.1. Dwell Time

It is interesting to note that although clearly recognising the visco-elastic nature of polymeric powders, Bigg[15] and Maeda *et al.*[16] do not make any mention of time effects under pressure. On the other hand Jayaraman *et al.*[18] and Crawford and Paul[11] found a significant increase in density and strength of the compacts if the maximum pressure was maintained for some time as shown in Figs 15 and 16. Subsequently Bigg[19] tried hold times of 15 s and 5 min but observed no effect and decided on a standard dwell time of 2 min.

4.2. Compaction Rate

Maeda *et al.*[16] did not give any details on the effect of loading rate but did report that under dynamic compaction conditions the polyethylene compact was much stronger than under static compaction conditions. Jayaraman[18] and Crawford and Paul[11] studied this effect for selected compaction rates and both reported increases in strength and density which continued to exhibit an upward trend at the highest rates which could be achieved on their test equipment (see Figs 17 and 18). Recently Crawford[27] carried out a detailed study of the effect of

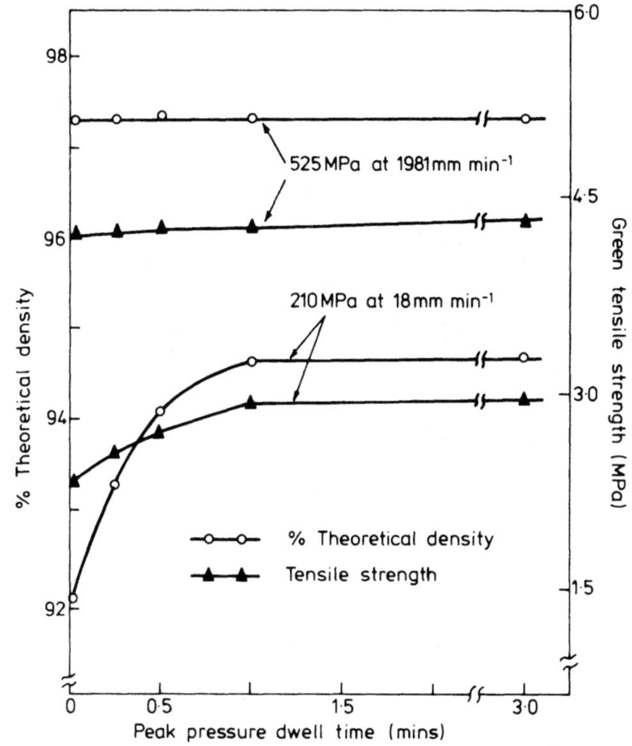

FIG. 15. Influence of peak presure dwell time for polystyrene (Jayaraman *et al.*[18]).

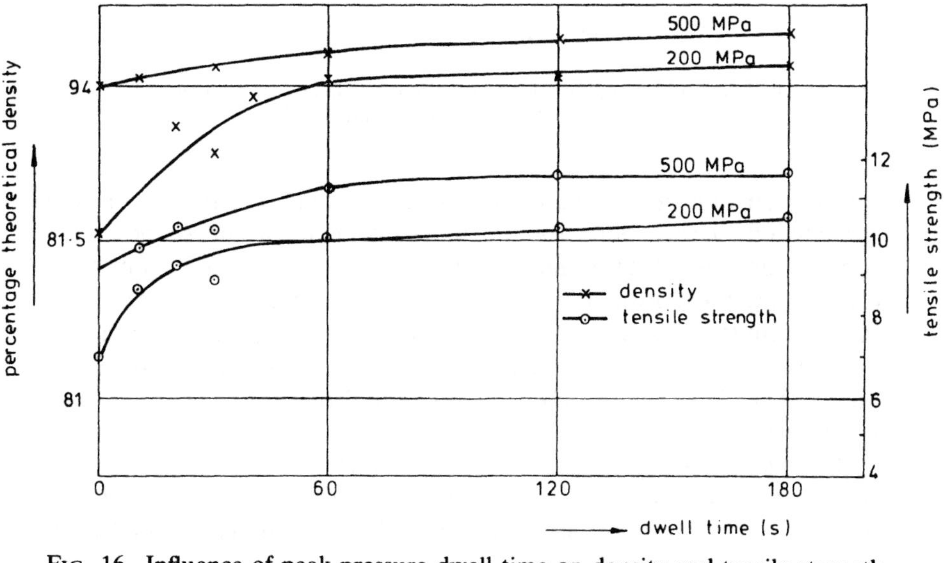

FIG. 16. Influence of peak pressure dwell time on density and tensile strength of Viclan PVdC copolymer (Crawford *et al.*[11]).

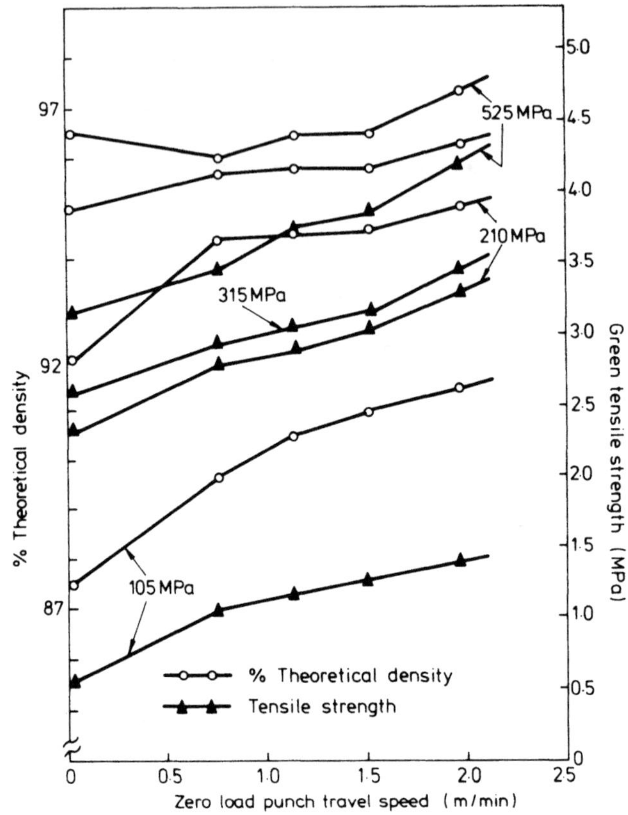

FIG. 17. Effect of pressing speed on density and tensile strength of polystyrene at various pressures (Jayaraman *et al.*[18]).

compaction rate on the forces transmitted through the powder during compaction. Using both force and stroke control for single-ended compaction it was found that for rates up to 8 MPa/s and 7·5 mm/s the transmitted force ratio is independent of the compaction rate. Thus the transmitted force may be estimated from eqn (1) where the 'material constant' K is independent of rate. At compaction rates higher than those quoted above it was found that the value of K, and hence the transmitted force, increased.

Blazynski *et al.*[37] have studied the effects of very high rates of compaction by using pressure waves generated by explosives. It was found that insofar as density is concerned no advantage accrued from

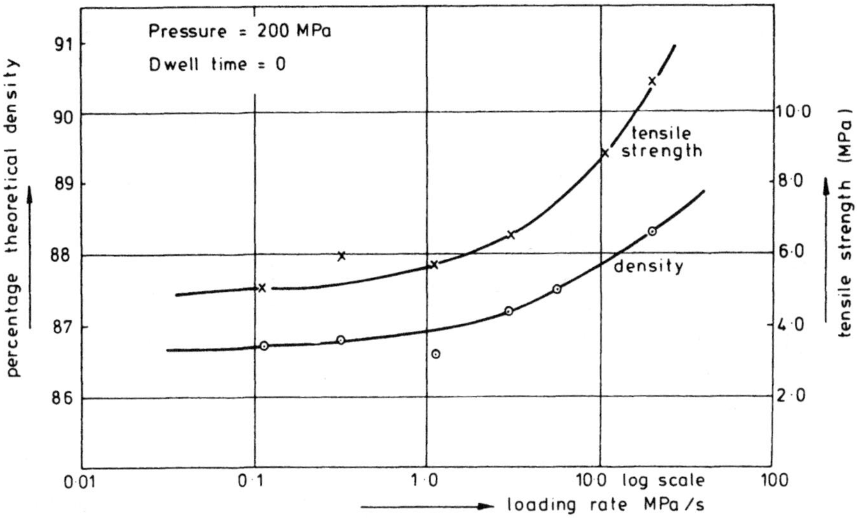

FIG. 18. Effect of loading rate on tensile strength and density of Viclan PVdC copolymers (Crawford *et al.*[11]).

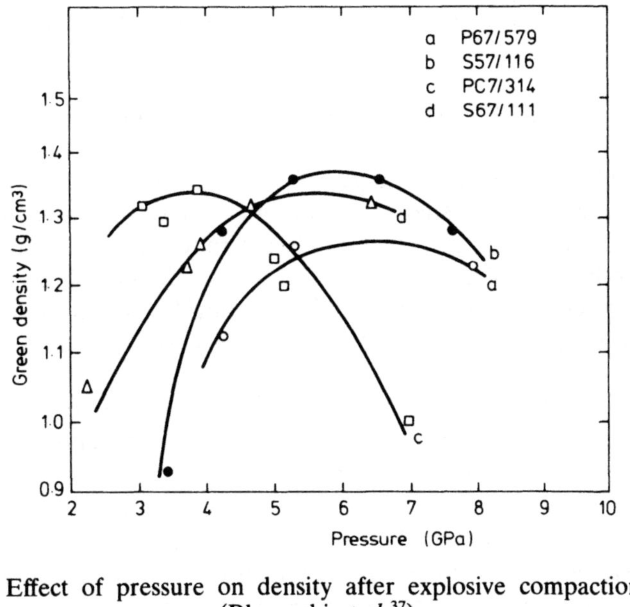

FIG. 19. Effect of pressure on density after explosive compaction of PVC (Blazynski *et al.*[37]).

TABLE 1
DENSITY AND STRENGTH VALUES FOR STATICALLY AND EXPLOSIVELY COMPACTED PVC

Material	*Static*		*Explosive*	
	Density (%)	*Compressive Strength (MPa)*	*Density (%)*	*Compressive Strength (MPa)*
(a) P67/579	95·0	29·5	85·0	38·9
(b) S57/116	97·0	23·3	97·0	62·2
(c) PC7/314	90·5	0·5	92·7	47·0
(d) S67/111	92·0	28·0	95·0	36·0

From Ref. 37.

the use of explosive compaction as opposed to static compaction. Figure 19 shows that for the four materials tested there was a peak in the density/pressure graph. The reduction in density at the very high pressures is associated with the mode of shock wave propagation and the higher degree of melting which occurs in these cases. However, it should be noted from Table 1 that explosive compaction produces a very large increase in the compressive strength of the compacts.

5. POWDER CHARACTERISTICS

As a result of the wide range of variables in polymeric powders (particle size, particle shape, crystallinity, molecular weight, etc.) it has proved difficult to define precisely the requirements to ensure good compactibility. Tables 2, 3 and 4 give the details of the materials successfully compacted by Maeda[16] and Bigg[19,20] respectively. Table 5 shows in decreasing order of compatibility the powders tested by Crawford *et al.*[25]

Maeda *et al.*[16] concluded from their work that the strength of their compacts increased as the molecular weight of the material increased and as the particle size of the powder decreased. Jayaraman *et al.*[18] working with polystyrene agreed with the conclusion on particle size but did not have enough data to comment on the molecular weight effect. Bigg *et al.*[20] restricted their conclusions about particle size to its effect on the density of the compact. In this regard they found that

TABLE 2
MATERIALS INVESTIGATED BY MAEDA *et al.*

	Material	Molecular weight ($\times 10^1$)	Granular size (μm)
MDPE[a]	Takasen Picw-38	8·9	7
	-14	6·4	7
	-15	6·0	8
	- 8	5·5	7
	-16	5·3	8
	-25	5·3	8
	-17	4·5	8
	- 6	3·0	9
	-44	2·7	10
	-33	2·2	10
	- 9	1·5	12
	-12	1·5	16
HDPE[b]	Hizex-2200P	4·0	50 ~ 80
	-1200P	3·0	60 ~ 90
	Million-Hizen-1	85·0	100 ~ 260
	-2	80·0	80 (Max.)
	-3	43·0	70 (Max.)
	-4	70·0	55 (Max.)
	PY-20 (Mitsubishi-Yuka)	3·0	(a) 39 (b) 39 ~ 88 (c) 88 ~ 127
	PX-80 (Mitsubishi-Yuka)	10·0	(a) 39 (b) 39 ~ 88 (c) 88 ~ 127
LDPE[c]	MS-30	2·5	70 ~ 150
	Flothene G801	4·0	60 ~ 100
PP[d]	TA-3 (Mitsubishi-Yuka)	27·0	(a) 39 (b) 39 ~ 88 (c) 88 ~ 127
PVC	135J (Nihon-Zeon)	7·5	1
	309M (Nihon-Zeon)	2·5	20
	110A (Nihon-Zeon)	2·5	100

From Ref. 16.
[a] Medium density polyethylene.
[b] High density polyethylene.
[c] Low density polyethylene.
[d] Polypropylene.

TABLE 3

MATERIALS INVESTIGATED BY BIGG

Polymer	*Type*	T_g(°C)	T_m(°C)	*Density*[a] g/cc	*Remarks*
Poly(amide–imide) A	High temperature	—	—	1·40	PAI-A is almost fully polymerised
Poly(amide–imide) B	High temperature	—	—	1·40	PAI-B is not fully polymerised
Melamine	Thermoset	—	—	1·48	Cures at 150°C and evolves water vapour
Poly(phenylene oxide)[b]	Thermoset	210	248	1·06	Degrades above T_m —
Polytetrafluoroethylene	Semi-crystalline	97	343	2·20	—
Polypropylene	Semi-crystalline	24	163	0·91	—
Nylon 11	Semi-crystalline	44	192	1·05	—
Polycrylamide	Glassy	—	—	1·05	Contains residual monomer; degrades at 150°C
Poly(vinyl alcohol)	Glassy	85	—	1·35	Degrades at 220°C
Poly(methyl methacrylate)	Glassy	100	—	1·28	—
Poly(vinyl chloride)	Glassy	83	—	1·35	Degrades at 170°C
Polysulphone	Glassy	184	—	1·24	—

From Ref. 19.
[a] Density achieved by conventional processing techniques.
[b] Unstabilised PPO oxidatively cross-links in air at 170°C.

TABLE 4

MATERIALS INVESTIGATED BY BIGG *et al.*

Polymer	*Type*	T_m(°C)	T_g(°C)	(g/cc)	*Particle size* (μm)
PAI[a]	Linear reactive	—	—	1·40	<175
PPO	Thermosetting	245	210	1·06	20–150
PP	Semi-crystalline	163	−18	0·91	<175
HDPE	Semi-crystalline	127	−120	0·938	~30
UHMWPE LS-501	Semi-crystalline	140	−120	0·96	60–300
UHMWPE LS-502	Semi-crystalline	140	−120	0·96	30–150
PTFE	Semi-crystalline	343	−112	2·20	Unknown

From Ref. 20.
[a] Poly(amide–imide).

TABLE 5
MATERIALS INVESTIGATED BY CRAWFORD *et al.*

Material	*Bulk density*[d] *(%)*	*Particle size (μm)*	*Particle shape*	*Work done (Nm)*	*% punch movement during first 5 MPa*
PVdC	21	37	Irregular	390	58
PTFE	20	<10	Irregular	85	84
UHMWPE	32	400	Regular (spherical)	350	77
PE[a](PP560)	44	100	Irregular	215	63
PES[b]	36	800	Irregular	480	51
PVC(P130/1)	42	3	Regular (spherical)	280	52
PE(BDH)	33	300	Smooth	149	76
PVC(S71/102)	37	150	Irregular	306	37
PS[c]	62	200	Spherical	310	10
Acrylic	60	75	Spherical	304	18
PP	60	250	Smooth	245	42

From Ref. 26.
[a] Polyethylene.
[b] Poly(ether sulphone).
[c] Polystyrene.
[d] Bulk density expressed as a percentage of solid density.

smaller powder particles gave a higher density in the 'green' (unsintered) compact and this in turn resulted in a higher strength after sintering.

Crawford *et al.*[25] concluded from their work that the materials which compacted very well exhibited a low bulk density, a small particle size and irregular shaped particles. The exception to these general rules was UHMWPE which compacted very well despite having relatively large, regular shaped particles. It could be that in this case the molecular weight effect referred to by Maeda *et al.*[16] was dominating the other parameters.

For the powders in Table 5 it was found that the compactability of the powders tended to decrease depending on the degree of departure from the three desirable features referred to above. For example, if only two of the desirable features were present the powder could still be pressed into a compact but it was weak and friable. Figure 20 shows particle sizes and shapes before and after compaction.

FIG. 20. Typical micrographs before and after compaction (Crawford.[25]). (a) Particles of PVdC before compaction; (b) micrograph of PVdC compact. (c) Particles of PVC (P130/1) before compaction; (d) after compaction. (e) Particles of poly(ether sulphone) before compaction; (f) after compaction.

The most undesirable particle shape was a sphere particularly if the particle size was large (>300 μm). This was the situation with the polypropylene powder tested in Table 5 which could not be compressed into a solid compact at pressures up to 800 MPa. In an attempt to improve the compactability of polypropylene and possibly validate the suspicions that particle size and shape have a major influence, some of the polypropylene powder was re-precipitated. This was done by dissolving the polymer in 1·5 litres of nearly boiling xylene containing antioxidant, an atmosphere of nitrogen being maintained in the vessel. This solution was then added slowly to well stirred methanol and the precipitate was filtered and washed thoroughly. The polymer was then dried *in vacuo* at 100°C. The resultant polypropylene powder had a very small particle size (<5 μm) and the particles were irregular. These features coupled with a much lower bulk density meant that the powder was well within the range of properties regarded as desirable and indeed the powder compacted extremely well.

The work by Crawford *et al.*[26] included a detailed investigation of the force transmissibility and deformation characteristics of their range of powders but although these were quite different in each case (see Figs 21 and 22) it was not possible to identify a feature which could be correlated with compactibility.

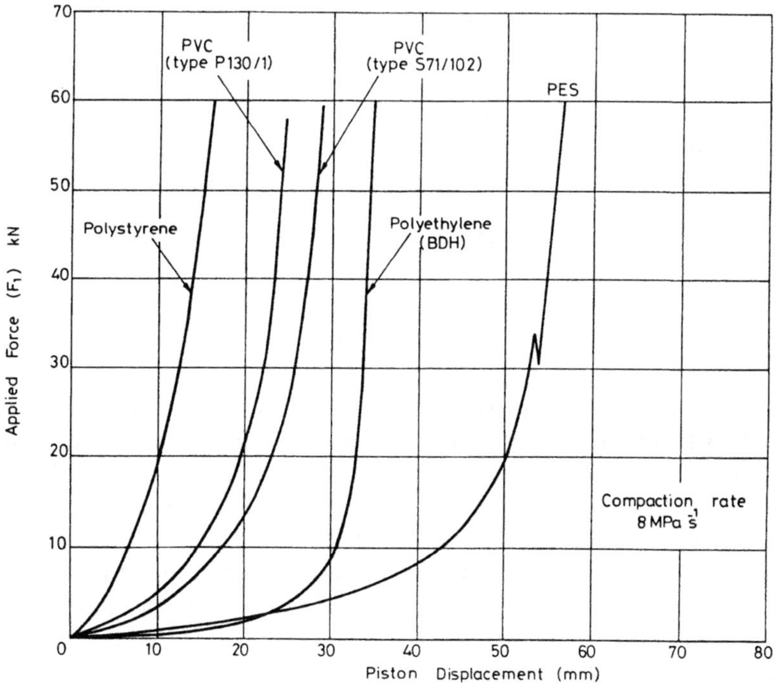

FIG. 21. Deformation characteristics for a range of plastic powders (Crawford *et al.*[25]).

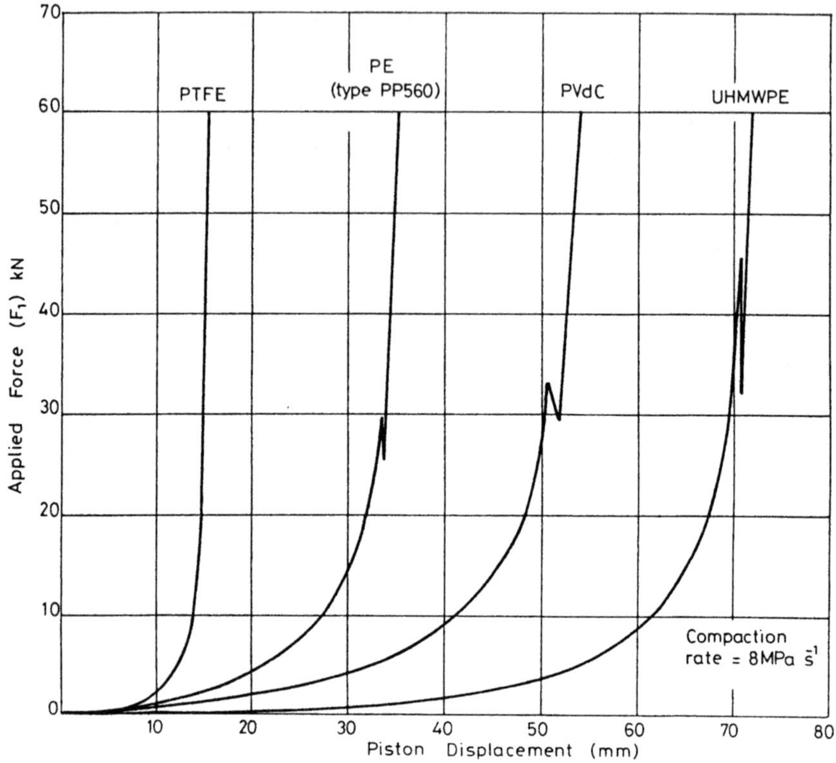

FIG. 22. Deformation characteristics for a range of plastic powders (Crawford *et al.*[25]).

6. SINTERING

Sintering, that is the process of heating the 'green' compact to a specified temperature for a specified time in order to permit structural changes to occur, is an essential part of powder metallurgical production methods. In a similar manner it is generally considered that in order to encourage polymeric particles to fuse into a homogeneous structure, a sintering operation is required after cold forming. However, the conditions of sintering have been found to be quite critical and the heat treatment will not be successful if the 'green' compact is not in the correct state. For example, Bigg[20] reported that samples having 'green' densities below 90% of the bulk density could not be sintered into useful parts.

6.1. Sintering Temperature

The sintering temperature is particularly important in that it must be sufficient to permit the viscous flow and surface tension forces to cause coalescence of individual particles. Bigg[19] found that for polyethylene the sintering temperature is critical to within 1°C, as shown in Fig. 23. After sintering at 128°C a tensile strength of 21 MPa was achieved at a test speed of 2·5 mm/min (the expected tensile strength of the material) but at 127°C a strength of only 12 MPa could be achieved. It is important to note that the ductility of the compact is also very dependent on sintering temperature but the optimum condition lags behind that for the strength. At a sintering temperature of 128°C (sufficient for optimum strength) the elongation at break, for polyethylene compacts, is less than 10%. Samples sintered at 130°C exhibited much greater ductility and did not break until strain values of 40–50% had been achieved. This is illustrated in Fig. 24.

Jayaraman[18] sintered polystyrene in the temperature range 150–173°C (50–73°C above the T_g) in order to optimise its strength. Crawford and Paul[11] sintered PVdC at temperatures between 100°C and 140°C and found that for optimum strength (and a three-fold increase over the green strength) the best temperature was 128°C. Above 130°C the polymer started to degrade and lose strength

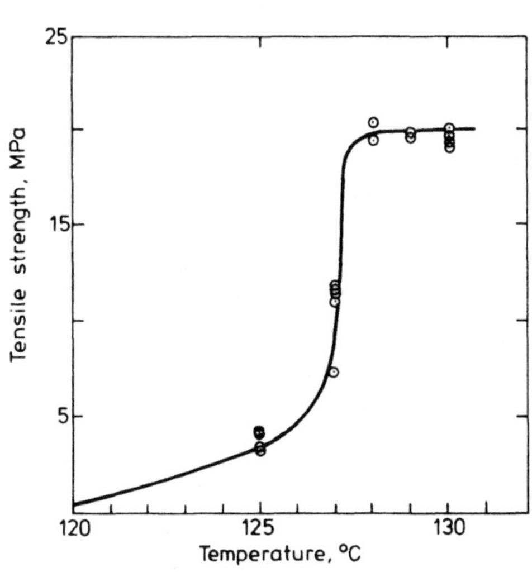

FIG. 23. Tensile strength versus sintering temperature for HDPE samples (Bigg *et al.*[20]).

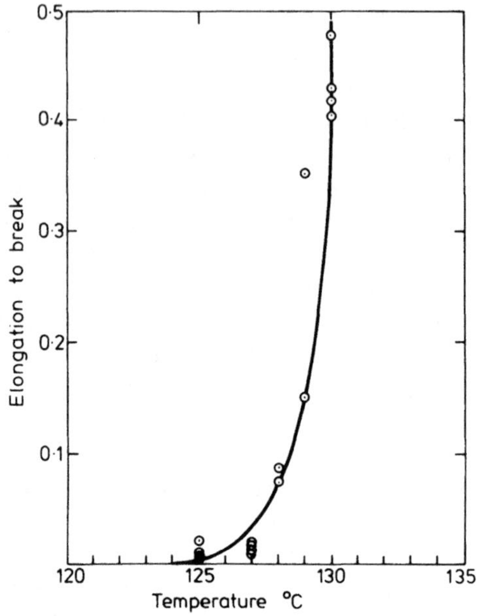

FIG. 24. Elongation to break versus sintering temperature for HDPE samples (Bigg *et al.*[20]).

although the elongation to break continues to show a dramatic improvement for temperatures up to 138°C—in agreement with the earlier observations by Bigg.

6.2. Sintering Time

Sintering time also has an influence on the properties of the compacted article. Figure 25 shows the complete picture for PVdC when a range of sintering times and temperatures was used. The optimum condition is 128°C for 1 h. A longer period at this temperature results in degradation of the polymer. Bigg *et al.*[20] used a sintering time of 1 h for all their materials although they did examine the effect of sintering time for PPO. The results are shown in Fig. 26 and indicate an optimum sintering time of 15 min. For times in excess of 20 min there was evidence of degradation of the sample.

6.3. Sintering Techniques

Maeda *et al.*[16] explored a range of sintering methods of which hot air ovens and immersion in boiling water proved to be best. The strength

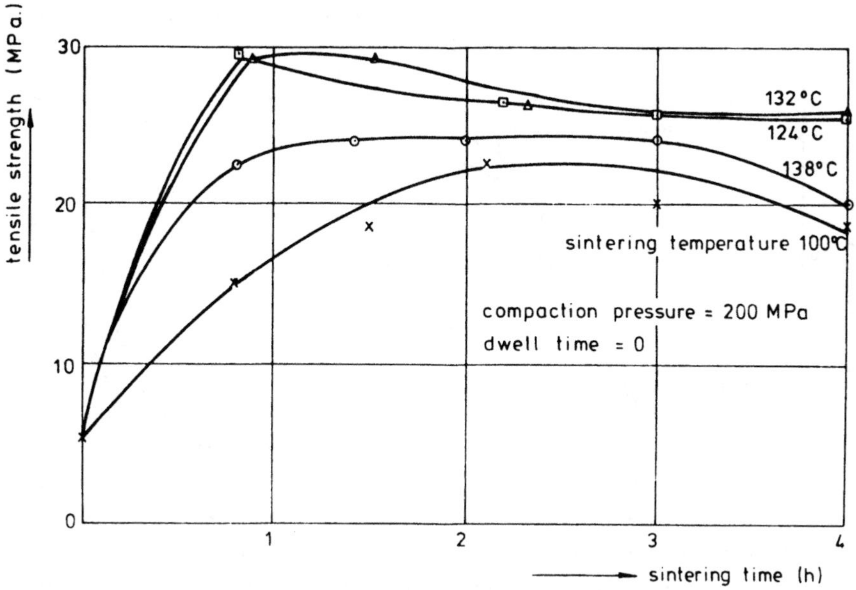

FIG. 25. Effect of sintering temperature and time on tensile strength of Viclan PVdC (Crawford *et al.*[11]).

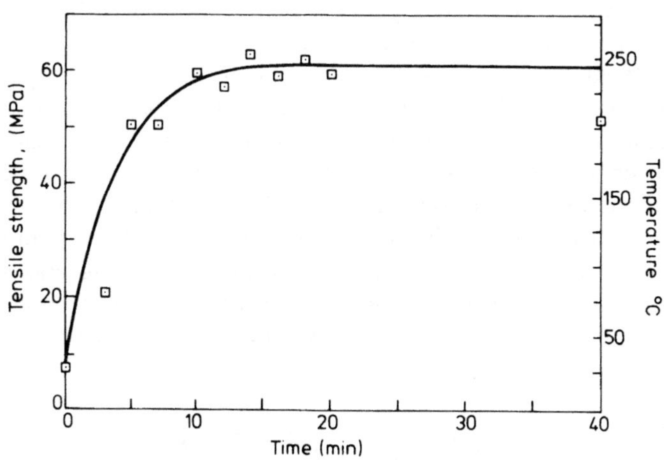

FIG. 26. Tensile strength versus sintering time for PPO sintered at 245°C (Bigg *et al.*[20]).

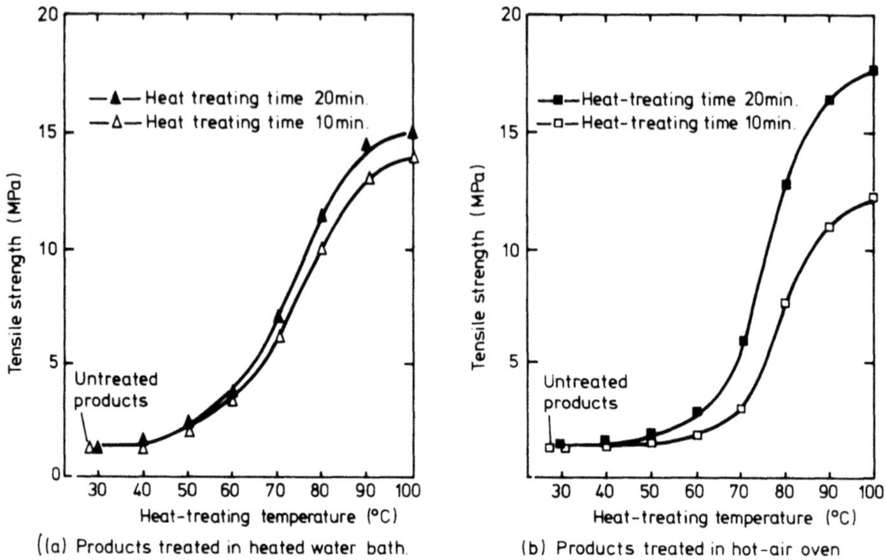

FIG. 27. Effects of heat-treating temperature on tensile strength of PVC (Maeda *et al.*[16]).

improvements achieved with polyethylene were quite good (a factor of about 2) but with PVC the results were quite startling—an increase of up to 15 times the green strength as shown in Fig. 27. The optimum sintering times were relatively short, being somewhere between 1 and 20 min depending on the section thickness and the material. It is evident from the range of results which have been reported on sintering trials that the optimum sintering conditions are very sensitive to the state of the 'green' compact and the nature (shape and size) of the powder particles. It is evident that polymeric powders of nominally the same material but obtained from different suppliers can have quite different characteristics. This may go some way to explaining why Bigg,[19] for example, had no success when compacting and sintering PVC whereas Maeda *et al.*[16] and Crawford *et al.*[25] were able to cold form this material quite successfully and achieve acceptable strength values.

7. EFFECTS OF FILLERS

The use of fillers in polymeric powder compacts has been very much exploratory and not very co-ordinated. A variety of fillers have been tried out to satisfy inquisitive minds and the objectives have not been clearly established. In the vast majority of cases the powders tested have been in the 'as-received' state which is often far from the ideal state for solid phase compaction. However, this has been a necessary step in order to establish the compaction characteristics of polymeric powders. The next step will be to follow the lead of the established manufacturing industries based on powder compaction (powder metallurgy, ceramics, pharmaceuticals, animal feedstuffs, etc.). In these industries the basic powder is combined with lubricants, binders and reinforcing fillers in order to improve the processability of the powder, enhance adhesion between the particles and maximise the load carrying capability of the moulded product.

To date the fillers which have been tried include fibres,[16,20] metal particles[16] and different polymeric particles.[16] In general the effect has been to reduce final strength values due to the incompatibility of the filler/powder combination at the sintering stage. However, Maeda[16] did report some success in relation to carefully chosen (compatible) combinations of dissimilar polymeric powders and achieved strength enhancement when metal particles were added to polymeric powders provided the 'green' compact was subjected to microwave sintering. Hot air or boiling water sintering resulted in a drastic drop in strength.

It is clear that this aspect of solid phase compaction of polymeric powders is still very much in its infancy but the identification and strategic use of appropriate fillers will be an essential step in the development of this process as a viable production method.

8. SPECIAL COMPACTION METHODS

8.1. Hot Compaction

The possibility of compacting polymeric powders in a heated die has been suggested by several workers.[16,38] The effect of moulding in a heated die is to remove the need for subsequent sintering. Figure 28 shows that pressing polyethylene powder at 100°C has the same effect as sintering a 'green' compact (produced cold under the same pressure) for 10 min at 100°C. The advantages of performing the compaction and the sintering in a single operation are readily apparent and work is currently being carried out to explore this possibility in greater detail.

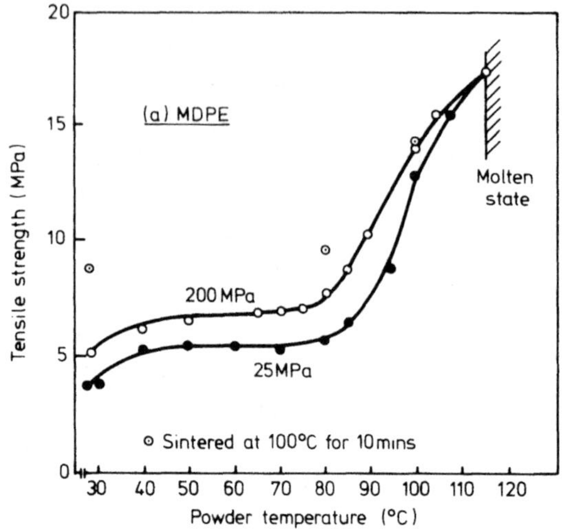

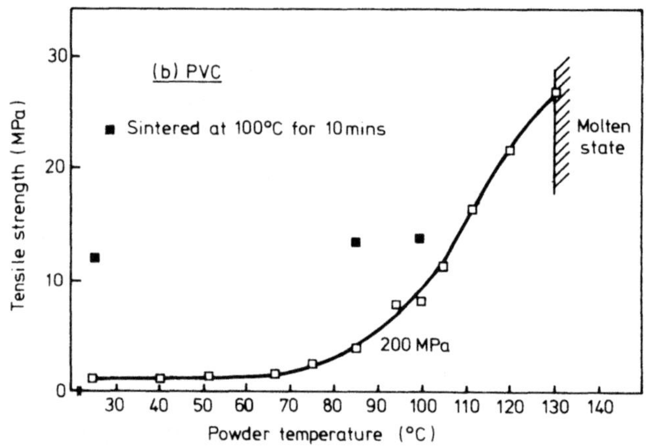

FIG. 28. Effects of powder temperature on tensile strength of statically hot compacted products of MDPE and PVC (Maeda *et al.*[16]).

8.2. Isostatic Compaction

The compaction of metallic and ceramic powders in flexible dies using fluid pressure is already well established. To date there have been no reports of this technique being used for polymeric powders although Crawford and Paul[11] have illustrated that the method is feasible. Very good quality, strong compacts have been moulded with excellent detail reproduction as shown in Fig. 29. Once again this is an area which is currently being explored.

FIG. 29. Articles produced by isostatic compaction of polymeric powders (except gear wheel).

8.3. Ultrasonic Moulding

There has been a small amount of interesting work on the possibility of using ultrasonic vibrations to mould thermoplastic powders. Fairbanks[39] and Maeda *et al.*[16] showed that ultrasound could be used to convert plastic powders into products which were as strong as conventional mouldings. Crawford *et al.*[40] recently showed that this technique could be used to mould polyproplyene powders which had proved impossible to mould by solid phase compaction. Once again more work is needed to perfect this technique because although the strengths are not far short of injection moulded material, the ductility is lower than would be expected—15% elongation to break was the best that could be achieved. However, despite this, the possibility of moulding using ultrasonic vibrations is exciting since relatively thick mouldings could be produced with a cycle time of under 4 s.

9. CONCLUDING REMARKS

As a manufacturing method the solid phase compaction of polymeric powders is still at the development stage but already it has demonstrated considerable potential. It is evident that there are several ways in which it can be put to good advantage.

(a) Granulation.
(b) Manufacture of materials with special properties.
(c) Manufacture of articles.

Future work is likely to be primarily concerned with the optimisation of particle characteristics (e.g. size and distribution). This will be an important step towards good quality compacts but the biggest breakthrough will be when additives (e.g. lubricants and binders) are used to enhance the compactability of the powder and improve the properties of the end-product. The final piece in the puzzle, as far as component manufacture is concerned, will be the determination of the inter-relationships between post-moulding dimensional changes, compaction variables and die geometry.

REFERENCES

1. Fischmeister, H. F., *Proc. I. Mech. E.*, **196,** 1982, 105–21.
2. Jones, W. D., *Fundamental Principles of Powder Metallurgy,* Arnold, London, 1960.
3. Hausner, H. H., *J. Metals,* **9,** 1957, 331.
4. Brown, G. T., *Chartered Mech. Engng,* **27**(7), 1980, 49.
5. James, P. J., *Chartered Mech. Engng,* **26**(7), 1979, 67.
6. Schwartz, E. G. and Weinstein, A. S., *J. Am. ceram. Soc.*, **48,** 1965, 346.
7. Strijbos, S., *Science of Ceramics,* Vol. 8, British Ceramics Society, Stoke-on-Trent, 1976.
8. Train, D., *J. Pharm. Pharmac.*, **8,** 1956, 276.
9. Train, D. and Lewis, C. J., *Trans. Inst. Chem. Engng,* **40,** 1962, 18.
10. Hausner, H. H., *Powder Metallurgy Processing,* 7th Powtech Conference, Basel, Switzerland, March 1980.
11. Crawford, R. J. and Paul, D., *J. Mat. Sci.*, **14,** 1979, 2693–702.
12. Maeda, T., *Japans Plastic Age,* October 1972, 35–42.
13. Palmer, H. K. and Rowe, R. C., *Powder Technology,* **9,** 1974, 181–6.
14. Palmer, H. K. and Rowe, R. C., *ibid.*, 225–30.
15. Bigg, D. M., *SPE Tech. Papers,* **21,** 1975, 472–6.
16. Maeda, T. and Matsuoka, S., *J. Faculty of Eng.*, Univ. of Tokyo, **23,** 1975, 191–226.
17. Matsuoka, S. and Maeda, T., *Modern Plastics,* March 1975, 58–62.
18. Jayaraman, G. S., Wallace, J. F., Geil, P. H. and Baer, E., *Polym. Eng. Sci.*, **16,** 1976, 529–36.
19. Bigg, D. M., *Polym. Eng. Sci.*, **17,** 1977, 691–9.

20. Bigg, D. M. and Epstein, M. M., in *Science and Technology of Polymer Processing,* N. P. Suh and N-H. Sung, Eds, MIT Press, Cambridge, Massachusetts, August 1979, 897–919.
21. Halldin, G. W. and Kamel, J. L., *Polym. Eng. Sci.,* **17,** 1977, 21–6.
22. Kulkami, K., *Polym. Eng. Sci.,* **19,** 1979, 474–81.
23. Crawford, R. J., Paul, D. and Adeebnia, Y., *Euro. Polym. J.,* **16,** 1980, 401–5.
24. Crawford, R. J. and Paul, D., *Euro. Polym. J.,* **17,** 1981, 1023–8.
25. Crawford, R. J. and Paul, D., *J. Mat. Sci.,* **17,** 1982, 2267–80.
26. Paul, D. W., *Solid Phase Compaction of Polymeric Powders,* PhD Thesis, Queen's University, Belfast, 1981.
27. Crawford, R. J., *Polym. Eng. Sci.,* **22,** 1982, 300–6.
28. Crawford, R. J. and Sprevak, D., *Euro. Polym. J.,* **20,** 1984, 441–6.
29. Crawford, R. J., Paul, D. and Sprevak, D., *Polymer,* **23,** 1982, 123–30.
30. Dean, T. A., *Plastics and Rubber International,* **3,** 1978, 109–11.
31. Train, D., *Trans. Inst. Chem. Engs.,* **35,** 1977, 258–64.
32. McLeod, H. and Marshall, K., *Proc. I.C.C.,* Brighton, UK, 1977.
33. Takada, T., Kuramoto, M. and Kuno, H., *Powder Technology,* **14,** 1976, 51–6.
34. Seelig, R. P. and Wulff, J., *Trans. Am. Inst. Mining Engs.,* **171,** 1947, 492–9.
35. Duwez, P. and Zwell, L., *Metals Trans.,* 1949, 137–45.
36. Strijbos, S., Oudermans, G. J. and Klein-Wassink, R. J., *Powder Technology,* **18,** 1977, 187–94.
37. Blazynski, T. Z., Hegozy, A. A. and El-Sobky, H., *8th HERE Conf.,* San Antonio, Texas, 1984, 113–20.
38. Dean, T. A., *Proc. 18th Int. MTDR Conf.,* 1976, 275–81.
39. Fairbanks, H. V., *Ultrasonics,* **12,** 1974, 22–4.
40. Crawford, R. J. and Paul, D. W., *Ultrasonics,* **19,** 1981, 23–7.

INDEX

GPSR Compliance

The European Union's (EU) General Product Safety Regulation (GPSR) is a set of rules that requires consumer products to be safe and our obligations to ensure this.

If you have any concerns about our products, you can contact us on ProductSafety@springernature.com

In case Publisher is established outside the EU, the EU authorized representative is:

Springer Nature Customer Service Center GmbH
Europaplatz 3
69115 Heidelberg, Germany

Batch number: 10607212

Printed by Printforce, the Netherlands